BASIC CONCEPTS
OF
SOIL SCIENCE

BASIC CONCEPTS OF SOIL SCIENCE

(SECOND EDITION)

A K Kolay

Publishing Globally

NEW AGE INTERNATIONAL (P) LIMITED, PUBLISHERS

New Delhi • Bangalore • Chennai • Cochin • Guwahati

Hyderabad • Kolkata • Lucknow • Mumbai

Visit us at **www.newagepublishers.com**

Published by New Age International (P) Ltd., Publishers
First Edition: 1993
Second Edition: 2000
Reprint: 2013

BRANCHES

- **Bangalore** 37/10, 8th Cross (Near Hanuman Temple), Azad Nagar, Chamarajpet Bangalore-560 018, Tel.: (080) 26756823, Telefax: 26756820 **E-mail: bangalore@newagepublishers.com**
- **Chennai** 26, Damodaran Street, T. Nagar, Chennai-600 017, Tel.: (044) 24353401 Telefax: 24351463, **E-mail: chennai@newagepublishers.com**
- **Cochin** CC-39/1016, Carrier Station Road, Ernakulam South, Cochin-682 016 Tel.: (0484) 2377004, Telefax: 4051303 **E-mail: cochin@newagepublishers.com**
- **Guwahati** Hemsen Complex, Mohd. Shah Road, Paltan Bazar, Near Starline Hotel Guwahati-781 008, Tel.: (0361) 2513881, Telefax: 2543669 **E-mail: guwahati@newagepublishers.com**
- **Hyderabad** 105, 1st Floor, Madhiray Kaveri Tower, 3-2-19, Azam Jahi Road Nimboliadda, Hyderabad-500 027, Tel.: (040) 24652456 Telefax: 24652457, **E-mail: hyderabad@newagepublishers.com**
- **Kolkata** RDB Chambers (Formerly Lotus Cinema) 106A, 1st Floor S.N. Banerjee Road, Kolkata-700 014, Tel.: (033) 22273773 Telefax: 22275247, **E-mail: kolkata@newagepublishers.com**
- **Lucknow** 16-A, Jopling Road, Lucknow-226 001, Tel.: (0522) 2209578, 4045297 Telefax: 2204098, **E-mail: lucknow@newagepublishers.com**
- **Mumbai** 142C, Victor House, Ground Floor, N.M. Joshi Marg, Lower Parel Mumbai-400 013, Tel.: (022) 24927869, Telefax: 24915415 **E-mail: mumbai@newagepublishers.com**
- **New Delhi** 22, Golden House, Daryaganj, New Delhi-110 002 Tel.: (011) 23262368, 23262370, Telefax: 43551305 **E-mail: sales@newagepublishers.com**

ISBN: 978-81-224-1199-7

₹ 250.00

C-12-10-6639

Printed in India at Mohanlal Printers, Delhi.
Typeset at Spectrum Media.

PUBLISHING FOR ONE WORLD
NEW AGE INTERNATIONAL (P) LIMITED, PUBLISHERS
7/30 A, Daryaganj, New Delhi - 110002
Visit us at **www.newagepublishers.com**

PREFACE TO THE SECOND EDITION

The author of an elementary book in soil science must be scientifically correct in presenting his subject to the students and extension workers at the grass roots level. He must explain soil science to beginners who may be students or extension workers, in a language simple enough to be readily understood. This is not an easy job because the average agricultural students commonly possess rudimentary knowledge of basic sciences. This book should also cover the entire syllabus of the course which is currently being taught to students. These problems have been taken into account. Thereafter some portions of chapters 3, 4, 6 and 7 have been altered / supplemented with new information. This book may also be very useful to the extension workers at the grass roots level since a thorough knowledge of basic soil science facilitates the optimum use of the soil resources to increase yield of crops at the farmer's field.

A. K. KOLAY

PREFACE TO THE FIRST EDITION

National progress is dependent upon the rapid development of agriculture. Agricultural production is mainly dependent upon the maintenance and improvement of soil productivity. Farmers should be educated to use lands according to their capability, and to adopt proper soil conservation measures. Students of agriculture today, will be the architects of the agricultural development of the country in the future, when they will serve the agricultural departments of the different states of India. They must therefore be well trained in different agricultural subjects especially soil science. The maintenance and improvement of soil productivity for sustained agricultural production necesitates the acquisition of thorough knowledge of soil science. Therefore, the teaching of soil science is atleast one of the most important factors for the agricultural development of the country because tillage of the soil is involved in every kind of farming.

The teacher must understand and know his pupils at least as much as if not more than his subject. The average undergraduate students of Indian Agricultural Universities and Colleges are usually not mature enough to thoroughly assimillate their subjects from books written by foreign authors. Perhaps this factor has prompted the teachers to write books in basic science. But few teachers have written textbook in soil science especially for beginners. So a book on basic soil science should be simple enough to be understood by the average agricultural students who usually possess a rudimentary knowledge of basic sciences. This book should also cover the entire syllabus of at least a few core courses in soil science in a manner that would create an interest in soil science in them. Keeping all these basic principles of teaching in mind, the author, who at present is teaching soil science at the School of Agricultural Sciences and Rural Development under the North-Eastern Hill University, has ventured to write this book which is expected to cover the entire syllabus of a few core courses in soil science of the Indian Agricultural Universities and colleges. It covers all important and main aspects of soil science. The author sincerely believes that this book will build up basic knowledge of soil science in the minds of the undergraduate students.

A.K. KOLAY

CONTENTS

1
THE EARTH

Agriculture is mainly dependent on the soil and the climate. The soil covers portions of the surface of the earth. It is an intimate mixture of organic and inorganic materials, water and air. Loose powdery soils usually support crop growth. They have originated from disintegration and decomposition of rocks at the surface of the earth by the action of atmospheric agents like heat, water carbon dioxide, nitrogen oxides, sulphur dioxide etc. Rocks are the original components of the earth's crust. So a thorough knowledge of rocks and the earth's crust is essential for understanding the origin of different soil and the causes of variation of properties of different soils. The nature of the earth's crust varies from place to place in India and abroad. The crust is the outermost layer of the earth. A brief account of the origin of the earth itself would be appropriate, followed by a study of the origin of India and the nature of the earth's crust in the first chapter.

1.1. ORIGIN OF THE EARTH

Gas and dust particles remained scattered in the form of clouds called nebula in the infinite spaces before the birth of universe. These clouds were thicker at some places. The gas and dust particles attracted eath other much more at these places when the cloud of gas and dust particles contracted and broke down into sub-units of clouds which began to rotate around the centre. The bulk of the material was concentrated at the centre, to form the stars. The sun was formed this way. Some dust and gaseous matter began to move around the centre in the form of a disc. G. Kuiper proposed in 1951 that the sun was the nucleus of a rotating disc of cloudy material. When this disc was rotating at a tremendous speed, dust particles and gases collided and united with another to from the heavenly bodies which now include nine planets, thirty one satellites, millions of asteroids, meteors and numerous comets. One of the nine planets is our earth.

After birth, the earth began to revolve around the sun. It gradually cooled down by radiating heat. Gases first cooled down at the surface of the earth to form a liquid which went down deep into the earth, where it is again vapourised and came up to the surface of the earth. So the central heat was transferred to the surface of the earth. The gaseous masses again cooled down to from a liquid which went into the earth, was vapourized and returned to the earth's surface. These processes continued till the liquid accumulated at the centre of the earth. Radioactive elements disintegrated to generate more heat that slowed down the process of cooling of the earth, but ultimately the gaseous earth was converted to the molten state when the materials constituting the earth were arranged in concentric layers according to their densities within the earth. Heavier elements like iron and nickel went down and accumulated at core, which is the central portion of the earth of approximate radius of 4328 kilometre, whereas the relatively lighter elements like silicon, aluminium, magnesium etc. remained in the upper concentric layers called mantle and crust respectively. Thereafter the earth cooled down and solidified to form the outer hard mass called the lithosphere when it was surrounded by methane that constituted the atmosphere. Nitrogen, which now occurs in the atmosphere, was present as ammonium compounds in the lithosphere. Later, some of it was driven out the atmosphere, where it existed mainly as ammonia. When plants grew, they photosynthesised to enrich the atmosphere with oxygen. Simultaneously the ammonia was oxidised to nitrogen gas. Later gases like oxygen water vapour and oxides of nitrogen, sulphur phosphorus etc. were forced out from the interior of the earth to the atmosphere. Most of the water vapour condensed to from the hydrosphere.

The lithosphere consists mainly of hard inorganic materials called rocks. The molten mass of the earth cooled down after the birth of the earth to form the igneous rocks which were gradually disintegrated and decomposed by the atmospheric agents like heat, water, vapour and gases during the millions of years when the products of this disintegration and decomposition were deposited in layers that were cemented to form the sedimentary rocks.

The period between the solidification of the gaseous material to form the lithosphere and its present from has been divided into six eras. Whenever each era ended, some sort of geological revolution took place when the climate, rocks and living organisms abruptly changed, the land and the sea were re-distributed, volcanoes erupted and sea beds were elevated to form mountain ranges. These six areas have been named as (*i*) Azoic, (*ii*) Proterozoic, (*iii*) Palaezoic, (*iv*) Mesozoic or Secondary, (*v*) Tertiary or Cenozoic or Kainozoic, and (*vi*) Quarternary. No living organisms existed on the surface of the earth during the Azoic era. The first living organisms appeared on the surface of the earth during proterozoic era. Palaezoic was the era of ancient life. Mesozoic was the era of middle life. Tertiary was the era of new life and quarternary was the era of recent life. Each era has been divided into periods. Whenever a period of any era ended, major disturbances occurred on the surface of the earth, but they were not as drastic as those in the eras. Periods have been divided into epochs. Each epoch has been divided into ages which are the smallest sub-division in the geological time scale. Rocks which were formed in a particular era have been placed in a group of rocks. They occur between two zones of removal or deposition of materials which are known as unconformities that separate the younger rocks from the older ones. A system of rocks include those rocks that were developed during a particular period when those genera of plants and animals requiring specific climatic c lition for their survival lived on the earth. Geological conditions of the earth did not change within a particular period. A series of rocks includes those rocks developed in a particular epoch. Rocks that were formed during a particular age have been put in the stage of rocks, which in turn may be divided into zone of rocks.

Every kind of living organisms require certain specific conditions favourable to their growth. These conditions may be found within a particular geological period, epoch or age. Whenever any geological period ended, environments which were favourable for the growth of specific groups of living organisms, *i.e.* plants and/or animals, changed. So they died and remained buried in the sedimentary rocks on which they lived. These preserved bodies of animals and plants or their impressions on sedimentary rocks, are known as fossils. So every system of sedimentary rocks contains specific fossils that enable us to identify that particular system of rocks. Similarly certain living organisms lived during certain epochs and ages, remaining buried in the series and stages of sedimentary rocks are identified on the basis of those specific fossils possessed by them. A stage of sedimentary rocks may contain different fossils in different levels. So a set of sedimentary beds of stages of sedimentary rocks have been divided into 'zones' of sedimentary rocks.

Rocks of the Proteozoic and Azoic eras have been classified basically on field evidence like the order of superimposition of rocks, intensities of metamorphism of rocks, marked unconformities and changes in lithological appearances, although sedimentary rocks that were formed during later eras have also been classified on the basis of these criteria.

European geologists have classified rocks into groups and systems on the basis of unconformities and fossil content of rocks which have been found in Europe, as mentioned below. This classification has been used as a scale more or less for the entire world and is therefore called the standard stratigraphical scale as presented in Table 1.1.

TABLE 1.1 : Standard Stratigraphical Scale

Group of rocks	*System of rocks*	*Time of formation in million of years*
Quarternary	Recent	0.025
	Pleistocene	1
	Unconformity	
Tertiary or Kainozoic or Cenozoic	Pliocene	8
	Miocene	25
	Oligocene	40
	Eocene	65
	Unconformity	
Secondary or Mesozoic	Cretaceous	130
	Jurassic	170
	Triassic	230
	Unconformity	

Primary or Paleozoic	Permain	280
	Carboniferous	320
	Devonian	390
	Silurian	420
	Orodovician	490
	Cambrian	550
	Unconformity	
Proterozoic	Precambrian	2225
	Eparchaean Unconformity	
	Archaean	3600
	Earth's Crust	

No unconformity occurs between the Quarternary and the Tertiary groups of rocks and between the Tertiary and Mesozoic groups, in India. The rocks of the upper carboniferous system had been separated from the lower carboniferous system by an unconformity. So all the rocks formed within about 320 million years from the present have been placed in the Aryan group of rocks. Similarly, an unconformity exists between the cambrian and the Vindhyan system. So all rocks that originated between 320 to 550 million years ago have been placed in the Dravidian group of rocks. Geologically rocks have been classified in India as shown in Table 1.2.

TABLE 1.2 : Geological Classification of Rocks in India

Groups of Rocks	*System of Rocks*	*Time of formation in million of years*
Aryan	Recent	0.025
	Pleistocene	1
	Pliocene	8
	Miocene	25
	Oligocene	40
	Eocene	65
	Cretaceous	130
	Jurassic	175
	Triassic	230
	Permian	280
	Upper Carboniferious	310
	Unconformity	
Dravidian	Lower carboniferous	330
	Devonian	390
	Ordoviacian	490
	Cambrian	550
	Unconformity	
Purana	Vindhyan	1500
	Cuddapah	
	Eparchaean Unconformity	
	Dharwar	
Archaean	Archaean	3600
	Earth's Crust	

1.2. EARTH'S CRUST

Geophysical studies of the earth have indicated that it consists of a number of nearly concentric layers of different properties. The boundary between these layers where properties of the matter abruptly changes is known as discontinuity. Velocity of earthquake waves suddenly changes whenever they meet a discontinuity. Mohorovicic has discovered a seismic discontinuity called Moho which has been recognized by sudden increase of the velocity of the compressional shock waves from 6.5 to 8 kms/sec. Moho has been accepted as the base of the earth's crust.

The earth's crust, approximately 50 kms in thickness is the outer portion of the lithosphere between the ground surface and the Moho. The crust has been divided into two sub-zones called sial and sima. The sial is a heterogeneous mixture of rocks. The sima is a homogeneous plastic or semi-plastic concentric layer that behaves like a solid, the sial floats on the sima, which in turn floats on the lower concentric layer called the mantle. The sial contains about 65-75 per cen silica. Aluminium is the next important element in the sial, represented by the most common rocks like granite and rhyolite. Silica decreases to about 50-60 per cent in the sima where aluminium has largely been replaced by magnesium with minor quantities iron. Sima is represented by the most common rocks like basalt and gabbro at the upper level and by olivine rich rocks at the lower level. There two layers of the earth's crust are called sial and sima respectively because their chief constituents are silica and alumina and silica and magnesia respectively. Their average densities are 2.7 gms/cc and 2.9 gms/cc respectively.

Besides oxygen (approximately 47%), silicon is the most important element (approximately 28%) found in the sial which also contain about 8 per cent aluminium, 5 per cent iron, 3.5 per cent calcium, and 2.6 per cent each of sodium, magnesium and potassium. Phosphorus (approximately 0.09%) and sulphur (approx. 0.07%) are the most important non-metals that occur in the earth's crust. Ninety elements have been detected in the earth's crust, mostly in the combined forms.

The earth has a tendency to approach an isostatic (Iso means same, static means state of rest) equilibrium. The upper crust (sial) has been gradually degraded; materials from the elevated areas like mountains were gradually transferred to the depressed areas by gravity, water, ice and wind. So the isostatic equilibrium of the crust was upset. The mantle rock below the crust behaves like a plastic flow and exerts pressure on the crust. So the regions from where material had been removed, were gradually elevated, to restore isostatic equilibrium. The surface of the earth folded. This process of deformation of the earth's crust is known as Eperiogeny. The force which combats degradation of the earth's surface is known as Diatrophism, resulting in folds, faults and fractures on the earth's crust. Folds usually occur in belts or systems where many folds exist, together. They are irregularly shaped in unconsolidated sediment and on steep slopes, and in metamorphic rocks where flowage has been a prominent part of the folding process.

A portion of the earth's crust was subjected to stress, stretched and pulled under tension. When the elastic limit of the earth's crust was exceeded, rocks in the crust were at first deformed and eventually ruptured. Fractures are breaks in the earth's crust, along which materials have not been displaced. A fault is a break fracture in the materials of the earth's crust along which materials had been displaced. Faults occurred in rocks exposed on the surface of the earth. (*see* Figs. 1.1, 1.2 and 1.3) The fractures have been found in most consolidated igneous, sedimentary and metamorphic rocks.

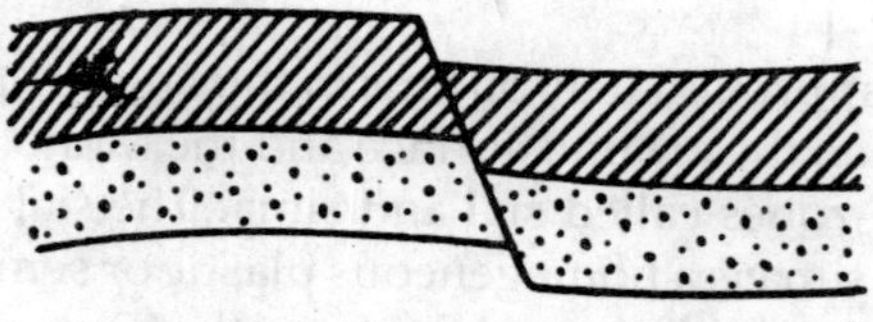

Fig. 1.1. Normal fault.

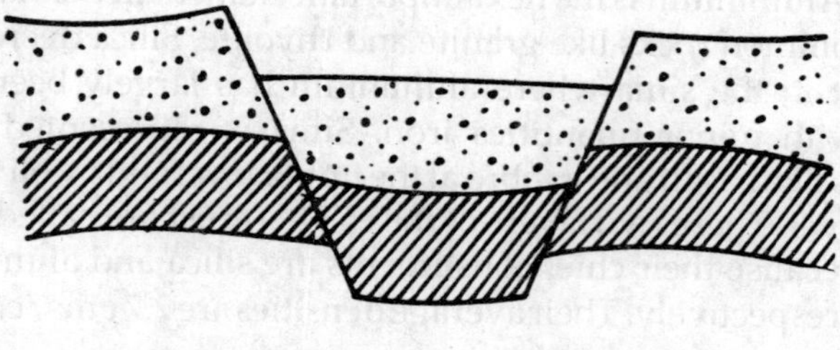

Fig. 1.2. Through or trench fault.

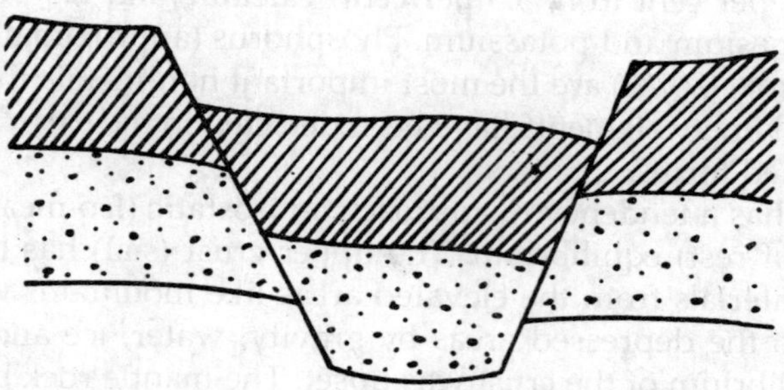

Fig. 1.3. Ridge fault or horst.

1.3. FORMATION OF INDIA

The earth was one continuous land mass called Pangea (meaning one land mass in Greek). Mountains were formed at the end of the Azoic era as a consequence of intense earth movements: the oldest mountains like the Aravalli mountains, were probably formed then. The lithosphere crumpled at many places to form the elevated areas and basins which were filled with sediments during the

Cuddapah period. These areas are, presently included in the present Andhra Pradesh, Orissa, Madhya Pradesh and Rajasthan. The lithosphere was subjected to broad uplifting and downwarping during the Vindhyan period when sediments derived from the elevated areas were deposited in the valleys in the large areas that now extend from South Bihar to South-Western Utter Pradesh, and from Rajasthan to Andhra Pradesh. A large number of trough faults were formed in areas which are now in the Damodar valley in Bihar and West Bengal, the Narmada valley in Madhya Pradesh and the Godavari valley in Andhra Pradesh. Sediments were deposited in these low laying regions during the carboniferous period, when plants growing there died and remained burried under fresh deposition of sediments. They were ultimately converted to coal seams, presently occurring in the Gondwana coalfields of Bihar, West Bengal and Madhya Pradesh.

An elongated trough like depression called the geosyncline was formed on the lithosphere during the Jurassic period. It has been named the Tethys sea. It virtually separated the lithosphere into two super continents. The northern supercontinent included Laurasia, Eurasia and Angaraland whereas the southern one which was called Gondawana land, included India, Australia, Africa and South America.

The region now known as Gujarat and western Rajasthan was flooded by an arm of the Tethys ocean. A large area, from Pondicherry to Cauvery valley and the southern parts of Assam plateau, was flooded by the sea during the Cretaceous period when the Indian sub-continent broke loose from Africa and started moving towards the north. This northernly movement of the Indian shield, thrust the sediments accumulated in Tethys ocean against the Asian mainland *i.e.* the northern supercontinent. First, the Tethys ocean was furrowed into a series of parallel ridges and basins during the upper cretaceous period. Sediments were also being deposited in the basin during the same period. The Himalayan mountain ranges were formed as a consequence of stupendous earth movement during the upper Eocene to middle Miocene period. A long narrow trough called foredeep had been gradually forming at the base of the Himalayan mountain ranges. Later sediments were deposited in it to form the Siwalik system of rocks at the Himalayan foot-hills.

When the Indian penninsular shield was moving towards the north, its northern border was thrust below the sediments of the Tethys ocean. So a long depression as formed, which has since been filled up by sediments derived especially from the lofty chains of the Himalayan mountain ranges to form the Indo-Gangetic Alluvial plains which are at present in Punjab, Haryana, Uttar Pradesh, North Bihar, West Bengal and Assam.

Normal faults developed in the Indian peninsula, one of which ran along the western coast and the other along the eastern coast. Large blocks of land sank beneath the Arabian sea one side and the Bay of Bengal on the other. The existing Indian peninsula is therefore a horst.

Volcanic eruptions continued intermittently up to the Eocene period when the basaltic lava flows (trap rocks) covered vast areas which are now in Maharashtra, Madhya Pradesh and Gujarat. The Indian peninsula was tilted in the south-west and southern direction during the post Miocene period. The eastern coast of India took its present shape in the beginning of the cretaceous period, but on the west, there was still some land to the west of Bombay stretching westwards, which is now part of the Arabian sea. Most of Bengal formed the shelf area of a marine basin where marine sediments were gradually deposited from the Cretaceous period to the Eocene period in areas which now form part of the districts of Midnapur, Burdwan and Murshidabad. Later, Gangetic alluvium was deposited during the Pliocene period and the pleistocene period in areas which now include southern Bengal.

Kerala was formed from the geological upheaval of the Arabian sea bed at the south-western foot of the western Ghat mountains and subsequent sedimentation on this new elevated sea bed as described below.

1.4. FORMATION OF KERALA

The Arabian sea extended right up to the foot of the western Ghat mountains, where the remains of marine animals were deposited along the south-western foot of the western Ghat mountains, for millions of years. This area underwent subsequent upheaval to form the present Kerala. This geological upheaval resulted in the formation of numerous shallow lakes and back waters and the Palghat gap nearly 30 km wide in the western Ghat mountain range which appeared thrown back and heaped up at Palghat as if some overwhelming deluge had burst through, sweeping them left and right, Geologist think that an ancient river flowed through the Palghat gap to the Arabian sea.

Kerala was flooded by the sea during the Miocene period after which ferruginous sandy and clayey materials were intermittently deposited at and around Varkala and other places along the Travancore-Cochin Malabar sea coast from Nagarcoat to Cannanore in which forest plants grew, died and remained buried under fresh deposition of sediments. These buried plants were carbonised. Sediments containing carbonised plant materials, presently known as the Varkala sandstone which was overlaid by compact grew coloured Quilon limestone containing fossils.

Mud banks occur along the sea coast from the mouth of the Kotta river to Cape Comorin, the most remarkable being those at Pantalayini-Kollam, Calicut and Alleuppy. The mud is peculiar. It is dark green in colour, and clayey and oily to touch. These mud banks are carried along littoral currents. But at times, their smooth surface is disturbed by mud volcanoes from which mud and water come bubbling up from below, often bringing with them dead fish.

The product of weathering of rocks constituting the western Ghat mountains have been carried by numerous rivers draining them and deposited along their paths and at their mouths. So the sea-beds have been raised to form the low-lying inland areas. For example, the low-lying areas of Kuttanad are believed to have once been part of the sea and later filled up by the silt carried by the Pumpa and other rivers.

SELECTED READINGS

1. Chaitanya, Krishan, (1972), *Kerala*, National Book Trust of India, New Delhi-110006.
2. Datta, A.K. (1984), *An Introduction to Physical Geology*. 15th Ed., Kalyani Publishers, 24 Daryaganj, New Delhi-110002.
3. Daji, J.A. (1968), *A Textbook of Soil Science*, Media Promoters and Publishere Pvt. Ltd., Bombay.
4. Krishnan, M.S. (1982), *Geology of India & Burma*, CBS Publishers and Distributors, 485 Bhola Nath Nagar, Shahdara, Delhi-110032.
5. Mehdiratta, R.C. (1985), *Geology of India, Pakistan, Bangladesh and Burma*, Revised by Dr. Chandha; Atma Ram & Sons, Delhi.
6. Sinha, R.K. (1975), *An Introduction to Geology*, Vikas Publishing House, Pvt. Ltd., 5 D aganj, Ansari Road, New Delhi-110002.
7. Spenser. E.W. (1962), *Basic Concept of Physical Geology*, Oxford and IBH Publishing Co., New Delhi.

2
SOIL FORMING MINERALS AND ROCKS

The earth's crust may be regarded as the vast laboratory of nature where varieties of minerals have been produced during several geological periods. Nearly 2000 minerals have been detected in the earth's crust but only a few constitute soil forming rocks. These rocks were disintegrated to form the mineral which were decomposed to form the soil. The nature of the soil basically depends upon the nature of minerals present in them: Rocks are identified on the basis of minerals they contain. So the minerals should be studied before studying the soil forming rocks.

2.1. DEFINITION AND BROAD CLASSIFICATION OF MINERALS

A mineral is a naturally occurring inorganic material that has fairly definite internal structure, composition and properties. Soil forming minerals have been broadly grouped as primary and secondary minerals. Primary minerals originate in the parent rock. Some of them remain unchanged during the course of soil formation. They are usually present in the sand fraction. Secondary minerals have formed from the alteration and decomposition of primary minerals and subsequent recombination of the products of their decomposition. They are also called clay minerals because they are the chief constituent of clay.

2.2. GENERAL CONSTITUTION OF MINERALS

Ions systematically occur to make the basic building blocks of crystals of minerals under natural conditions. Most of the minerals are made up of silica tetrahedra (singular-tetrahedron) and aluminium octahedra (singular-octahedron). A few of them are built up of silica tetrahedra only.

One silicon ion occupies the central space between four oxygen ions that are closely packed to form a silica tetrahedron. A tetrahedron is a geometrical figure that has four surfaces as shown diagramatically in Fig. 2.1B. All the basal oxygens of one silica tetrahedron have been shared by silicon ions of the neighbouring silica tetrahedra to make a continuous silica tetrahedral layer of minerals. One aluminium or magnesium ion occupies the central space between six hydroxyl ions that one closely packed to form an aluminium or magnesium octahedron which has eight surfaces. The aluminium octahedron is called a gibbsite. The magnesium octahedron is called a brucite. Gibbsite has been diagramatically represented in Fig. 2.1A. All the hydroxyl ions of one aluminium/magnesium octahedron have been shared by the aluminium/magnesium ions of the neighbouring octahedra to make a continuous sheet of aluminium/magnesium octahedral layer of minerals. These silica tetrahedra and aluminium/magnesium octahedra are joined together to form minerals, when the top oxygen of the silica tetrahedron occupies the position of one of the hydroxyl ions of the aluminium/magnesium octahedron. One aluminium octahedron has been shown to be joined with one silica tetrahedron from below in Fig. 2.1C and from below and top in Fig. 2.1D. Some silicon ions are replaced from the tetrahedral position of some minerals by aluminium ions. Some

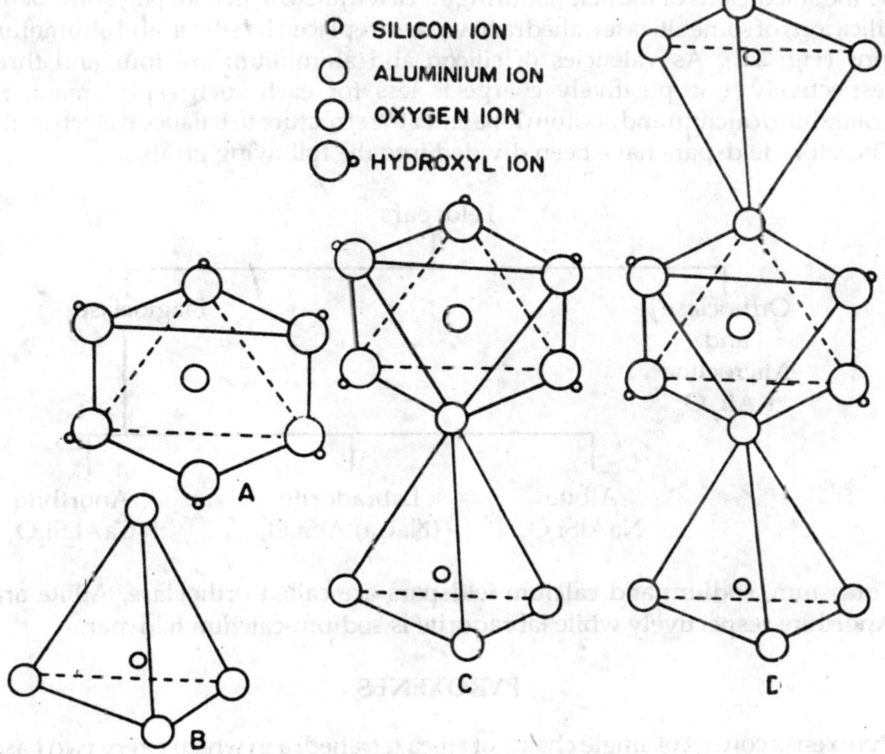

Fig. 2.1. Structures of silica tetrahedron and aluminium, octahedron and how they are joined togeher to form the minerals. A. One aluminium ions equidistant from six hydroxyl ions to form one aluminium octahedron (Gibbite). B. One silicon ion is equidistant from four oxygen ions to form a silica tetrahedron. C. One aluminium octahedron is joined with one silica tetrahedron at the bottom when the top oxygen if the silica tetrahedron occupies the position of one of the hydroxyl ions of the aluminium octahedron. D. Similarly, the aluminium, octahedron is joined by two silica tatrahedron, one at the top and other at the bottom.

aluminium ions are replaced from the octahedral position by magnesium, ferrous, ferric and zinc ions. Minerals become negatively changed as a consequence of the above process and therefore attract positively charged cations.

2.3. IMPORTANT SOIL FORMING PRIMARY MINERALS

A few very important soil forming primary minerals which are the constituent of important soil forming rocks are briefly described below:

FELDSPARS

Feldspars consist of a continuous three-dimensional framework of silica tetrahedra where all the four oxygens of one silica tetrahedron have been shared by the silicon ions of the neighbouring silica tetrahedra. In feldspar, some of the silica ions of some silica tetrahedra have been replaced by silicon and aluminium ions (Fig. 2.2). As valencies of silicon and aluminium are four and three respectively, one positively charge is less for each such replacement. So potassium, calcium and sodium ions enter the structure to balance it electrically. Therefore feldspars have been divided into the following groups.

Feldspars
- Orthoclase and Microcline $KAli_3 O_8$
- Plagioclase
  - Albite $NaAlSi_3O_8$
  - Labradorite (NaCa) $AlSi_3O_8$
  - Anorthite $CaAl_2Si_2O_8$

Potassium, sodium and calcium feldspars are called orthoclase, Albite and Anorthite respectively while labradorite is sodium-calcium feldspar.

PYROXENES

Pyroxenes consist of single chains of silica tetrahedra in which every two basal oxygen ions of each silica tetrahedron are shared by silicon ions of the neighbouring silica tetrahedra as shown in Fig. 2.3A. The unsatisfied negative charges of the third basal oxygen ions and the top oxygen ions are satisfied by mostly by calcium, magnesium and ferrous ions which connect one pyroxene chain with another. A dark coloured mineral called Augite of formula $Ca_2(AlFe)_4$ $(Mg\ Fe)_4\ Si_6\ O_{24}$, belongs to the pyroxene group of minerals which also include hypersthene and enstatite.

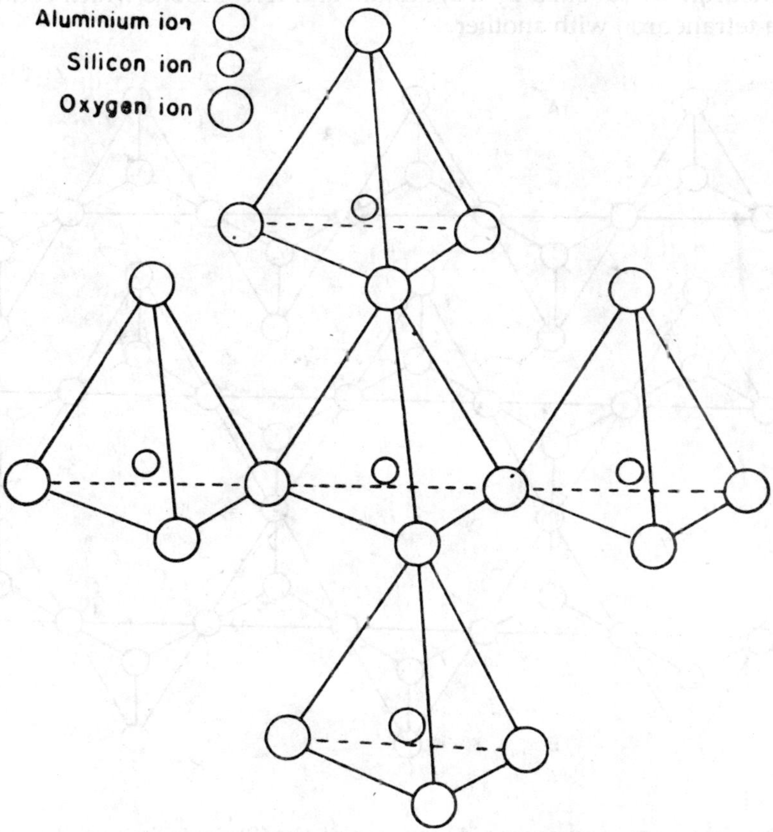

Fig. 2.2. Structure of Feldspar. All the oxygen ions of the central tetrahedron are common to it and the neighbouring tetrahedra. The silicon of the bottom tetrahedron has been replaced by the aluminium ion.

AMPHIBOLES

Amphiboles consist of a double chain of silica tetrahedra where every two and every three basal oxygen ions of each silica tetrahedron are alternately shared by silicon ions of the neighbouring silica tetrahedra as shown in Fig. 2.3B. The unsatisfied negative charge of the third basal oxygen ions and the top oxygen ions are satisfied mostly by calcium, magnesium and ferrous ions which connect one amphibole chain with the another. A dark coloured mineral, horoblende of formula $Ca_2Al_2Mg_2Fe_3Si_6O_{22}(OH)_2$ belongs to the amphibole group.

OLIVINE

Olivine consist of independent silica tetrahedra in which oxygen ions of one silica tetrahedron are not shared by the silicon ions of the neighbouring silica tetrahedra. The unsatisfied negative charge of the oxygen ions of one silica tetrahedron are satisfied by magnesium and ferrous ions which connect one silica tetrahedron with another.

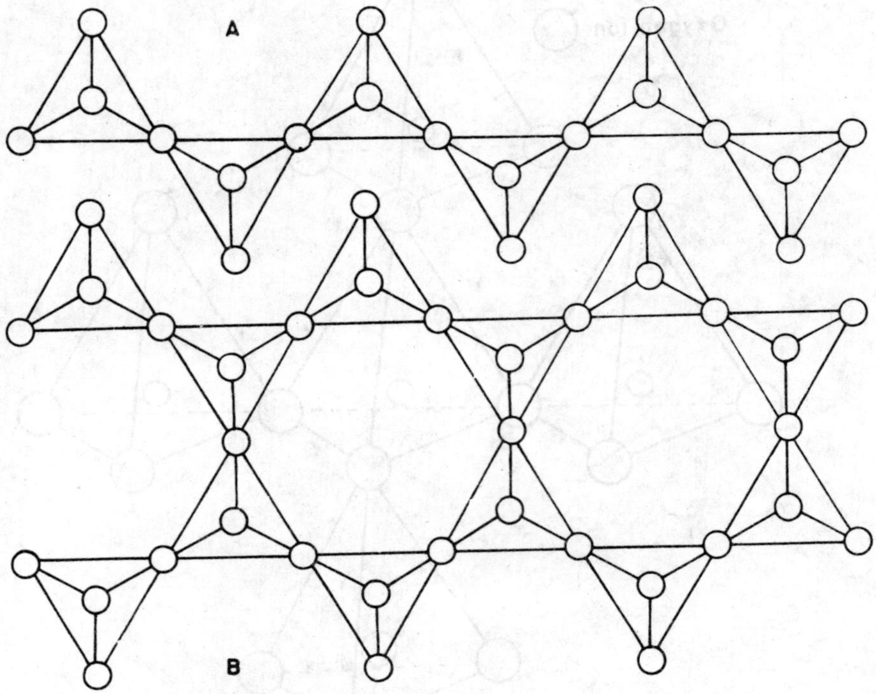

Fig. 2.3. Structures of pyroxenes (A) and amphiboles (B) single chain of silica tetrahedra of pyroxenes (A) and double silica chain of amphiboles (B). The silicon ions have been covered by the top oxygen ions and therefore silicon ions are not seen.

MICA

In the case of mica, one aluminium octahedral layer has been joined with two silica tetrahedral layers, one at the top and the other at the bottom as shown in Fig. 2.4A and B when the top oxygen ions of the silica tetrahedral layers replaces one of the hydroxyl ions of the aluminium octahedral layers. So these oxygen ions are common to tetrahedral and octahedral layer as shown in Fig. 2.4A and B. One-fourth of the silicon ions have been replaced from the tetrahedral position by aluminium ions. The valency of Si^{4+} is 4 and valency of Al^{3+} is 3. Hence there is one positive charge less for replacement of each silicon ion with one aluminium ion. Thus potassium ion occurs between the sheets of mica to balance it electrically. When two-thirds of the octahedral positions are occupied by aluminium ions and one-third remains empty as shown in Fig. 2.4B, the mica is called white mica, muscovite or dioctahedral mica. When all the octahedral positions are occupied by magnesium and/or ferrous ions, it is called black mica, biotite or trioctahedral mica as shown in Fig. 2.4A.

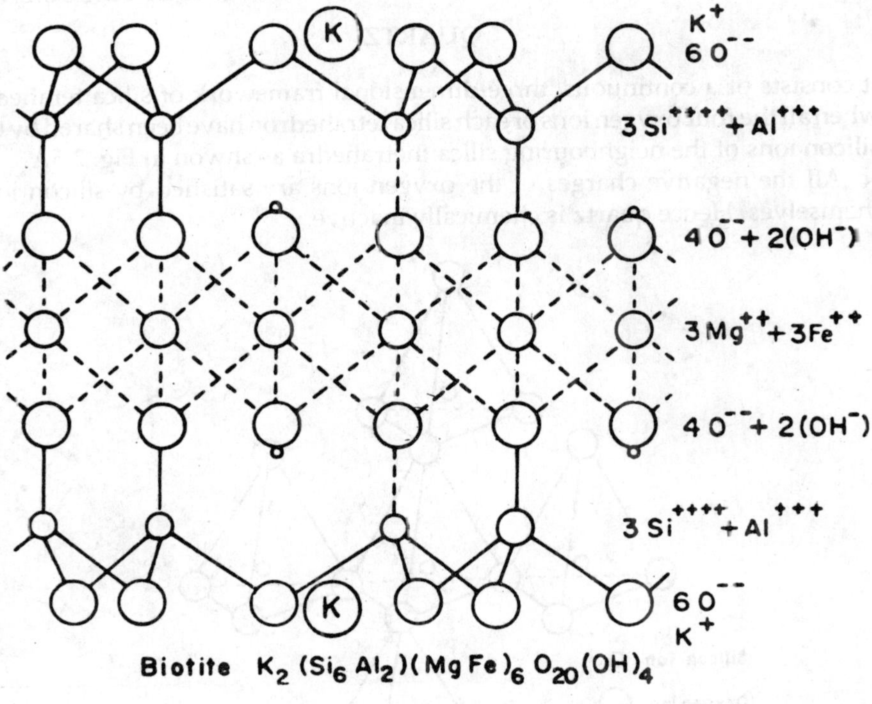

Fig. 2.4A. Structure of biotite mica.

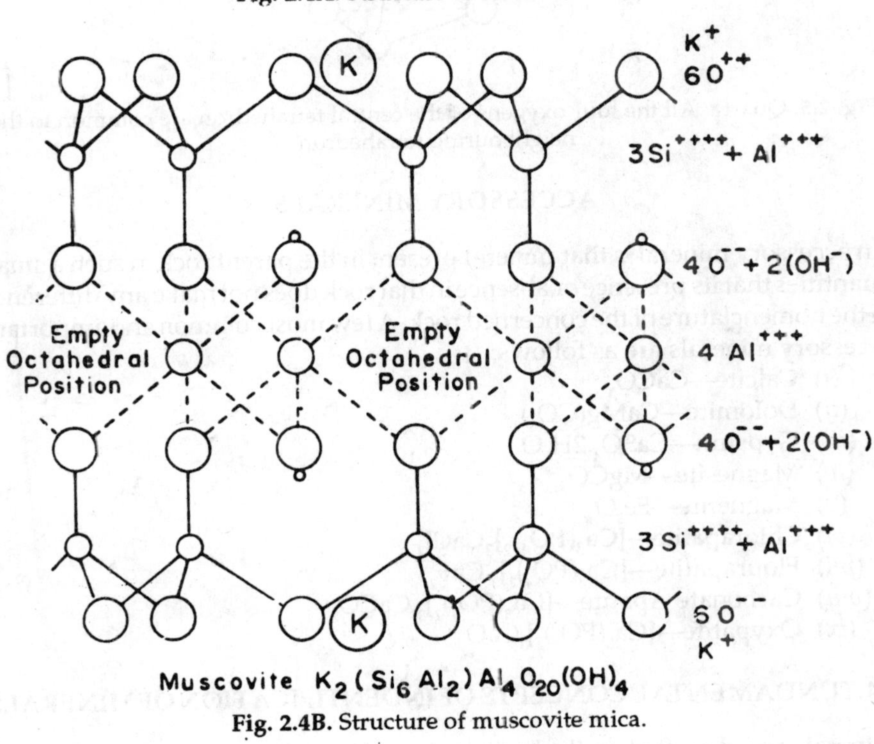

Fig. 2.4B. Structure of muscovite mica.

QUARTZ

It consists of a continuous, three-dimensional framework of silica terahedra when all the four oxygen ions of each silica tetrahedron have been shared by the silicon ions of the neighbouring silica thetrahedra as shwon in Fig. 2.5A.

All the negative charges of the oxygen ions are satisfied by silicon ions themselves. Hence quartz is chemically inactive.

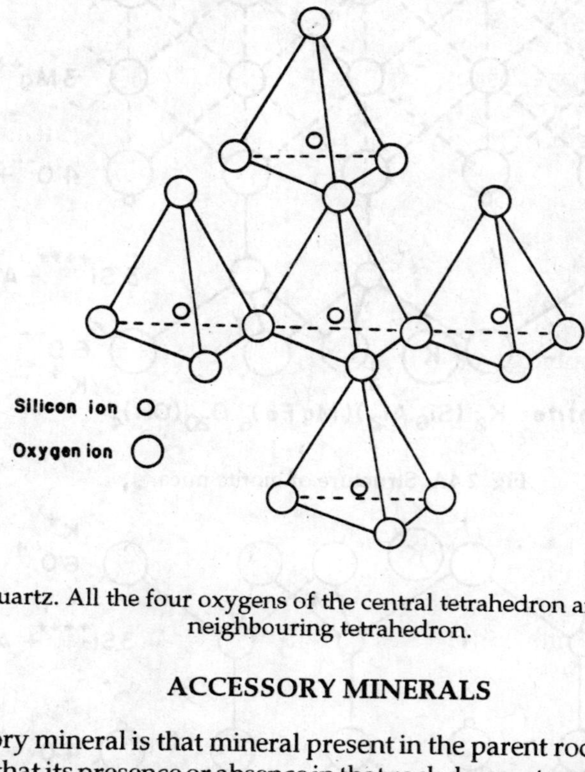

Fig. 2.5. Quartz. All the four oxygens of the central tetrahedron are common to the neighbouring tetrahedron.

ACCESSORY MINERALS

An accessory mineral is that mineral present in the parent rock in such a small quantities that its presence or absence in that rock does not make any difference in the nomenclature of the concerned rock. A few most common and important accessory minerals are as follows:

(*i*) Calcite—$CaCO_3$
(*ii*) Dolomite—$CaMg(CO_3)_2$
(*iii*) Gypsum—$CaSO_4.2H_2O$
(*iv*) Magnesite—$MgCO_3$
(*v*) Magnetite—Fe_3O_4
(*vi*) Chlorapatite—$[Ca_3(PO_4)_2]_3CaCl_2$
(*vii*) Flourapatite—$[Ca_3(PO_4)_2]_3.CaF_2$
(*viii*) Carbonate Apatite—$[Ca_3(PO_4)_2]_3.CaCO_3$
(*ix*) Oxypatite—$[Ca_3(PO_4)_2]_3CaO$

2.4. FUNDAMENTAL CONCEPTS OF INDENTIFICATION OF MINERALS

Minerals are identified on the basis of their physical properties, and chemical composition. They are also examined as thin polished sections under the petrographic microscope. The following characters of minerals are noted in the laboratory.

1. *Nature of the mineral crystal*: A crystal is a homogeneous body bound by smooth plane surface. A symmetry line or axis of a crystal is an imaginery line through a crystal, around which the crystal may be rotated when two or more identical views of the crystal are seen in one rotation. Crystals of a mineral may be one of the following three forms: (*i*) Euhedral crystals with perfectly developed faces; (*ii*) Subhedral crystals with imperfectly developed faces, and (*iii*) Anhedral crystals with no developed faces.

The axes of the crystal may be either of the following:

(*i*) Isometric or cubic crystal with three mutually perpendicular axes of equal length.
(*ii*) Hexagonal crystal with four axes, three of which are of equal length intersecting each other at 120° angle, with a fourth axis of different length, perpendicular to the plane of the first three axes.
(*iii*) Tetragonal crystals with three mutually perpendicular axes, two horizontal axes of equal length and the vertical axis of a different lengths.
(*iv*) Orthorhombic crystals with three mutually perpendicular axes of different lengths.
(*v*) Triclinic crystals with three axes of different lengths. They are inclined at an angle with respect to one another.
(*vi*) Monoclinic crystals with three axes of different lengths. One of them is incline a with respect to the other two.

2. *Form*: The degree of the development of the crystals of the minerals may be as follows:

(*i*) Crystallised minerals with well formed big crystals of a regular shape.
(*ii*) Crystalline minerals lacking an outward regular form although they possess a distinct internal molecular structure.
(*iii*) Crypto-crystalline where the crystals of the mineral can only be seen in the thin section under the microscope.
(*iv*) Amorphous: Crystals are totally undeveloped.

3. Specific Gravity of a mineral is the ratio of the weight of a given volume of the mineral to the weight of an equal volume of water.

4. Cleavage of minerals is the property of breaking along definite planes as described below:

(*i*) Perfect or distinct cleavage where the broken surface is very smooth.
(*ii*) Imperfect or indistinct cleavage where the broken surface is not smooth.

5. The hardness of a minerals is a measure of its ability to scratch another mineral, which has been placed lower in the scale of hardness than the concerned mineral. Mohs' scale of hardness of different is as follows:

(*i*) Talc	(*ii*) Gypsum	(*iii*) Calcite
(*iv*) Flourite	(*v*) Apatite	(*vi*) Feldspar
(*vii*) Quartz	(*viii*) Topaz	(*ix*) Corundum
(*x*) Diamond		

Only the fresh surface of the mineral should be used for scratching.

6. The lustre of a mineral is the appearance of its surface in the reflected light as described below:

(*i*) *Adamantine*: The surface is like that of a diamond.
(*ii*) *Vitreous*: The surface is like that of a glass.
(*iii*) *Resinous*: The surface is like that of the yellow resin.
(*iv*) *Greasy*: The surface is like that of the grease.
(*v*) *Pearly*: The surface is like that of the pearl.
(*vi*) *Silky*: The surface is like that of the silk.

7. *Colour*: Minerals may be different colours.

8. *Streak*: The mineral is rubbed against an unglazed porcelain plate. The colour of the powder of the mineral that is obtained by rubbing it against the unglazed porcelain plate is known as the streak of the mineral.

9. Fracture is the nature of the surface produced when the mineral breaks in a direction other than the cleavage plane. The broken pieces may exhibit:

(*i*) concave surface when the fracture is called conchoidal.
(*ii*) irregular surface when the fracture is called uneven.
(*iii*) somewhat smooth surface when the fracture is called even.
(*iv*) very much irregular surface when the fracture is called hackly
(*v*) splinter like parts when the fracture is called splintery.

10. Twining occurs when the crystals of the mineral have grown together or they penetrated each other.

11. *Miscellaneous physical properties*: A few minerals may be identified by their magnetic properties. Some may flash different colours whenever they are turned. Some have the property of fluorescence, which means they emit light from within when exposed to ultra violet light. Some may be identified by their taste.

2.5. DESCRIPTION AND COMPOSITION OF ROCK/SOIL FORMING MINERALS

1. *Pyroxene*: The pyroxene group includes: (*i*) Enstatite, (*ii*) Hypersthene, (*iii*) Augite, (*iv*) Diopside and (*v*) Rhodonite. Enstatite has orthorhombic crystal. Augite and Diopside have monoclinic crystals. Rhodonite has triclinic crystals. They may be of massive (heavy or bulky) lamillar (thin leaf like sheets) or granular forms. They have two sets of cleavages, nearly at right angles to each other, and an uneven fracture. They are silicates of calcium, magnesium, iron, aluminium and sodium. The formula of augite is (CaNa) (MgFeAlTi) Al_2O_6. Its colour varies from dark green to black. Its streak is whitish grey. It is of vitreous lustre. Its hardness is six. Its density varies from 3.2 to 3.5. Enstatite ($MgSi_2O_6$) is of light green colour, white streak and vitreous lustre. Its density varies from 3.2 to 3.9.

2. *Amphibole*: The amphibole group includes Hornblende, Actinolite, Tremolite, Anthophyllite and Aenigmatite. Hornblende, Actinolite and Tremolite possesses a monoclinic crystal system and Aenigmatite possesses a triclinic crystal system. They are usually massive (bulky and heavy) but may also be fibrous or granular.

These minerals have a perfect cleavage (two sets inclined at angles 124° and 56°) and an uneven fracture of black, green or greyish colour. The Hardness is 5 to 6 and specific gravity is 2.9 to 3.5. They are similar to pyroxenes in composition by they contain a little water. Formula of Hornblende is $Ca(MgFe)_4$ $Al(SiAl) O_{22}(OH)$.

3. *Feldspars*: Feldspars include potash feldspars or orthoclase and microcline and soda-lime feldspars or plagioclase feldspar which includes sodium feldspar or albite and calcium feldspar or anorthite.

(*i*) Orthoclase and microcline. Orthoclase possesses monoclinic crystals and microcline processes a triclinic crystal system. The colour of orthoclase is white, pink or gray. Microcline is cream green coloured. Both are massive (bulky and heavy) in form and have an uneven fracture, and are uncoloured. The lustre of both is vitreous. The specific gravity and the hardness of both are 6 and 2.55 respectively. Their formula is $KAlSi_3O_8$.

(*ii*) Plagioclase feldspars are all twined. Albite ($NaAlSi_3O_8$), labradorite or andesine $NaCa(AlSi_2O_8)$ and anorthite ($CaAl_2Si_3O_8$) belong to plagioclase feldspar. They are of triclinic crystal ststem, massive (bulky and heavy) form, perfect cleavage (two sets are nearly at right angles to each other) uneven fracture, uncoloured streak and vitreous lustre 6 to 6.5 hardness and of specific gravity of 2.60 to 2.76.

4. *Mica*: Mica includes muscovite and biotite, both of which are of monoclinic crystal system, perfect cleavage, flaky or scaly form and of pearly lustre. Fracture is not obtainable in both the cases due to good cleavage.

(*i*) Muscovite is colourless, or of grayish or pinkish colour, uncoloured streak, 2 to 2.25 hardness, with a specific gravity varying from 2.7 to 3.0. It is of pearly to vitreous lustre. Its formula is $K_2(LiAl_2)(Al_4)O_{20}(OH)_4$.

(*ii*) Biotite is black or dark green in colour, with an uncoloured streak. It is often grayish or brownish due to decompostion. Hardness is 2.5 to 3.0. It specific gravity varies from 2.7 to 3.1. The formula $K(Si_6Al_2)(MgFe)_6O_{20}(OH)_4$. Both muscovite and biotite are transparent but the former is of pearly to vitreous lustre whereas the latter is of vitreous to a pearly lustre.

5. *Olivine*: It is green or yellowish green is colour, yet lacks colour in the streak. It is of massive form, as an imperfect cleavage, conchoidal fracture and vitreous lusture, with a hardness of 6.5 to 7.0 and specific gravity of 3 to 3.4. The formula is $(MgFe)_2SiO_4$.

6. *Calcite*: Calcite commonly occurs as hexagonal crystals but sometimes occurs in fibrous, granular or nodular (aggregate of big rounded bodies) form. White or colourless calcite has a perfect cleavage, vitreous lustre, the hardness is 3 the specific gravity 2.7 and the formula $CaCO_3$.

7. ***Dolomite***: It has a hexagonal crystal system and granular form, uneven to conchoidal fracture. It is of usually white or sometimes pinkish and has a white streak. The hardness is 3.5 to 4.0 The specific gravity is 2.8 and the formula is $CaCO_3.MgCO_3$.

8. *Gypsum*: It usually occurs in the monoclinic crystal form, but may also occur in the granular, fibrous or massive form. It has a perfect cleavage and

uneven fracture. It is white greyish white or pinkish white colour, white streak and vitreous lustre. The hardness and the specific gravity of gypsum are 2 and 2.3 respectively. The formula is $CaSO_4.2H_2O$.

9. *Kaolin or Kaolinite*: Monoclinic crystals have a perfect cleavage, and uneven fracture. It easily breaks down to powder. It is of white, greyish white or brownish white colour, with a white streak and dull lustre. the specific gravity is 3.2 and the hardness is 5.

10. *Tourmaline*: It occurs in the prismatic form. Hexagonal crystals are striated length wise. It has an imperfect cleavage, uneven to somewhat conchoidal. It is black, green, red or brown has an uncoloured streak and a vitreous lustre. The hardness and specifies gravities are 7.5 and 3.0 respectively. It is complex borosilicate of aluminium with a little magnesium, iron and alkali metals.

11. *Garnet*: Isometric crystals of garnet occur in the granular form. They are very heavy with a conchoidal fracture and no cleavage. They are black or green and have an uncoloured streak and vitreous lustre. The hardness and the specific gravity are 7.0 and from 3.2 to 4.3 respectively. Formula is $(CaFeMgAl)SiO_4$.

12. *Bauxite*: It is amorphous and occurs in the granular, earthy, concretionery, colitic (aggregate of bodies resembling fish roe) or pisolitic (aggregate of pea like masses from with no cleavage and sub-conchoidal fracture. It is greyish white or brownish white in colour with a dull or earthy lustre. The hardness and the specific gravities are 2 and 2.5 respectively. The formula is $Al_2O_3.2H_2O$.

13. *Chlorite*: It is micaceous in appearance due to its perfect cleavage. The crystal are hexagonal tabular and often bent. Green chlorite occurs as a scally coating on other minerals. The formula is $Mg_6(OH)_{12}(Si_6Al_2)(Mg_3Fe_3)O_{20}(OH)_4$.

14. *Serpentine*: It occurs as fine granular masses sometimes fibrous. It is dark green and has greasy, waxy or silky lustre and hardness of 2.5 to 3.0. The formula is $Mg_6(OH)_8Si_4O_{10}$.

15. *Talc*: It is pearly, greasy or waxy lustre and soapy or greasy feel. Its of white, sea green or grey colour. The formula $Mg_3(OH)_2Si_4O_{10}$.

16. *Chalcopyrite*: Tetragonal crystal of chalcopyrite commonly occur in the massive form with indistinct cleavage and uneven fracture. It is of golden yellow colour, has a greenish black streak and metallic lustre. The hardness and the specific gravity are 3.5 to 4 and 4.2 respectively. The formula is $CuFeS_2$.

17. *Malachite*: Crystals of Malachite are very rare. It usually occurs in massive, botryoidal (aggregate of grape like masses) fibrous or encrusting form, with an indistinct cleavage and conchoidal fracture. It is green, with a pale green streak, an often silky or otherwise dull lustre. The hardness is 3.5 to 4.0, the specific gravity is 4.0 and the formula is $CuCO_3.Ca(OH)_2$.

18. *Azurite*: Monoclinic crystals of azurite are very fare. It is usually of massive and earthy form with an indistinct cleavage and conchoidal fracture, deep sky blue colour and pale blue streak and vitreous to adamantive lustre. The hardness is 3.5 to 4. The specific gravity is 3.8. The formula $2CuCO_3.Cu(OH)_2$.

19. *Pyrite*: Brass yellow coloured pyrite occurs in straited cubic crystals, massive, fibrous, reinform (kidney-shaped) or granular form. The crystal

system is isomeric. It has greenish black streak, metallic lustre 6 to 6.5, and hardness and specific gravity are both 5. The formula is FeS_2.

20. *Hematite*: Hexagonal crystals of hematite commonly occur in the tabular form. It also occurs in reinform, botryoidal, scally or foliaceous form. It has an indistinct cleavage and uneven fracture. It is of steel and metallic lustre. Its hardness varies from 5.5. to 6.5 while its specific gravity is 5.2. The formula is Fe_2O_3.

21. *Limonite*: Amorphous limonite occur in mammilary (aggregate of masses bigger than that of grapes) or botryoidal (aggregate of grape like masses or fibrous or concretionary or earthly or massive form. The cleavage is absent but the fracture is uneven. It is brown, has a yellowish brown streak and dull or silky lustre. The hardness and the specific gravity are 5.5 and 4.0 respectively. The formula is $2Fe_2O_3.3H_2O$

22. *Magnetite*: Isomeric crystals of magnetite are usually or octahedral form, but they may also occur in massive or granular form. It is magnetic with an indistinct cleavage and uneven fracture. Colour and streak and both black iorn. It has a metallic to dull lustre. The hardness is 5.5 to 6.5 and the specific gravity is 5.1. The formula is Fe_3O_4.

23. *Magnesite*: It is a hexagonal crystal system and massive compact or earth form with perfect cleavage and conchoidal fracture colour is white or greyish white or brownish white. Streak is white. It is white. It is of vitreous 4.0 to 4.5 hardness and a specific gravity of 3. The formula is $MgCO_3$.

24. *Ilmenite*: A hexagonal crystal system of ilemenite occur tabulars massive, compact, granular or loose form as sand with indistinct cleavage and conchoidal fracture. It is of iron black colour, black or brownish black streak and sub-metallic lustre. The hardness and the specific gravity varies from 5 to 6 and from 4.5 to 5.0 respectively.

25. *Montmorillonite* is greyish yellow in colour, with a white streak and dull lustre. Its hardness varies from 2.0 to 2.5 and its specific gravity varies from 2.0 to 2.7.

26. *Glauconite* is of green to brown in colour, greenish grey in streak, earthy to dull lustre. It is opaque. Its hardness and specific gravity are about 2.5 and 2.65 respectively.

2.6. DEFINITION OF ROCKS

The earth's crust is covered by very hard solid materials called rocks which usually consist of minerals in different amounts and combinations. Ore is that rock which contains large amounts of industrially useful metals. Most of the rocks are made up of large crystals of different minerals. A few rocks contain loose grains of the same mineral. Rocks are identified on the basis of certain characteristics like colour, cleavage, fracture, mineralogical composition, structure, density etc. which however, are not as rigid as in the case of minerals.

2.7. CLASSIFICATION OF ROCKS

Rocks have been broadly classified as igneous, sedimentary and metamorphic rocks on the basis of their origin.

All rocks that originated from cooling and subsequent solidification of the molten material called magma, are known as igneous rocks (igneous means born from fire) which have subdivided into two sub-classes intrusive rocks and extrusive or volcanic rocks. Magma solidified below the surface of the earth to form the intrusive rocks magma solidified at a very great depth inside the earth to form the plutonic rocks. Magma solidified on the surface of the eartn to form extrusive or volcanic rocks. Both extrusive and intrusive rocks have also been grouped on the basis of silica contents as acidic intermediate, basis and ultra basic rocks.

Acidic rocks contain more than	66% Silica
Intermediate rocks contain	52 to 66% Silica
Basic rocks contain	45 to 52% Silica
Ultrabasic rocks contain less than	45% Silica

Existing rocks were disintegrated and decomposed by the force of nature *i.e.* flowing water, glaciers, frost, atmospheric gases. The product of disintegration and decomposition of older frocks were deposited in layers that were recemented to form the sedimentary rocks.

Metamorphic rocks are those rocks which have been changed or metamorphosed by high temperature and pressure and chemical action magnetic fluid after their original formation when they were buried very deep below the surface of the earth. Igneous rocks were metamorphosed to form the orthometamorphic rock whereas sedimentary rocks were metamorphosed to form the parametamorphic rocks.

2.8. ORIGIN OF IGNEOUS ROCKS

The earth cooled down some 4600 million years ago to form molten silicate material called magma, that contained varying amounts of water and highly reactive fluids. The magma rapidly cooled down at the surface of the earth to form extrusive igneous rocks of fine texture because the molecules of the minerals did not get enough time to form large crystals of minerals with will defined faces. Hence an insulating mat of extrusive igneous rock was formed at the surface of the earth. Below this pressure was exerted and therefore the magma slowly cooled down to form intrusive or abyssal igneous rocks of coarse texture because the gases could not escape and the minerals got enough time to arrange themselves to form large crystals of minerals with well defined faces. The intrusive rocks formed due to the cooling down and solidification of the magma within the vertical cracks between the existing rocks below the surface of the earth are called the dykes. Similarly, intrusive rocks formed within the horizontal cracks and narrow and irregular cracks are known as sills and veins respectively. The magma cooled at a very great depth below the surface of the earth to form the plutonic igneous rock.

Igneous rocks were buried under high pressure and temperature very deep in the earth. When the pressure was decreased, the hot solid rock melted to form the magma, which was forced out to the surface of the earth in the form of lava which cooled down rapidly to form volcanic or extrusive rocks of a fine texture.

Extrusive and intrusive rocks have not been clearly demarcated. The hypabyssal rocks are intermediate in origin between the extrusive and the intrusive rocks. They are of intermediate texture.

2.9. ORIGIN OF SEDIMENTARY ROCKS

The term sediment means to anything that settles down in water or wind. Sediments originated by mechanical, chemical and organic processes. In the mechanical process, rocks were gradually disintegrated to smaller pieces and ultimately to the constituent minerals, which were carried away by running water or a blowing wind elsewhere, where they were deposited in layers according to the densities, which means that the heavier particles settled down first followed, by lighter particles. Then these layers were cemented or lithified at a relatively lower temperature and pressure to form the sedimentary rocks. Lithification is a process that converts newly deposited sedimentary layers into compact sedimentary rocks by the combined effects of pressure, heat and cementing action. The sedimentary rocks of mechanical origin are more or less similar to the original igneous rocks in mineralogical and chemical composition, although some proportion of easily decomposable minerals of the igneous rocks had dissolved, increasing the proportion of resistant minerals in the sedimentary rocks. For example, quartz is the chief constituent of both granite, and igneous rock and sandstone, a sedimentary rocks. Granite contains about 66 per cent silica whereas sandstone contain about 79 per cent silica.

In the chemical process, minerals were gradually dissolved in water and carried away elsewhere where the solution became concentrated by evaporation when the minerals crystallised, and settled down in layers that were cemented or lithified to form sedimentary rocks e.g. limestone.

In the organic process, certain kinds of marine plants and animals and their excreta were deposited at the bottom of the ocean bed in layer which were lithified to form sedimentary rocks e.g. chalk.

2.10. ORIGIN OF METAMORPHIC ROCKS

Rocks were buried several thousand feet deep below the surface of the earth where they were subjected to a high temperature and the chemical action of magmatic fluids. Consequently, existing minerals decomposed to form new minerals that arranged themselves parallel to the plane of application of high pressure. This parallelism in texture is known as schistosity. New kinds of rocks were formed. These were known as schistose rock of metamorphic rock. The banding produced in the schistose rock is known as foliation. During the course of formation of metamorphic rocks, less stable minerals were gradually converted to more stable minerals. Metamorphic rocks might have undergone several cycles of metamorphism each time, making new adjustments in the constituent minerals and texture until they finally became stable although the total chemical composition might not have changed. For example, a bed of limestone or dolomite might have changed into marble, shale into phyllite and slate to mica schist depending upon the intensity of heat and pressure.

2.11. SOIL FORMING IGNEOUS ROCKS

Hot molten matter must contract on cooling and solidifying. So all igneous rock have cracks or joints. The fabric of an igneous rocks is determined by the physical state and arrangement of the material that solidified from the magma to form the rock. An igneous rocks in which minerals form an interlocking system is a crystalline igneous rock. The texture of igneous rocks means the sizes of grains or crystals of minerals present in them. If they can be seen with the naked eye, the rock is said to have a coarse medium or fine grained texture depending on the size of their crystals. If the mineral crystal of the rock can only be seen with the help of a microscope, then the concerned rock possesses the aphinitic texture. Some igneous rock possess a porphyritic texture, which means that large perfectly developed crystals called euthedral crystals, are enclosed within a ground mass of very fine grains.

In the past, lighter coloured igneous rocks containing more than 66 per cent silica and relatively lower amount of basic elements were called acidic rocks, and darker igneous rocks containing 45-52 per cent silica and relatively higher amount of basic elements were called basic rocks because the former weathered to form relatively coarser textured soils poorer in basic elements and acidic in reaction and the latter weathered to form relatively finer textured soils, richer in basic elements and alkaline in reaction. Today, acidic rocks are also called felsic rocks and basic rocks are also called mafic rocks. A few most important soil forming igneous rocks are as follows:

(*i*) *Granite* is a coarse to medium textured plutonic or intrusive acidic rock. It mainly consists of orthoclase and quartz, and minor amounts of biotite or muscovite and hornblende, rarely augite. The minerals occur as interlocking irregular crystals but linear and graphic fabrics are sometimes found. As orthoclase and quartz is the most abundant mineral in granite, the rock is white light gray or pink like the minerals.

(*ii*) *Rhyolite* is made up of the same minerals as granite but has a fine, glassy or porphyritic texture, Rhyolite is usually light, may be white, light grey or various shades of red. A characteristic textural feature is a streaked pattern known as flow banding which results from the concentration of coloured material or glass in layers during the flow of highly viscous lava.

(*iii*) *Syenite and Trachytes* are almost entirely composed of orthoclase and a small amount of dark minerals like hornblende, augite and biotite. So their colour is blackish grey or blackish pink. Syenite is a coarse to medium textured equiangular plutonic rock, but trachyte is a fine textured volcanic rock.

(*iv*) *Diorite* is a coarse to medium textured dark grey, plutonic rock whose chief constituent is plagioclase feldspar. It also contain hornblende or biotite. These dark minerals may be as abundant as plagioclase.

(*v*) *Andesite*: It is a grey to greyish black fine grained volcanic rock of porphyritic texture. It contains plagioclase, augite, hornblende and biotite.

(*vi*) *Gabbro* is a coarse textured plutonic basic rock which consist of coarsegrained inter growth of cystals of anorthite, hornblende or augite. Many gabbro varieties contain olivine and are therefore coloured greenish black. Typical gabbro contains either more augite and hornblende than anorthite or equal amounts of augite and / or hornblende and anorthite and is therefore dark grey in colour.

(*vii*) *Anorthosite* consists almost entirely of anorthite, the interlocking coarse textured and therefore rendering it white or grey in colour. It is a volcanic basic rock.

(*viii*) *Pyroxenite* is a coarse textured plutonic ultrabasic rock that consists almost entirely of pyroxene, and is therefore black in colour.

(*ix*) *Hornblendite* is a coarse textured plutonic ultrabasic rock consisting almost entirely of hornblende. It is black in colour.

(*x*) *Peridotite* is coarse, medium textured ultra basic equiangular rock that consists almost entirely of olivine and a little pyroxene and is therefore greyish green in colour.

(*xi*) *Dunite* is a coarse textured plutonic ultrabasic rock that consists of only olivine and therefore is green in colour.

(*xii*) *Basalt* is a volcanic basic igneous rock that consists of anorthite, augite and hornblende. It is fine grained to aphinitic in texture and coal black to dark grey in colour. Many varieties also contain olivine. Basalt is commonly frothy and cellular and is filled with numerous holes called vesicles where gas bubbles were trapped in the molten lava when it solidified. Such a structure is known as secoriaceous.

2.12. SOIL FORMING SEDIMENTARY ROCKS

One of the outstanding characteristics of sedimentary rocks is their arrangement into layers where sediments of different textured and composition were deposited when the velocity and direction of the stream and the wind changed. Sedimentary rocks usually contain fossils. Layers may be inclined or horizontal. The most abundant minerals in the sedimentary rock are quartz, kaolin and calcite, which are white or colourness when pure. Sedimentary rocks may also be of different shades of yellow, brown, red, blue, green, grey and black, Red and yellow colours are due to hematite and limonite respectively. The blue green colour is due to the presence of copper oxide. The deep black colour is due to the presence of manganese oxide. Most shades of grey and black, very common in sedimentary rocks, especially shales, are due to the admixing of various amounts of carbon derived from organic matter in the sediment.

Concretions are formed by the precipitation of minor inorganic constituents from solution during or after deposition of sediments. While moving through sedimentary beds the ground water may dissolve a minor constituent of a large mass of rocks and redeposit the accumulated material in a concentrated form around certain nuclei to form concretions, calcite, limonite, pyrite and other concretions occur in sandstone and shales. Concretions vary from a fraction of an inch to several feet in diameter.

A few most common soil forming sedimentary rocks are described below:

(*i*) *Conglomerate and Brecia*. Conglomerate consists of a cemented mixture of rounded particles of gravels and sands. Gravels are usually composed of quartz, but may also contain other minerals. But brecia consists of cemented angular gravel and sand particles.

(*ii*) *Sandstone*: Sand particles have been cemented together to form sandstone. Sand particles are chiefly quartz grains but other minerals like calcite, gypsum, limomite, haematite, magnetic, olivine, garnet etc. may occur as impurities. The cement is usually silica or calcium carbonate in the white, lighter coloured rocks while iron oxide is responsible for the cementing of sand grains in rocks with a yellowish, brownish or reddish colour. Sandstone which has unequally sized quartz grains is called grit. The massive and somewhat impure variety of quartz is called Flint. Chert is an impure dark, flint-like rock.

(*iii*) *Shales*: Mostly Kaolinites usually contaminated with impurities like silt, iron oxide, organic material and others, have been cemented together to form the layered beds called shales or mudstone. Most shale usually split along the bedding planes. Red, brown and yellow varieties contain iron oxide while black varieties contain organic matter.

(*iv*) *Siltstone*: Siltstones are an intermediate between the sandstone and shales because they are all gradations in texture between sandstone and shales. They consist of fine sand, silt and some clay and include such varieties as shaly sandstone and sandy shales.

(*v*) *Limestone*: Limestones mainly consist of calcite but some impurities like clay, iron oxide, fine sand, silt etc. may also occur in some. Most limestones consist of materials that were once the skeletal parts of living organism. Limestone effervesces in dilute hydrochlorine acid.

(*vi*) *Chalk*: Chalk is a fine grained, soft and porous variety of limestone. Mixtures of microscopic shales of forminifera, (marine protozoa that had been enclosed within the perforated calcareous shale), pulverized fragments of other shale and other calcareous matter have been cemented together usually with calcareous cements to form the chalk.

(*vii*) *Marl*: It is an impure variety of limestone containing calcite and clayey material as impurities.

(*viii*) *Dolomite*: It contains considerable amounts of magnesium carbonate and calcium carbonate.

2.13. SOIL FORMING METAMORPHIC ROCKS

Metamorphic rocks may be foliated or unfoliated. Foliated metamorphic rocks have a tendency to break along parallel planes. Unfoliated metamorphic rocks break down irregularly. Foliated metamorphic rocks are made up of leaf like layers and their constituent minerals are flattened and oriented parallel with the folia. There are three kinds of foliation. In some rocks like gneiss, lighter and dark coloured minerals occur in alternate layers which give a banded appearance to the rocks. This kind of banded appearance is not found in some other kinds of rocks like schistoses because they contain only one mineral. Fine grained foliated rocks like slate are said to have a slaty cleavage because they can be split in smooth thin plates.

The foliated rocks are often folded, meshed and injected with igneous materials and crumpled into intricate and complex patterns due to movements of rocks under high temperature and pressure. Some of the minerals which usually occur in igneous and sedimentary rocks also occur in metamorphic rocks. For example, quartz occurs in igneous, sedimentary and metamorphic rocks. Feldspar, pyroxenes, micas, and hornblende commonly occur in igneous and metamorphic rocks. Similarly, calcite and dolomite occur in both limestone and marble.

A few common soil forming metamorphic rocks are as follows:

(*i*) *Slate*: Slate is the most perfectly foliated metamorphic rock. It is of such a fine texture that the crystals of minerals cannot be seen with the naked eye. It has a slaty cleavage *i.e.* it can be broken along closely spaced parallel planes. It is usually black or grey coloured, but red coloured slates have also been found.

(*ii*) *Phyllite*: Phyllite is intermediate in the size of grains between slate and schist. It is coarser than slate and finer than schist. It has glimmering lustre which is due to the presence of small flakes of lustrous muscovite mica. Probably most phyllites are formed from metamorphosis of shales.

(*iii*) *Schist*: Slate grade into schist with increasing grain size. The slaty minerals of schist can be recognised with the naked eye whereas those of slate cannot be recognised with the naked eye. All schists include tabular, flaky or even fibrous minerals. The individual folia lamination of schist are regularly spaced at a distance of up to about half a centimetre. This is the only features by which schists can be identified from more closely layered gneisses. Mineral grains of schist are larger than those of phyllite. Sometimes one band may be composed of mica, hornblended, chlorite etc. and the adjacent band composed of quartz. So they are called mica schist, hornblende, schist, chlorite schist etc. Generally schists contain more than 50 per cent of the flaky minerals.

(*iv*) *Gneiss*: Gneiss is a coarse grained banded metamorphic rock. The bands are commonly folded and twisted in a complex manner. It consists of alternate layers or bands of dark coloured minerals like hornblende, biotite and light coloured minerals quartz and feldspar.

(*v*) *Quartzite*: The sandstone had been metamorphosed to form the quartzite. The pore spaces which separated sand grains from each other have been filled up with newly crystallised quartz. The non-foliated quartzite may be stained red, yellow or other colours due to the presence of impurities.

(*vi*) *Marble*: Marble is a non-foliated coarse to fine textured metamorphic rock which consists mainly of calcite and/or dolomite. Pure marble is white in colour, but may be of different colours due to presence of impurities.

2.14. ROCK SYSTEMS OF INDIA

India may be divide into three well marked regions of distinguishing characters as mentioned below:

1. Peninsular Shield (Geologically old parts of the earth's crust is known as the shield) is the triangular plateau like region located south of the Indus and the Ganges river systems, the major part of which projects out in the Indian Ocean.
2. The Extra-Peninsular area, consisting of the Himalayan region.
3. The Indo-Gangetic alluvial plain, located between the Peninsular and the

Extra Peninsular area from Punjab through Uttar Pradesh, Bihar and West Bengal to Assam.

The Archaean groups of rocks covers more than half the Peninsular shield. The rest of the land area except some part of the coastal region, has been covered by the Purana, the Gondwana and the Deccan Traps. Some Mesozoic and Tertiary sediments have been found in coastal region. Marine sedimentary rocks predominate in the extra Peninsula, although the Archaean groups of rocks occurs in parts of the Himalayas. The Indo-Gangetic alluvial plain is made up of pleistocene and recent deposits of clay and silt. The rock systems of India have been briefly discussed below:

ARCHEAEAN GROUPS OF ROCKS

The Archaean groups of rocks comprises of the Archaean System of rocks and the Dharwar System of rocks.

ARCHAEAN SYSTEM OF ROCKS

The Archaean Systems of rocks lies below the oldest unconformity. (unconformities are surfaces of removal or deposition that separate the younger strata of rocks from the older strata of rocks), known as the Eparchaean Unconformity. As they are the oldest systems of rocks they occupy the cores of mountains like the Himalayas or are exposed in the most ancient segment of the earth's crust such as the Indian Peninsula. Some of the them might have been formed from the cooling and subsequent solidification of the molten material of the earth. Others may have been formed from the cementation of the sediments that were deposited in the primitive oceans of high chemical potency. Some of them might have been formed from the metamorphosis of varieties of intruded magma and extruded masses of lava. The Archaean Systems of rocks are called the basement complex because of their complex nature.

The Archaean system of rocks are widely distributed in the Indian Peninsula including Karnataka, Andhra Pradesh, Tamil Nadu, Eastern Ghat Mountains, Orissa, Madhya Pradesh, West Bengal, Assam, Bihar and Rajasthan and the Central Himalayan ranges, Assam and Meghalaya.

The complex Archaean system of rocks of different origin are crystalline in nature, some parts of which were twisted and foliated as they were extremely metamorphosed. Different kinds of gneisses and schists are the most common types of rocks recognised in the Archaean System of rocks. The gneiss consists of different amounts of white or pink orthoclase, a little plagioclase, small grains of bluish quartz, muscovite, biotite, hornblende and the accessory minerals like tourmaline, epidote, chlorite, garnet, magnetite etc. The gneiss is extremely foliated and coarsely banded in texture. At many places they are so coarsely banded that the rock resembles a granite or some other plutonic rocks.

The relatively finer textured schist is very foliated. It composition is similar to gneiss, except that it contains a little less orthoclase. Phyllites, granulites marbles, dolomites, slates with graphite talc-schists, and quartzites have been found interbedded in the gneiss and the schists. The Archean Systems include the following kinds of rocks:

(*i*) The Peninsular Gneiss was also called the Bengal gneiss because it was first studied in the Midnapur district of West Bengal. These finely foliated rocks vary in composition from granite to granodiorite or even amphibolites. They also contain plenty of accessory minerals like epidote, apatite, tourmaline, garnet, magnetite etc. They are also called dome gneiss because they have weathered to form the dome shaped mass.

(*ii*) Charnockites are bluish gray to dark coloured red rock of a medium to coarse grained texture. They consist of hypersthene, enstatite, blue coloured quartz, microcline, plagioclase, hornblende, augite and biotite. They also contain accessory minerals like zircon, magnetite, ilmenite, graphite and garnet.

(*iii*) Bundelkhand gneisses are so coarse grained that they look like granite. The typical Bundelkhand Gneiss is a pink to reddish brown, medium grained non-foliated and non-porphyritic granite that consists of quartz, orthoclase, microcline and a little ferro-mangnesian minerals. The ferromagnesian minerals, biotite and green hornblende have been more or less altered to epidote and calcite. Accesory minerals are usually rare. The Archaean system of rocks do not contain any fossil. They were formed during the Azoic era when no life existed on the earth. Some geologist think that some life existed during the Azoic era because some calcareous rocks occur in the Archaean System of rocks. No fossils have been found in the Archaean Systems of rock because they were destroyed by intense metamorphism.

DHARWAR SYSTEM OF ROCKS

The Archaean system of rocks, gneisses and schists were disintegrated and decomposed. The products of disintegration and decomposition were deposited in layers over the Archaean gneisses and schists. These layers were later on cemented to form sedimentary rocks that were profusely injected with molten magma from below when they were metamorphosed to form the Dharwar System of rocks in South India, the Champaner System of rocks in Gujarat and the Aravalli System of rocks in Rajasthan. Hence the Dharwar system of rocks mainly comprises of metamorphic rocks and a small proportion of sedimentary rocks. The Dharwar system of rocks are mainly schists, phyllite and slate. The schists include hornblende schist chlorite-schists etc., which are interbedded with quartzites, and marbles. The Dharwar System of rocks are intruded by plutonic rocks like nepheline syenites, tourmaline granites, dunites etc.

The Dharwar system of rocks does not contain fossils since they were formed in the Azoic era. This system of rocks, first studied in Dharwar in Karnataka, occur in Dharwar in Karnataka, Chotanagpur in Bihar, Assam and Rajasthan.

PURANA GROUPS OF ROCKS

The Cuddapah and the Vindhyan systems of rocks were grouped together under the Purana groups of rocks by Sir T.H. Holland in 1904. The Purana groups of rocks which are formed during the Proterozoic era, do not contain fossils.

CUDDAPAH SYSTEM OF ROCKS

The name 'Cuddapah' originated from the Cuddapah districts of Andhra Pradesh where the Cuddapah system of rocks were first studied. Intense earth movement occurred at the end of the Azoic era when the primitive seas gradually receded and low-laying areas were elevated to form mountain ranges. Materials were removed from the elevated areas and deposited in the valleys in layers which were cemented to form sedimentary rocks. Again the earth's surface was crumpled to form elevated areas and basins which were gradually filled up with sediments derived from the elevated areas. The layers of sediment were cemented to form the sedimentary rocks. These sedimentary rocks were folded and metamorphosed to different degrees to form the Cuddapah system of rocks. They mainly occur in the Cuddapah district of Andhra Pradesh, the Chhatisgarh district of Madhya Pradesh, and Sundargarh district of Orissa. The Cuddapah System of rocks are more folded in Rajasthan where it is known as Delhi system, than elsewhere. The Cuddapah system of rocks comprises of quartzite, indurated sandstone, slates, shales and limestone.

VINDHYAN SYSTEM OF ROCKS

The earth's crust was subjected to broad uplifting and down warping after the Cuddapah period. Sediments were derived from the elevated areas and deposited in the valleys. These layers of sediments were cemented and metamorphosed to form the Vindhyan system of rocks which are almost horizontal and consists of sandstones, limestones and shales. Vindhyan system of rocks had not been metamorphosed in any part. Shales have retained their laminated character and not been converted to slates. Similarly limestone had not been converted to marble. Lower Vindhyan rocks are mostly calcareous and argillaceous of marine origin. Their junction with older Aravalli rocks is faulty.

The Vindhyan System of rocks have been widely distributed. They cover a large area that extends from Dehri-on-Sone in Bihar to Gwalior and Hoshangabad and from Chittorgarh to Agra. Lower Vindhyan rocks occur in the valley of the Sone, Bhima and Godavari which the upper Vindhyan rocks occur at the north of the Narmada valley.

GONDWANA GROUPS OF ROCKS

The name GONDWANA was derived from the kingdom of the Gonds, a great ancient tribe who still inhabit Madhya Pradesh, where these rocks were first studied by H.B. Medlicott in 1872. The Gondwana groups of rocks occur in the Narmada valley in Madhya Pradesh, Sone valley and Rajmahal Hills in Bihar, Damodar valley in Bihar and West Bengal, Mahanadi valley in Orissa, Godavari valley in Andhra Pradesh, Himalayan Foot Hills of Nepal, Bhutan, Assam, Kashmir, east coast of India between Cuttack and Cape Commorin, Saurashtra and Kutch.

Large number of trough faults were formed in the Damodar valley in Bihar and West Bengal, Narmada valley in Madhya Pradesh, Mahanadi valley in Orissa and Godavari valley in Andhra, towards the end of the middle carboniferous period. Sediments were deposited in these basins till the end of the Jurassic period. There basins were tilted to one side or the other. Plants grew on these sediments, died and were buried under fresh deposits of sediments. The plants were converted to coal seams. The entire thickness of the continental deposits subsided and cemented to form the Gondwana groups of rocks. Gondwana coal fields contain dolerite and basalts. The lower Gondwana system of rocks contain Glossopteris flora and include Talchir, Damuda (from Damodar) and Panchet series of rocks. The upper Gondwana system of rocks which contain ptilophyllom flora, include the Mahadeva, Jabalpur, Rajmahal and Umia series of rocks.

The Talchir series of rocks consist of greenish or brownish sandstone that lies over aeranaceous (sandy), micaceous or calcareous shales, conglomerates, grits and coal seams. The panchet series consists of greenish, buff and brownish sandstones and shales in the lower part and greenish micaceous and felsphatic sandstones and shales in the upper part.

The Mahadeva series of rocks consist of sandstones and reddish shales. Rajmahal series comprises of white and light coloured clays and massive soft sandstones. Umia series of rocks consist of sandstones that contains plant fossils.

DECCAN TRAP ROCKS

Volcanoes erupted intermittently towards the end of the cretaceous period to eject huge masses of molten materials called lava from the interior of the earth. The lava covered about two lakhs square miles of area in the present Maharashtra, Madhya Pradesh, Andhra Pradesh and Kutch where it cooled down and formed step like rocks called the Deccan Trap rocks, which, however are not composed of volcanic materials from top to bottom since they had been interculated by sedimentary rocks called the inter-trappean beds. Almond shaped cavities occur in the Deccan Trap rocks. These cavities had been wholly or partially filled up with secondary materials like calcite, zeolite, chalcedony etc. The Deccan Trap rocks had been horizontally laid over the entire area in western and south-western India. The Deccan Trap rock is a dark basic volcanic rock of mainly basaltic composition on the surface. It is thus called plateau basalt. It's dark colour is due to the dark colour of basalt and magnetite. The lava solidified below the surface of the earth to dolerite, of a coarse texture.

The Deccan Trap rock are of porphyritic texture. The large crystals called the phenocysts are made up of plagioclase and the ground mass consists of augite, plagioclase, olivine and magnetite. The inter-trappean beds are mostly composed of dark coloured cherty or siliceous rocks that usually contain fossils.

MESOZOIC GROUP OF ROCKS

The Mesozoic group of rocks comprises of Triassic, Jurassic and Cretaceous system of rocks.

The Triassic system of rocks occurs in Spiti and Kumaon in Himachal Pradesh, Kashmir and Sikkim, and mainly comprise of limestones and shales that contain fossils.

The Jurassic system of rocks consist of shale, limestone and quartzite that cover large areas in Kutch and Rajasthan. Black lustrous micacious shales containing abundant amounts of calcareous concretion occur at Spiti in Himachal Pradesh. The Spiti shales were intercalated with sandstones.

The Cretaceous system of rocks mainly comprises of varieties of sedimentary rocks that contain fossils especially those of foraminifera (a species of marine protozoa that had been enclosed within a perforated calcareous shale). For example, most sandstones, shale and quartzites occurring in Spiti and Kumaon in Uttar Pradesh and Kashmir contain fossils. In Maharashtra, Ahmednagar, sandstones consists of thick horizontally bedded pink, red and brown coloured sandstone, shale and conglomerate that contain plant fossils. Sandstone occurs in the lower and organic limestone occurs in the upper layer in Gwalior and Kathiawar. In Tiruchirapalli districts, the Cretaceous system of rocks consists of fine silts and calcareous shales and sandy clays containing ferruginous phosphatic and calcareous nodules. They contain coral and other fossils.

TERTIARY GROUPS OF ROCKS

The Tertiary groups of rocks are abundant in the Extra Peninsular regions from Kashmir along the Himalayan Foot Hills to Assam. They also occur from the Gulf of Cambay to Kutch in Gujrat, Rajasthan and the Coastal areas of Orissa, Tamil Nadu and Kerala. They mainly consist of various kinds of sedimentary rocks that include varieties of shale, limestone, sandstone, conglomerate etc. most of which contain different kinds of fossils including those of foraminifera.

PLEISTOCENE AND RECENT GEOLOGICAL FORMATION

During the Pleistocene and recent periods, mainly unconsolidated sediments of clay, silt and sand have been deposited. The closure of the Pleistocene period was marked by laterisation. The agriculturally important geological formation during the recent period are as follows :

1. *Indo-Gangetic alluvium*: Alluvium is the product of weathering of rocks constituting the mountains that have been brought by the great river systems the Indus, the Ganges, the Brahmaputra, the Narmada, the Godavari and the Mahanadi, in various degrees of fineness and deposited them as they traverse the plains. The Indo-Gengetic Alluvium has filled up a depression between the Himalayan mountain ranges and the Peninsular shield.

2. *Loess* : Large areas in Haryana, Punjab and Kashmir have been covered by wind blown dust called Loess which is fine grained buff or grey coloured fine sandy to clayey in nature.

3. *Desert Sands*: The Thar desert in western and south-western Rajasthan is covered by several metres of sand which is constantly being shifted by winds blowing from the south-west direction. The sub-soil water varies in quality from place to place. It can be used for irrigation in eastern sides. It is saline is semi-desert areas. Rainfall is very low because the Aravalli mountains run parallel to the direction of the south-west monsoon wind.

4. *Laterite*: Laterite is extensively distributed in the Peninsular India *i.e.* Kerala, Madhya Pradesh, Maharashtra, Karnataka, Tamil Nadu and Orissa.

Laterite is a porous, pitted clay like rock with red yellow, brown grey and mottled colours. It consists mainly of oxides of iron and aluminium. It is very soft when freshly dug up, but becomes hard like bricks, when it comes in contact with moist air.

2.15. ROCKS OF KARNATAKA

Almost the whole of Karnataka is made up of the Archaean groups of rocks, although Vindhyan system and Deccan trap rocks occur in small areas. The Archaean groups of rocks comprises of Archaean system and Dharwar system of rocks.

The Peninsular gneiss, and the Charnokite, the champion gneiss and the Closepet granite belong to the Archaean system of rocks. A heterogeneous mixture of different kinds of granitic rocks profusely penetrated the schistose rocks after the latter were folded, crumpled and metamorphosed to from the Peninsular gneiss which therefore included varieties of rocks like granite, granodiorite, gneissic granite and banded or composite gneisses, the granite constituent of which show distinct sign of intrusion. The banded gneisses are made up of white bands of quartz feldspar alternating with dark bands of hornblende, biotite and accessory minerals.

Charnokite is a bluish-grey to dark coloured medium to coarse grained rocks, composed of hypersthene, enstatite, blue coloured quartz microline, plagioclase, hornblende, augite and biotite. It also contains accessory minerals like zircon, magnetite, ilmentie, garnet and graphite.

Champion gneiss has been considered to be the oldest gneiss in Karnataka. It has been named after the Champion reef in the Kolar gold-field; the rocks are dull grey to pink in colour, and greasy looking comprising of quartz, oligoclase, andesine and muscovite with apatite and magnetite as accessory minerals.

Clasepet granites which are extensively exposed near Clasepet and Channapetna, are grey or pink coloured intrusive biotite-granite of a coarse texture.

The Dharwar system of rocks of Karnataka has been divided into five geographical groups from west to east. Hornblende schist and thin bands of haematite quartzite occur in the west. Banded ferrugineous rocks and magniferous rocks occur in the west-central area *i.e.* Shimoga and Babubudan belt. Lots of igneous and banded ferrugineous rocks and limestone occur in the Chitaldrug-Chiknaya Kanhalli and Naga Mangala belts of the central area.

Varieties of rocks like Quartz-magnetite-granulities, silliminite-quartzite, cordierite-hypersthene-gneisses, cordierite-mica-gneisses, cummington-schists, pyroxene-geneisses are included in the east central group. A belt of schist rocks occurs in the east.

The Vindhyan system of rocks which occurs in the Gulbarga and Bijapur districts comprises of lower, middle and upper divisions. The lower division consist of sandstone or green and purple shales. The middle division consists of creamy grey, blusih and buff coloured limestone. The upper division consist of black, blue, buff and purple shales, Sandstone and limestone occur at the bottom and top respectively of the upper Vindhyan system.

Deccan Traps that comprise of basalts and dolerites occur in small areas in Hassan, Kadur and Mysore districts. Cappings of laterites have been found over the Deccan Trap rocks.

2.16. ROCKS OF KERALA

Different kinds of gneiss belonging to the Archaean system of rocks have been found in south Malabar. Charnokite containing orthopyroxene occurs in Travancore, Magnesia rich pyroxenite has been found at Neyyoor in South Travancore. The Miocene system of rocks that comprises of fossiliferous limestone sand and clay, occur at Padappakara near Quilon.

Warkalli (Varkala) sandstone that occurs along the coast of Kerala also belongs to the Miocene system of rocks. They are ferruginous sandstones that have bveen penetrated by clays. They gradually change from gery to dark grey clay and carbonaceous sandstone. They occur just over the Quilon limestone which is a very compact and hard fossiliferous limestone. It contains high amounts of calcium and low amounts of magnesium and 6 to 15 per cent silica Varkala sandstones have been laterised to depth of 10 to 13 metres. Felspathic gneisses which occur below the Vaskala sandstone at some places have been converted to Kaoline to a depth of about ten metres or more. Kaoline occurs at Kundara near Quilon and several other places in Malabar. The laterite that occurs in Malabar is red, yellow, grey and mottled porous clay-like material that mainly consists of oxides of iron and aluminium. It is very soft when freshly dug and becomes very hard on exposure to the atmosphere.

2.17. ROCKS OF TAMIL NADU

Mesozoic (Cretaceous system) and Tertiary groups of rocks occur in the coastal regions of Tamil Nadu. The major areas of this state have been covered by Archaean groups of rocks, as follows :

Several bands of schistone rocks are widely distributed in Salem and Coimbatore districts where geologists have found a few bands of magnetite quartzite, formed from banded haematite-quartzite. Older grey gneisses and younger pink gneisses occur in Arcot, Tiruchirampalli and Madurai. Crystalline limestone occurs whithin the gneisses at many places.

Several places in Salem district have been covered by Olivine rocks containing magnesite. Magnesia rich pyroxenite occur at the Tamil Nadu-Kerala border. Ultra basic igneous rocks have been found near Ten Mundiyanur about 20 km south-west of Tiruvannamalai. These ultra basic igneous rocks which are serpentinised dunite and peridotites with veins of magnesite, are flanked by pyroxenite on one side by corundum-feldspar rocks associated with granite and pegamatite on the other meta-anorthositic gneisses and metagabbro have been found near Pattalur in the Salem district. The anorthosite of Tiruchirapalli district comprises of quartzite, amphibolite and crystalline limestone. Alkali rocks which include nepheline-syenite and its varieties containing hornblende and biotite, aegirine-augite, feldspar and feldspar corundum rock, have been found at and around Sivamalai.

The Gondwana groups of rocks occurring near Madras comprise of two stages. The lower stage is called Sriperumbudur and consists of white to pink clay, shales and felspathic sandstone and fossils. The upper stage is called the Satyavedu stage and consists of purple mottled ferruginous sandstone which contains fragmentary plant fossils. The Gondwana groups of rocks which cover a small area near Uttatur village in Tiruchirapalli district are micaceous shales, grey sandstone and grits containing calcareous concretion. They contain fossils. The Gondwana groups of rocks have been found near Sivaganga in the Ramnad district, where they are made up of boulders and conglomerates at the lower part and micaceous sandstone and alternating grits and shales in the upper part. They contain plant fossils.

The Cretaceous system of rocks that occur in Tiruchirappalli district, comprises of Uttatur, Trichinoploy, Ariyalur and Niniyur stages. The Uttatur stage consists of fine silt, calcareous shales and sandy clays containing ferruginous, phosphatic and calcareous nodules. Dark grey to yellow somewhat sandy limestone occurs at the base of the Uttatur beds. It contains coral and other fossils the Tiruchirappalli stage comprises of sandstone, grits, calcareous grits, shales and bands of shell limestone. The Ariyalur stage is made up of grey to light brown arigillaceous (clayey) and white sandstones. Fossils occur at the lower part of the Ariyalur stage. The Niniyur stage consists of grey brown ochreous sandstones and shales which contain fragments of flint and chert that contains fossils.

The Tertiary groups of rocks cover the coastal strip of South Arcot, Tanjore and Ramanathapuram districts. They are sandstone and shales that contain fossils. A wide stretch of somewhat ferruginous and laterised sandstone called Cuddalore sandstone occurs intermittently from Rameshwaram through Padu Kottah, Tanjore, Cuddalore, Pondichery and Madras.

2.18. ROCKS OF ANDHRA PRADESH

The important rocks of Andhra Pradesh include the Archaean groups of rocks, Cuddapah system of rocks, Gondwana groups of rocks, Deccan Traps and rocks formed during the Pleistocene period.

The Dharwar system of rocks, found in Karimnagar and Warangal district, includes hornblende-talc, chlorite and mica schists and quartzites. Other important Archaean groups of rocks of these districts include grey gneisses and pink gneisses. The gray gneiss which corresponds to Peninsular gneiss, consists of light bands of quartz and feldspar, and dark bands of mica and hornblenbe. Pink gneiss comprises of quartz, microcline and orthoclase, acid plagioclase, some hornblende and epidote.

Four kinds of gneisses have been identified in the Nellore and Guntur districts, of which two are schistose and two massive. The schistose gneisses comprises of quartz, mica, hornblende and talc schists and quartz magnetite rocks. This massive gneisses comprises of grey, sometimes prophyrite gneiss a red granitoid gniess. The banded grey gneiss varies in composition and contain streaks and bands of micaceous gneiss and charnokite. This belongs to the groups of Peninsular gneisses. Pegmatites and quartz veins have intruded into the schists and young granitic red gneisses.

The sedimentary rocks of the Cuddapah system occur on the gneissic and schistose rocks belonging to the Archaean group. The Cuddapah system of rocks, consists of the Papaghmi, Cheyair, Nallamalai and Kistna series. The Pagni series occurs on the gneisses and schists of the Archaean groups of rocks with a profound unconformity called eparchean unconformity, and consists of the lower Gulcheru stage and the upper Vempalle stage. The Gulcheru stage consists of conglomerates, grits and sandstone; the Vempalle stage consists of grey to buff coloured limestone that contain chert, chalcedony and shales. The Cheyair series includes lower Pulivendla stage and upper Tadpatri stage. The Pulivendla stage consists of quartzite, conglomerates, and sandstones. The Tadpatri stage consists of slaty shale that contains thin beds of siliceous limestone, chert jasper and intrusive basic silts. The Nallamalai series consists of lower Bairenkonda quartzite and upper Cumbum shales, that includes shales and slates. The Kistna series consist of quartzite.

The Gondwana groups of rocks occurring between Rajahmundry and Vijaywada consists of Golapilli sandstone below, Raghavapuram shales in the middle and the Tirupati sandstone above. All these stage contain fossils. The Gondwana groups of rocks occurring near Budavada in the Ongole district comprises of three beds. Buff coloured Budavada sandstone occurs at the lowest bed. The middle bed consists of purplish variegated shales. The topmost bed consists of brown and red unfossiliferous sandstone.

Deccan Traps which mainly comprise of basalt and dolerite, occur in Rajahmundry. The Godavari alluvium is mainly composed of brown clay and sandy silts, with nodules of Kankar and beds of gravel.

2.19. ROCKS OF MAHARASHTRA

Major areas of Maharashtra have been covered by the Deccan Trap rocks, formed by the solidification of molten lava. This lava formed step-like terraces. Deccan trap rocks have been called plateau basalt by H.S. Washington because they consist mainly of basalts on the surface and dolerities and gabbros below the surface of the earth. They are generally a dark grey to dark greenish colour. Brown and purple tints have also been seen. Buff to cream coloured, more acidic (Trachytic) types have been seen near Bombay. The non-vesicular basaltic types are hard compact and medium to fine grained, and break with a fracture. The vesicular types are comparatively soft and break more easily. The Archaean groups occur in Nagpur and Bhandara district. They include the Sakoi series in the north and the Sauser series in the south. The Sakoi series consist of chlorite and sericite schists and hematitic iron ore. The Sauser series comprises of calo-granulites, marbles, garniteferous schists and manganese silicates and gondites.

The cretaceous system of rocks include Ahmednagar sandstone which comprise of thick horizontally bedded sandstone shales and conglomerates of pink, red and brown colour. It contains plant fossils.

The Godavari alluvium of Maharashtra consists mainly of brown clay and sandy silts and nodules of kankar and beds of gravel.

2.20. ROCKS OF MADHYA PRADESH

The Archaean groups of rocks occurs in the districts of Raipur, Durg, Bilaspur, and Balaghat, Chhindwara and Jabalpur.

The Archaean groups of rocks of Raipur district comprise of quartzite, phyllites, micas schist and banded haematite-quartzite. Volcanic agglomerates and epidorite covered the schistose rocks of Durg district. Schistose rocks also occur in Bastar, Bilaspur and Balaghat district. Highly metamorphosed schistose rocks which occur in the Chhindwara district comprises of varieties of metamorphic rocks like different kinds of marble, schists, quartzites etc. The Archaean groups of rocks of Jabalpur district include conglomerates, phyllites, mica schists, calcitic and dolomitic marbles and banded ferruginous rocks.

The Vindhyan system of rocks have been found in the districts of Rewa, Gwalior, Bastar, Raipur, Durg and Bilaspur, sandstone and shale occur at Rewa and Gwalior. Shale, quartzite, conglomerate and limestone have been found in Bastar district. Shaly limestone, calcareous shale and felspathic sandstone occurs at Raipur, Durg and Bilaspur districts.

The Gondwana groups of rocks have been found at Rewa, Panchmarhi and Jabalpur. The Parsora stage of the Panchet series which occur at south Rewa, consist of medium grained sandstone with micaceous and ferruginous bands. The Mahadeva series which occurs at Panchmarhi includes Panchmarhi, Denwa and Bagra stages. The Panchmarhi stage comprises of red and buff coloured sandstones with some red clays at the top and the bottom. The Denwa stage which occurs just above the Panchmarhi stage, comprises of pale brownish or greenish yellow, bright mottled red clays. It contains bands of white and yellow sandstone which are often calcareous. The Bagra stage consists of pebble-beds and conglomerates with red jasper in a matrix of red sandy clay. Jabalpur series consists of white and light coloured clays and massive soft sandstones. Some of the shales are carbonaceous.

The cretaceous system of rocks which have been found in Gwalior, are made up two portions. The lower portion is called Nimar sandstone, comprising of basal conglomerate and layers of sandstone and shale above. Organic limestone occur at the upper portion. These rock systems contain fossils.

2.21. ROCKS OF RAJASTHAN

The Archaean groups of rocks of Rajasthan comprises of Bundelkhand gneiss, banded gneissic complex and the Aravalli system of rocks, Erinpura granite and Malani igneous suite.

The Bundelkhand gneiss which occur in the Berach valley between Chittor and Bhilwara is a pink to reddish coloured, medium grained non-foliated and non-porphyuritic granite that consists of quartz, orthoclase and minor amounts of microclinic, buotite and green hornblende. Bundelkhand gneiss have become foliated near the junction of Berach and Began rivers.

The banded gneissic complex of Rajasthan usually consists of alternating bands of biotite gneiss and granite. Several varieties of banded gneissic complexes have been found in Rajasthan. These banded gneissic complexes consist of biotite and chlorite schists in south Mewar. They grade into granite-gneiss or even into unfoliated granite at many places. They also contain some hornblende-schist and epidorite. The banded gneissic complex of central and north Mewar and Ajmer consists of dark schists and granite ferous granulites which are profusely penetrated by bio-tiegranite. Other varieties include a grained and somewhat foliated porphyritic granite and grey, fine, slightly foliated granitie gneiss.

The Aravalli system of rocks are dominantly argillaceous (clayey) in composition and are metamorphosed and folded in an increasing degree from east to west. They consist of a basal and upper beds. The basal bed of arkose and grity quartzites lies over the Bundelkhand gneiss or the banded gneissic complex; shales and phyllites occur over the arkose and gritty quartzite. Some altered basic volcanies are associated with the shales and phyllite at some places. Limestone and quartzite occur at certain places. The Aravalli system of rocks of eastern Mewar include Binota shales which are brown and dive shales that contain ferruginous and clay concretions. The Aravalli system of rocks, found near Ranthambhor and Sawai Madhopur, include reddish sandstones and quartzite. The quartzites have been interbedded with shales and dolerite at Ranthambhor.

The Raialo series of rocks, found in Alwar, Jaipur around Railo, Ajmer and ewar Makrana, Ras and Godwar, North of Udaipur city, and through athdwara, Rajsamand and Kankroli into Par-Benara and Jahazpur hills, mainly consist of limestone (about 600 metre thick) that lies over thin basal beds of sandstone and conglomerates.

The Erinpura granite which has been found in eastern Mewar, Palanpur, Idar, Sirohi, Beawar, Jaipur and Alwar and Mt. Abu generally consists of biotite and granite though its pegamtite contains muscovite and tourmaline. Reddish malani rhyolite containing pink orthoclase oligoclase and corroded quartz crystals covers a larger area in Jodhpur.

The Delhi System of rocks includes 'Calc-gneisses', 'Calc-schists', phyllite and biotite-schists of Mewar and Ajmer, Alwar and Ajabagrah Series of Alwar and Jaipur, Hornstone breccia and Kushalgarh limestone of Alwar and Jiran sandstone and shales and grits of Chittor and Nimbahera. The calc-schists were originally thin bedded biotite limestones, part of which had metamorphosed to form the schistose rocks when tremolite disposed, biotite and feldspar were developed. They, show dark and white bands, Calc-gneisses are mainly dark banded, siliceous limestone Alwar series consisting of compact quartzite, conglomerate and grits. The Ajabgrah series is dominantly argillaceous in nature and contain minor amounts of siliceous limestone, calcareous silt and ferruginous quartzites. Hornstone breccia consists of angular particles of quartz in a fine grained dark matrix of ferruginous and siliceous matter. Kushalgarh, limestones are dominantly limestones with dust-like inclusions forming a banded structure. Jiran sandstones are compact, hard, pale grey quatzite which are some time ferruginous and mottled with purple stains. The Vindhyan system of rocks occurring in Bundi, Kota, Dholwar Bharatpur, Jaipur and Bikaner, consists of fine grained, soft red sandstone with white specks. The Vindhyan system of rocks have also been found in Karauli and Chittor-Jhalrapatan area. Breccia have been found over limestone, which occurs above sandstone and conglomerate, that lies over phyllities at Karauli. Grits and conglomerate occur above shale at the Chittor-Jhalrapatan area.

The Jurassic system of rocks which have been found in Jaisalmer comprises of different kinds of sedimentary rocks. The earliest formation is Lathi beds of sandstone and fossils in the lower part and limestone in upper part. The next formation was Jaisalmer limestone that consists of organic limestone and calcareous sandstone, followed by a Baisákhi bed of sandstone and gypseous shales. This was followed by Bedser beds of ferruginous sandstone and grits The succeding bed are Parihar sandstone of unfossiliferrous and felspathic sandstone and quartzite. Abur bed of limestone, grit and shales containing fossils and the topmost Bandah and Khuila bed of dense siliceous limestones containing fossils.

A large area in western and south-western Rajasthan is covered by several meters of sand.

2.22. ROCKS OF GUJARAT

The Champaner series of rocks belong to the Archaean Groups of rocks, and includes quartzite, conglomerates, slates and limestone.

The Gondwana system of rocks, which has been found in Kutch consists of sandstone and shales containing fossils. A large area in Kutch has been covered by the Jurassic system of rocks, which mainly comprises of four series. The patcham series consists of clastic sandstone and yellow limestones that contains fossils. The Chari series consists of organic limestones that contains fossils. The Katrol series consist, of different types of sandstone and shales that contain fossils. The Umia series including sandstone, conglomerates, shales and fossils.

The Deccan Trap rocks, extensively found in Kathiawar and Kutch, originated from basaltic lava flows. They mainly consist of dolerite and basalt but other rocks like olivine-gabbro, andesite, rhyolite, felsite etc. have also been found interbedded with them. The Eocene system of rocks of Kutch, lying over the Deccan Trap rocks, comprises of the lowermost Madh series, the middle Kakdi series and the upper Berwali series. The Kakdi series includes a lower layer of carboniferous shales with lignite and ferruginous and gypsiferous clays and upper layers containing fossils. The upper Berwali series consists of limestone that lies over clay containing fossils. The Eocene system of rocks occurring between Surat and Broach, comprises of basal bed of impure limestone containing fossils, and an upper bed of sandstone, shales, conglomerates etc. in which fossils are present at some places.

The Oligocene and Miocene system of rocks have also been found in Kutch. The Oligocene system of rocks comprises of organic limestone containing fossils. The lower part of the Miocene system comprises of organic limestone and the upper part comprises of clays and siltstone containing fossils.

2.23. ROCKS OF UTTAR PRADESH

Most of the plain areas of Uttar Pradesh are made up of Indo-Gangetic alluvium, brought down from the Himalayas by the Ganges river system and deposited in the plain areas. The older alluvium called Bhangar is darker clay, rich in calcereous kankar. It was formed during the middle to upper pleistocene period. The newer alluvium called khadar is a light coloured, sandy and gravely material, almost free from lime kankar. It was formed during the upper Pleistocene to recent period.

The Siwalik system of rocks occur at the Himalayan foot hills in Haridvar and Dehra Dun. They are composed of sandstone, grits, conglomerates, pseudoconglomerates, clay and silts that contain fossils. They are often folded and faulted and lie at steep angles against other formation.

A large out-crop of gneissic granite occurs in Bundelkhand. It is mainly metasomatic granite which had assimilated older rocks like quartzite, micaschist, talc-schists amphibolites etc.

Schistose rocks have been found in the Mirzapur district.

2.24. ROCKS OF BIHAR

Iron-ore series rocks include basic lava, shales, banded haematite-quartzite and purple and grey limestone and occur at the northy of a thrust zone in the Singhbhum district. They are called iron-ore series because they contain iron-ore. They lie over hornblende-schist, quartzite, micaceous and chlorite schist. Metamorphosed shales and sandy shales occur within the thrust zone. Highly metamorphosed mica-schist, hornblende-schist and chlorite schist occur at the north of the thrust zone in the Singhbhum district belongs to the Dharwar system of rocks.

The Archaean system of rocks includes Chota Nagpur granite-gneiss and Singhbhum granite. Chota Nagpur granite gnesis is generally coarse, prophyritic and contains quartz, microcline, orthoclase, oligoclase, bitoite and little apatite. It occurs in Santhal Paraganas, Hazaribagh, Ranchi and Palamau districts. This Chota Nagpur granite gneiss weathers in Dhanbad, Ranchi and Hazaribagh districts to form dome-gneiss or Bengal gneiss.

Various types of schistose rocks occur in the Gaya and Hazaribagh districts. The Gondwana system of rocks occurs in the Damodar valley in south Bihar. This mainly comprises of sandstone and coal seams. Conglomerate, sandstone and limestone occur at Chaibasa in Singhbhum district.

The Gondwana system of rocks have been found in the coal fields of Bihar. Pebbly grits and sandstone with intercalated coal seams occur in the Giridih. Daltonganj, Umaria, Mahapani and Shahpur field, while fawn coloured sandstone and grits and beds of shale occur in the Jharia coal field. The sandstone contains more or less decomposed feldspar. Coal seams are often present in the Gondwana system of rocks, which occurs in south Bihar at the Bengal border and comprises of basaltic lava flow which had been intercalated with carbonaceous shale that contains plant fossils. Sedimentary rocks belonging to the Kolhan series of the Cuddapah system have been found over the iron ore series in the south Singhbhum district. They comprise of unmetamorphosed conglomerate, purple sandstone, shale and limestone.

Sedimentary rocks belonging to the Semri series of the lower Vindhyan system of rocks have been found west of Sasaram. They comprise of several beds. The lowermost bed comprises of basal conglomerate and limestone overlaid by shales, sandstone and tuffs that in turn are overlaid by olive shales, fawn limestone and gluconitic sandstone.

Indo-Gangetic alluvium occur in north Bihar and at the north of the Chota Nagpur plateau. It comprises of Pleistocene and Recent deposit of clay and sand.

2.25. ROCKS OF ORISSA

Gondite rocks that belong to the Archaean group have found at the central part of an anticline in the Sundargrah district. They consist of quartzspessarite, rhodonite, winchite and manganese ore. These oldest rocks are overlaid by carbonaceous quartzites, phyllites, dolimitic and calcitic marbles, and carbonaceous shales which had been intercalated with phyllites and mica schist. The carbonaceous phyllites are slaty at certain places. The marble contains a large reserve of good limestone and dolomite. A shear zone occurs above them. Raghunathapalli conglomerate occurs above this shear zone.

The Cuddapah system (Kolhan series) occurs only in the Keonjhar district. It consists of basal conglomerate, purple sandstone, shale and limestone. The Gondwana groups of rocks occurs in the Mahanadi valley. The Telcher series of this group comprises of several beds rocks. The lowermost bed is of boulders overlaid greenish arenaceous (sandy), micaceous or calcareous prismatic fragments of shales and greenish brown sandstone.

Yellowish brown limestone containing abundant oyster shells have been found at Molia near Baripeda in the Mayurbhanj district. They belong to the Miocene system of the Tertiary groups of rocks.

2.26. ROCKS OF WEST BENGAL

The Bengal gneiss of the Archaean groups of rocks occurs in the Midnapore district. It consists, mainly of gneiss and schist. Anorthite belonging to the Archaean group occurs at the south of the Raniganj coal field. Its composition varies from almost pure plagioclase rocks to anorthosite gabbro and diorite. Its northern boundary is faulted against the Gondwana groups of rocks.

The Gondwana Groups of rocks occur at Raniganj and comprise of sandstone, shales and coal seams. They belong to the Raniganj series of the Damuda system. Shales, limestone, sand and calcareous sandstone of the Cretaceous system occur at Ghatal in Midnapore district. The Cretaceous system of rocks mainly trap wash and some white sandstone have also been found at Bolpur in the Birbhum district. Bengal formed the shelf areas of a marine basin where sediments were deposited during the Tertiary era. The Eocene system or rocks, found at Jalangi in the Murshidabad district, consist of esturine carbonaceous shale containing pyrite, lignite and resins. The Oligocene system of rocks comprises of sandstone and lignitic shales at Burdwan, calcareous shales, shaly limestone and glauconitic sandstone containing fossils of foraminifera at Mesmari and sand, silt and shale at Pandua.

The sediments deposited during the Plio-pleistocene period in the Ganges delta are represented by the Debagram and Ranaghat formation. The Debagram formation comprises of clay, calcareous shales and silts which are deltaic. The alternating marine deltaic Matla formation comprises of clay and silt. Clay and silt of Plio-pleistocene period have been found at Ranaghat.

Strips of laterised sandstone, felspathic grits and bluish grey mottled clay, known as the Durgapur beds, occur along the western border of the Raniganj coal field. These are covered by alluvium at some distance east of Durgapur. These sandstones also occur at Suri in the Birbhum district.

2.27. ROCKS OF ASSAM

Gneisses, schists and granites belonging to the Archaean system of rocks, occur in the Mikir hills. The Shillong series belonging to the Dharwar system also occur in at the same area. It comprises of quartzite, conglomerates, phyllites, chlorite-schists mica-schist and hornblende-schists. Limestone containing plant fossils that belong to the lower Gondwana system of rocks have been found in the Aka hills north of Tejpur and the Bharati valley.

A Tertiary group of rocks have been found in eastern and south eastern Assam. The Dissang series of rocks were developed in the Brahamputra valley during the upper cretaceous to the upper Eocene period. It consists of splintery dark grey shales intercalated with fine grained sandstone. The Oligocene and Miocene system of rocks of Assam include the Barail and the Surma series. The Barail series of the upper Assam include Tilak Parbat stage of 450 metres of carbonaceous shales and coal seams, the Baragoloi stage 3000 metres of sandstone, carbonaceous shales and coal seams and the Naogaon stage of 2400 metres of sandstone. The Barail series of the Surma valley and north Cachar includes the Renji stage of 900 metres of hard massive sandstone and very subordinate shales, the Jenam stage of 900 to 1200 metres of sandstones, alternating with shales and the Laisong stage of 2400 metres of hard, thin bedded sandstone and subordinate shales. The Surma series of the Surma valley include 900 to 1500 metres of the Boka Bil stage and 1200 to 2400 metres of the Bhuban stage. The Boka Bil stage consists of sandy shales, silts, sandstone. The Bhuban stage consists of shales, sandy shales, sandstones and conglomerates.

The Siwalik system of rocks of Assam which developed during the middle Miocene to the lower Pleistocene period, comprises of the Tipam series, the Dupi Tila series and the Dihing series. The Tipam series has been found from the Surma valley to upper Assam. In upper Assam, it includes a 760 metres thickness of Namsang beds of sandstone, grit and conglomerate, 909 to 1018 metres. Thickness of Girujan clay, composed of mottled clay and sandy clay subordinate sandstone, and a 909 to 2424 metres thickness of Tipam sandstone, composed of gritty ferruginous sandstone and some cônglomerate shale parting. The Dupi Tila series consists of about a 3000 metre thickness of clay and sandstone. The Dihing series consists of a great thickness of pebblebeds with subordinate sandstone and clay bands.

2.28. ROCKS OF MEGHALAYA

Meghalaya is a detached blocks of the Peninsular plateau. So it is mainly made up of the Archaean groups of rocks though some Tertiary groups of rocks have also been found.

The Archaean groups of rocks of the Khasi and Jaintia hills consist of banded, composite, and biotite-granite-gneisses. The granatic constituent consists of quartz, microperthite, some microcline, oligoclase and biotite and accessory minerals like garnet, apatite, zircon etc. The Gneisses are associated with garnet-quartzites.

The Shillong series which belongs to the Dharwar system of rocks, occurs above the gneissic complex. The Shillong series consists of quartzite, conglomerate, phyllite, chlorite-schist, mica-schist and hornblende-schist. Occasionally carbonaceous slates and ferruginous rocks also occur.

The Shillong series had been intruded upon by Khasi green stone that consist of epidorite, amphibolites and amphibole-schits. Mylliem granite is younger than Khasi greenstone. It is a homogeneous, fairly coarse biotitegranite containing prophyritic pink microclinic orthoclase some acid plagioclase, biotite and hornblende and accessory minerals like appatite, magnetite and zircon. The granite and gneisses are traversed by dykes of dolerite.

The marine cretaceous system of rocks occurs in the Garo, Khasi and Jaintia hills. They are dominantly sandy in nature although shales and carbonaceous materials are so occasionally present. They are composed of sandstone and some coaly layer in the Garo hills. The Mahadek and Langpur stages occur over the basal bed of conglomerate in the Khasi and the Jaintia hills. The Mahadek stage consists of a 230 metre thickness of hard, gritty, coarse glauconitic sandstone with a fossil ferous horizon near the top. This is overlaid by the Langper stage of impure limestone, calcareous shales and sandstone. The Tertiary groups of rocks of Meghalaya include the Jaintia series and the Kopili series. The Jaintia series consists of shales, sandstone and limestone, the Kopili series of alternate layers of shale and sandstone and fossiliferous limestone.

2.29. ROCKS OF NAGALAND

Nagaland is made up the Paleozoic, Mesozoic (Cretaceous system only) and Tertiary groups of rocks. Alluvium of the Pleistocene and the recent period occurs along the courses of rivers.

The Paleozoic groups of rock includes only the Nimi formation, composed of limestone, quartzite, phyllite, carbonaceous phyllite, schistose granite and quartz-serictie-schist. The Zepuhu formation belongs to the Cretaceous system of rocks. It is composed of amphibolite, serpentinite, peridotie, dunite, pyroxenite gabbrodiorite, basalt, and esite, glaucophane schist, basic schist etc. Which are intermingled with oceanic sediments slike chert, grey wacke, limestone, cherty quartzite and phyllite. The Tertiary groups of rocks comprises of the Disang series of the Eocene system, the Barail series of the Oligocene system the Surma and Tipam series of the Miocene system and the Dihang series of the Pliocene system. Each of these series of rocks are separated from the others by an unconformity. The Barail and the Surma series of rocks are dominantly by mollusic sediments. The former is composed of well bedded felspathic sandstone, carbonaceous shale and coal, while the latter is composed of felspathic sandstone, shale and clay. The succeeding Tipam series comprises of two layers, a lower one contain felspathic and ferruginous sandstone, and upper one contain shale, mottled clay, sandy clay and ferruginous clay (Girujam clay). The Disang series consists of pebble beds, gravels, thin clay and sandy.

The alluvium is composed of unassorted clay, coarse sand, gravel and boulder bed.

2.30. WEATHERING OF ROCKS AND MINERALS

"Rocks occurring within the earth remain in equilibrium with their underground environment. When the portion of the earth's crust is upheaved (gets lifted up from beneath) or the rock/soil overlying the rocks are naturally are removed, the rock thus exposed at the surface of the earth are in a state of disequilibrium with their new environment at the surface of the earth when the rocks are acted upon by the atmospheric agents *i.e.* heat, cold, rain, wind and chemical action of atmospheric gases. Consequently rocks are partially/wholly disintegrated and decomposed. All these changes which have been caused by the action of the atmospheric agents on rocks and minerals at or near the surface of the earth is called the 'Weathering' of rocks which may be mechanical/physical, chemical and biological. The weathering process may be mechanical or physical and chemical and biological.

Mechanical or physical weathering of rocks breaks the rocks down to smaller and smaller pieces without appreciably altering their chemical composition. But definite chemical changes takes place by the chemical weathering of rocks. Easily decomposable minerals are decomposed, the products of the decomposition may be washed away or may recombine with each other to form new minerals.

MECHANICAL OR PHYSICAL PROCESSES OF WEATHERING

Rocks heat up during the day and cool down during the night. They are composed of minerals which differ in their coeficient of expansion. Different minerals present in the rocks expand and contract at different rates, on being heated and cooled. Some expand and contract more than others. This differential expansion and contraction of minerals continues for long time. Consequently, cracks appear on the surface of the rocks. Ultimately, rocks break down to smaller and smaller pieces.

Rocks occurring inside the earth remain under tremendously high pressure of the rocks overlying them. When the overlying rocks are eroded away, or they are geologically upheaved to be exposed at the earth's surface, they are relieved of the high over burden pressure. Consequently they expand when the outer layer of the rock may separate from the rest of the rock masses. This process is called exfoliation. In cold weather, water freezes in the cracks of rocks to form ice when tremendous pressure (about 150 tons per square foot) is exerted upon the rock masses by the expanding ice. The cracks widen. Ice melts during the warm season and water again freezes to form ice during the next cold season, which again expands. So the width of the crack increases more due to the pressure of the expanding ice. Repetition of this process causes the rocks to gradually break down to smaller and smaller pieces.

Rocks and minerals were disintegrated by the abrasive action of glaciers as they moved under their own weight. Water, while it flowed over the rocks, dislodged smaller pieces of rocks and carried them away. Water loaded with such materials has tremendous cutting power. Rocks roll in running water and becomes rounded.

Wind, when it carries coarser particles, scrapes rocks and carries away the finer particles.

CHEMICAL PROCESS OF WEATHERING

Atmospheric gases like carbon dioxide, sulphur dioxide, nitrogen peroxide water vapour, and oxygen react with the minerals contained in the rocks at different temperatures, and changes them to new minerals by the following processes:

1. *Hydrolysis*: The reaction of the minerals with water is called hydrolysis of minerals as shown in the following equation.

$$KAlSi_3O_8 + H.OH \longrightarrow HAlSi_3O + KOH$$

Orthoclase or Microcline → Alumino Silicic Acid

$$2HAlSi_3O_8 + 8H.OH \longrightarrow Al_2O_3.3H_2O + 6H_2SiO_3$$

Alumino Silicic Acid → Bauxite + Silicic Acid

Bauxite and Silicic acid may combine with each other to form clay minerals like Kaolinite.

2. *Hydration*: Hydrogen and hydroxyl ions are rigidly attached with the crystals of the minerals to become an intergral part of them. This phenomenon is called hydration of minerals. For example, red haematite reacts with water to form yellow limonite

$$2Fe_2O_3 + 3H.OH \longrightarrow 2Fe_2O_3 + 3H_2O$$

Haematite (red) → Limonite (yellow)

when micas become hydrated some hydrogen and hydroxyl ions move in between plate-like units of mica expand it and make it more porous.

3. *Carbonation*: The process by which carbonic acid reacts with minerals and changes them to new minerals, as shown below is called carbonation:

$$\underset{\text{Orthoclase or Microcline}}{4KAlSiO_8} + 2H_2OCO_3 + 3H_2O \longrightarrow 2K_2CO_3 + \underset{\text{Kaolinite}}{Si_4Al_4(OH)_8} + 8SiO_2$$

$$\underset{\text{Spaingly Soluble}}{CaCO_3 + H_2O + CO_2} \rightleftharpoons \underset{\text{Soluble}}{Ca(HCO_3)_2}$$

4. *Solution* : Rain water gradually dissolves sparingly soluble minerals as shown above.

Heavy rain washes the calcium and the bicarbonate ions down, the above reversible reaction proceeds in the forward direction. So the sparingly soluble calcium carbonate is gradually dissolved in water.

5. *Oxidation*: The process by which electrons are lost from ions whose positive charge is increased is called oxidation. For example, ferrous iron loses an electron to become ferric iron

$$Fe^{++} \rightleftharpoons Fe^{++} + {}^{+}\text{electron}$$

A compound is said to be oxidized when it combines with oxygen as shown in the following equation

$$\underset{\text{Ferrous oxide}}{4FeO + O_2} \longrightarrow \underset{\text{Hematite}}{2Fe_2O_3}$$

Ferromagnesian minerals contain ferrous ions which are earily oxidized to the ferric ion. So some positive ions must go out of the crystal lattice of the mineral in order to maintain its electrical neutrality. So some position in the crystal lattice of the mineral becomes empty, facilitating the entry of ions into it and exist of some ions from it. Ferromagnesian minerals are rapidly decomposed this way:

$$\underset{\text{Olivine}}{6MgFeSiO_4} + 2HOH \longrightarrow \underset{\text{Serpentine}}{Mg_6(OH)_8Si_4O_{10}} + 6FeO + 2SiO_2$$

BIOLOGICAL PROCESS OF WEATHERING

Some lower forms of plants like mosses, algae and lichens (Sec. 8.4) grow even on bare rocks and respire producing carbon dioxide, which combines with water to form the carbonic acid that decomposes minerals contained in the rock to form the clay. These organisms are short lived. They die and their bodies quickly decay to add organic matter to rocks, on which another crop of microorganisms grow. Higher plants also grow within the crevices of rocks, widened by the pressure exerted by the growing roots on rock masses. They also add organic matter to the rocks which decomposed to form organic acids decomposed rock minerals to form more clay.

Big masses or rocks are also disintegrated by human activities. The physical process of weathering of soil mineral is accelerated by tillage operations. Application of lime and fertilizers to soils hastens the decomposition of the soil minerals. Animals e.g. moles and rats make burrows when the soil minerals become smaller by abrasion. Earthworms eat large amounts of organic matter along with fine clay. The earthworm casts are ideal environments for quicker decomposition of the soild minerals.

FACTORS AFFECTING WEATHERING OF ROCKS AND MINERALS

(i) *Climate*: The physical or mechanical process of weathering dominates over the chemical process under conditions of low rain fall of when the sizes of the minerals particles are decreased without much change in their chemical composition. But the minerals are decomposed by high rainfall and high rainfall conditions in humid regions.

(ii) *Physical characteristics*: Rocks composed of large crystals are easily disintegrated when the temperature frequently change. Hence physical or mechanical weathering dominates. But when the rocks are composed of small crystals of minerals, more surface area is available for a chemical reaction. Hence the chemical process of weathering dominates over the mechanical process of weathering.

(iii) *Chemical composition*: Dark coloured ferromagnesian minerals containing ferrous ions are easily decomposed, because ferrous ions are oxidized to ferric ions. So some ions must leave the crystal structure of the mineral to balance it electrically thus leaving empty spaces in the crystal which reduce the electrostatic forces that binds the different parts of the crystal together. They also become more susceptible to acid attack and dissolution. The crystal structure is gradually weakened and ultimately broken down.

SELECTED READINGS

1. Berry, L.G. Mason, B. and Dietrich, R.V. (1985): *Mineralogy*. CBS Publishers and Distributors, 485 Jain Bhawan, Bhola Nath Nagar, Shahdara, Delhi-110 032.
2. Brady, N.C. (1984), *Nature and Properties of Soil*, 9th Ed. Macmillan Publishing Co., New York.
3. Daji J.A. (1968), *A Text Book of Soil Science*, Modern Promoters Publishers Pvt. Ltd., Bombay.
4. Datta, A.K. (1984); *An Introduction to Physical Geology*, 15th Ed., Kalyani Publishers, 24 Daryaganj, New Delhi-110 002.
5. Directorate of Geology and Mining, Government of Nagaland (1978), *Status of Geological Work and Inventory of Mineral Discoveries in Nagaland*.
6. Donahue, R.L. (1977), *An Introduction to Soils and Plant Growth*, Prentice Hall, Englewood Cliffs, New Jersy 07632.
7. Krishnan, M.S. (1982), *Geology of India and Burma*, CBS Publishers and Distributors, 485, Jain Bhawan, Bhola Nath Nagar, Shahdara, Delhi-110 032.
8. Mehdiratta, RC (1985), *Geology of India, Pakistan, Banglade*[illegible] *a* Atma Ram & Sons, Delhi.
9. Read, H.H. (1984), *Rutley's Elements of Mineralogy*, 26th Ed., CBS Publishers & Distributors, 485, Jain Bhawan, Bhola Nath Nagar, Shahdara, Delhi 110 032.
10. Sinha, R.K. (1975), *An Introduction to Geology*, Vikas Publishing House Pvt. Ltd., 5 Ansari Road, New Delhi-110 002.
11. Spence, r.W.(1962), *Basic Concept of Physical Geology*, Oxford & IBH Publishing, Co., 66, Janpath New Delhi.
12. Whilten, D.G.A and Brooks, J.R.V. (1972). *Dictionary of Geology*, The Penguin, p. 393.

3
SOIL GENESIS

"Most of the Pedologist (The person who knows pedology that deals with the origin, formation and geographic distribution of the Soil as a natural body) have briefly defined the soil to be the outermost weathered layer of the earth's crust. J.S. Joffe who is one of the representative of the Russian School of Pedology has defined the soil to be a natural body which have been differentiated into horizons and which is usually unconsolidated, of variable depth, and which differ from the underlying material in morphological characteristics, physical properties and chemical composition and properties and biological characteristics. This definition is limited by the fact that distinct horizon do not occur in some soils. The United States Soil Conservation Service (1975) have defined the soil as "The soil is the collection of natural bodies occurring on the surface of the earth which either supports the growth of plants or is capable of doing so. Biswas and Mukherjee (1987) have given the undermentioned definition: "The soil is a dynamic natural body (i) which has been formed as the result of pedogenic processes taking place during and after weathering of rocks (ii) which are constituted of inorganic and organic substances, (iii) which possess definite physical, mineralogical, chemical and biological properties, (iv) which is of variable thickness and (v) which acts as a medium for growth of terrestrial plants."

3.1. SOIL VERSUS REGOLITH

The bed rock is commonly overlaid by some unconsolidated debris called the regolith of thickness varying from a few to hundreds of feet, which might have originated at the same site from the weathering of underlying bed rock or which might have been brought from elsewhere by the action of wind, water or ice only to be deposited on the bed rock or on other material overlying the bed rock.

The regolith tends to vary in composition from place to place. Its upper portion of a few metre thickness, tends to differ from its remaining lower portion. Plant roots grow in it and add organic matter to it in the form of sloughed off root hairs which are decomposed by microorganisms. Earth worms are also active in this upper few metres of the regolith which is therefore higher in organic matter and more porous than its remaining lower portion. As the upper few metres of the regolith is nearer the surface than its remaining lower portion, so the former has been more weathered than its remaining lower portion. Some of the products of weathering moves down the former which is therefore comprises of layers or horizons. This upper biochemically weathered portion of the regolith has been designated as the solum or true soil which is the product of both destructive and synthetic processes. Weathering of rocks and minerals and microbial decomposition of organic matter are examples of destructive processes whereas the formation of secondary minerals i.e. clay and the formation of certain characteristic layered patterns are synthetic in nature. These forces are responsible for the formation of a distinct entity called the soil which comprises of a large number of individual soils that possesses recognized characteristics and properties.

3.2. FORMATION OF SOIL

Rocks were slowly broken down to smaller and smaller pieces by physical weathering, then decomposed by chemical weathering to form the parent material, which may be defined as the weathered material from which the soil was synthesised. The parent material was further decomposed by the action of atmospheric gases like water vapour, carbon-dioxide, sulphur dioxide, nitric oxide etc. at different temperatures. Some of the primary minerals present in rocks were altered to form some secondary minerals. Some were completely decomposed and the products of decomposition recombined with each other to form some secondary minerals, which also included clay minerals. These clay minerals or simply the clay remained intimately mixed with primary minerals like quartz or sand to form some soil. Some lower plants like algae, moss, lichens (association of algae and fungi is called lichens) etc. began to grow on the bare rocks and on this thin layer of soil, respiring to produce carbon dioxide, which reacted with water to form carbonic acid that decomposed the primary minerals to form clay. Later on, higher plants began to grow and continued this process. These plants continued to add organic matter to the soil in the form of leaves, roots etc. which decomposed in the soil to form humus that combined with the clay to form the clay-humus complex. Hence the colour of the surface soil gradually darkened. This process continued till a darker layer of soil of about one foot in thickness developed. This is known as the A horizon. The parent material, which at this stage, was at a considerable distance from the surface, could not be subjected to the direct action of atmospheric agents like heat, moisture and gases etc., so it continued to decomposed at a slower rate forming a relatively compact layer called the B horizon. Some clay, and sesquioxide (Sum of the oxides of iron and aluminium is called sesquioxide, R_2O_3) were gradually washed down or eluviated from, the lower portion of the A horizon and were washed in or illuviated in the B horizon. So the lower eluviated portion of the A horizon became the E horizon. The E horizon is usually lighter in colour and texture may be and a little acidic in reaction.

In the earlier stages of soil development, soils were dominated by characteristics which they inherited from the parent material. They are dominated by the acquired characters at the later stages of soil development. For example, soils which were developed from the basic parent material were rich in basic elements and alkaline in reaction during the earlier stages of soil development. Later on these basic elements were gradually washed down by high rainfall and the soils ultimately become acidic in reaction at the later stages of their development.

3.3. SOIL PROFILE

We have been that as the parent material was gradually converted to soil, definite layers developed. These layers are called horizon. A soil profile (Fig. 3.1) is a vertical section of the earths crust that includes different layers called horizons, formed due to the action of soil forming factors and also the deeper layers that influences soil formation. A theoretical soil profile consists of the following horizon:

O Horizon—	Organic Horizon develops from dead plant materials on forest lands. This surface horizon occurs just above the mineral or inorganic horizon. It has been subdivided as follows:
Oi Horizon—	Consists of undercomposed plant material, the original form of which can be recognised.
Oe Horizon—	Consists of incompletely decomposed plant material.
Oa Horizon—	Consists of completely decomposed plant material, the original form of which cannot be recognised. It is black in colour.
A Horizon—	is the topmost mineral horizon where humus has been thoroughly mixed and chemically combined with clay to form clay-humus complex. so it is black in colour.
E Horizon—	is the horizon of eluviation from which some clay and sesquioxide and humus have been washed down or eluviated. So it is lighter in texture (i.e. relatively richer in sand) and colour.
EB Horizon—	Transitional layer between E and B. It is more like E than B.
BE Horizon—	Transitional layer between E and B. It is more like B than E
B Horizon—	is the horizon of illuviation. Some clay, sesquioxide humus etc. have been washed in or illuviated in the B horizon, which is relatively heavier in texture and darker in colour. Lime may accumulate in the B horizon in arid region.
C Horizon—	is the parent material which is the partially or full weathered material from which A, E and B horizon have been developed.
R Horizon—	is the Bed rock.

Solum is the true soil developed by the action of soil forming factors. It includes A, E and B horizon.

Regolith is the unconsolidated material lying about the bed rock. It includes A, E, B and C horizon.

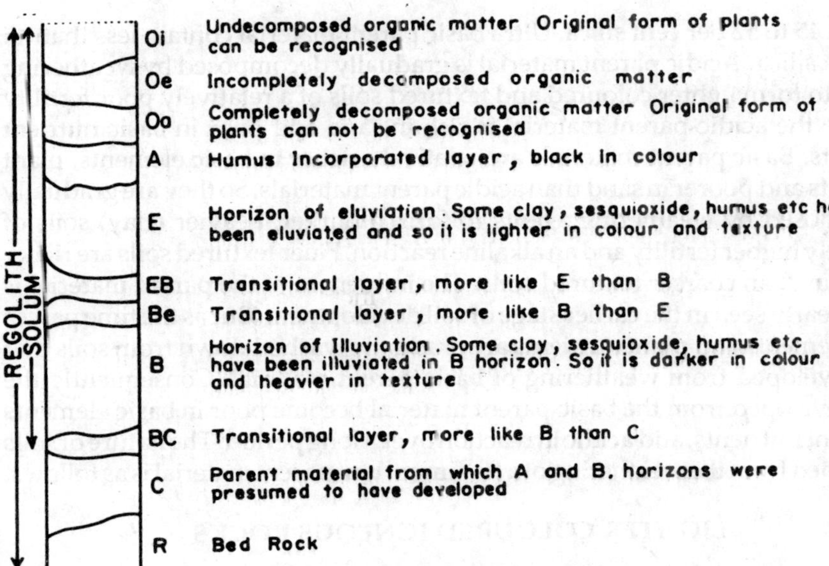

Fig. 3.1. Theoritical profile

Usually all the above mentioned horizons are not found in every soil profile because they may be immature or over influenced by poor-drainage, topography etc. Well drained mineral soils usually possess O and Oa if the land is under forest, A or E horizon, B and C horizons depending on soil development.

When farmers being to plough vigin soils, the original layered condition of the soil from 15-25 cms depth is destroyed. This layer has been thoroughly mixed up by ploughing and it includes A and E horizon when they are sufficiently deep. Otherwise it may include a little part of B horizon also. This mixed layer is called the furrow slice designated as Ap horizon.

3.4. DIFFERENCE BETWEEN FACTORS AND PROCESSES OF SOIL FORMATION

Soil forming factors are parent material, climate, vegitation topography and time, considered more or less independent of each other. But the simultaneous interaction of all these soil forming factors results in the particular soil forming process.

3.5. FACTORS OF SOIL FORMATION

Rocks are first broken down to smaller and smaller pieces which are further decomposed to form the parent material. Therefore a parent material may be defined as the unconsolidated, fully or partially weathered material from which the solum has been presumed to have developed. Parent materials have been classified on the basis of their silica content as acidic or Felsic, intermediate, basic or mafic and ultra basic or ultra mafic.

Acidic parent material contains more than 66 per cent silica. Intermediate parent material contains 52 per cent to 66 per cent silica. Basic parent material contain 45 to 52 per cent silica. Ultra basic parent material contain less than 45 per cent silica. Acidic parent material is gradually decomposed by weathering agents to form lighter coloured and textured soils of a relatively poor fertility because the acidic parent material is rich in sand and poor in basic nutrient elements. Basic parent materials are relatively richer in basic elements, plant nutrients and poorer in sand than acidic parent materials. So they are gradually decomposed by weathering agents to form a darker, heavier (clay) soils of relatively higher fertility and an alkaline reaction. Finer textured soils are richer in humus than coarser textured soils. The influences of the parent material is more clearly seen in the earlier stage of soil development. But as the time passes basic elements and plant nutrients are gradually washed down from soils that had developed from weathering of basic parent material. Consequently the soils developed from the basic parent material become poor in basic elements and plant nutrients, and acidic in reaction over a long period. The nature of soils developed from the weathering of a few important parent material is as follows.

LIGHTER COLOURED IGNEOUS ROCKS

The compositions of Rhyolite, and Granite (Sec 2.11) are almost similar, and therefore, they weather to form acidic parent material, which in turn decomposes to form friable, permeable sandy loam to loam soils of light yellow colour due to their low iron content. The primary minerals decomposed to form the secondary clay minerals like illite, vermiculite, montmorillonite etc. in drier climates, which may decompose to kaolinite when both rainfall and temperature increase. Therefore the soils which have developed in a dry climate may contain fair amounts of plant nutrients and are alkaline in reaction, but those which have developed in a warm humid climate are poor in plant nutrients and moderately to strongly acidic in reaction.

Both trachyte and syentite consists almost entirely of orthoclase. But they also contain low amounts of biotite and hornblende. Since orthoclase weathers very slowly to form clay, the soils developed from weathering of trachyte and syenite is medium in texture and fertility and natural in reaction.

DARK COLOURED IGNEOUS ROCKS

Andesite, diorite, basalt, gabbro, perdotite and dunite consists of feromagnesian and calcium bearing minerals, which rapidly weather to form dark coloured clayey soils rich in plant nutrients, basic elements and alkaline in reaction. Usually montmorillonite and allied mineral smectite (group) dominate the clay minerals. If the rainfall and temperature further increases, the montmorillonite decomposes to kaolinite.

SEDIMENTARY ROCKS

Sandstone mainly consists of quartz, but a little feldspar and mica are also present as impurities. In general, parent material formed from weathering of sandstone decomposes to form light coloured sandy soils, poor in plant nutrients and acidic in reaction. Shales are made up of clay minerals and smaller quantities of mica, feldspar, hornblende and quartz. Sometimes a little calcite may be present. Shales usually weather to from darker coloured, relatively impermeable clayey soils of a high base and plant nutrient status and alkaline in reaction in which usually illite and montmorillonite dominates the clay minerals. Soils are usually very fertile.

Limestone and dolstone contain more than 50 per cent calcite and dolomite respectively. They weather to form darker coloured clayey soils rich in plant nutrients and basic elements and alkaline in reaction in a cool dry climate. But the nature of soil formed from the weathering of limestone and dolstone in a warm humid climate depends upon the nature of the impurities present in them because calcium carbonate and calcium magnesium carbonates are washed down by high rainfall and the impurities are left behind. If the impurities are mainly clay, then dark coloured clayey soils fairly rich in plant nutrients and alkaline in reaction are formed. But if the impurities are quartz or iron oxide then light coloured, highly permeable sandy soils poor in plant nutrients and basic elements and acidic in reaction are formed.

METAMORPHIC ROCKS

Gnesis usually weathers to form soil which is dominated by smectite and hydrous mica groups of clay minerals in a cool humid climate. These soils contain a fair amount of plant nutrients and are neutral to slighly alkaline in reaction. But these clay minerals further decompose to kaolinite when the temperature and precipitation is increased. Hence highly permeable coarse textured acidic soils are formed in humid climate.

Soils developed from the weathering of mica schists tend to be silty, and are dominated by illite and vermiculite. Hornblende schist may weather to form the smectite group of clay minerals. Schists usually weather to form loamy to clayey soils, fairly rich in plant nutrients and alkaline in reaction. Recognisable horizons develop faster on sandy loam to loam soils which are high in easily weatherable minerals, are well drained and aerated and are located on moderate slopes. They develop slowly on sandy parent materials which are poor in easily weatherable minerals. They also develop slowly on highly clayey parent materials.

RELATION BETWEEN PARENT MATERIAL AND VEGETATION

If the parent material is rich in basic elements, it weathers the soil rich in basic elements and the vegetation growing on it is rich in basic elements. Parent materials like sandstones; quartzites etc. are poor in basic elements Coniferous narrow leaved plants usually grow on such acidic soils in cool humid climate.

CLIMATE AND CLIMATIC WATER BALANCE

Weather is the state of the atmosphere at any given time and place. Climate is the average weather. In Greek, climate means the angle of the suns rays striking the earth. Differential heating causes local variation in the earths temperature which cause air mass movement and induces precipitation.

The climate comprises of precipitation and temperature that affect various physical and chemical processes in soil. Climatic water balance is the difference between gain in water mainly through rainfall and loss of water due to evaporation. This water reached the surface of the earth and took part in soil formation. In arid region, less rain water reaches the ground surface, more water is evaporated and therefore, the climatic water balance decreases. Climatic water balance may be estimated by various means:

(*i*) Lang's Rain Factor = $\dfrac{\text{Mean annual rainfall in millimetres}}{\text{Mean annual temperature in °C}}$

(*ii*) Niederschlag Sathigungs deficit Quotient (NSQ of Meyer (1962)

$$\text{NSQ} = \frac{\text{Rainfall in Millimetre}}{\text{Absolute saturation deficit of air milimetre of mercury}}$$

(*iii*) Thornthwaite's (1931) Precipitation Effectiveness Index (P.E. Index)

$$\text{P.E. Index} = \sum_{n-1}^{n-12} 115 \left(\frac{P}{T-10}\right)$$

where P = Monthly rainfall in inches
T = Mean monthly temperature in degree F.
n = Cencerned month.

The climatic water balance effect, the decomposition of minerals the formation of clay and the growth of natural vegetation. The soluble products of decomposition are removed along with the percolating and the run off water.

As the climatic water balance increases, more natural vegetation grows and more organic matter is added to the soil. Hence the organic matter and nitrogen content of soil increase. Primary minerals decompose at increasing rates to form 2 : 1 types of clay minerals like smectite (montmorillonite) hydrous mica and vermiculite groups of clay minerals. Hence the percentages of clay in soils and their cation exchange capacities increase when the climatic water balance increase. However, excessive rainfall leaches the basic elements and silica down the soil profile to make soils acidic in reaction when 2 : 1 types of clay minerals decomposes to form the 1 : 1 group of clay minerals (Kaolinite) which are further decomposed to hydrous oxides of iron and aluminium, 2 : 1 types of clay minerals dominated the clay mineralogy of soils of relatively arid regions which are rich in basic elements and alkaline reaction. A calcium carbonate layer occurs near the surface of the soils if the climate is extremely arid. The depth of this calcium carbonate layer increases with an increase in rainfall.

When rainfall is constant and the temperature is increased, less vegetation grows thus less organic matter is added to the soil. Organic matter rapidly decompose. Hence the humus and nitrogen content of soils decreases. The increase in temperature also increases the rate of weathering of primary minerals to increase the clay content and cation exchange capacity of young soils. But as the clay decomposes, the cation exchange capacity of old soils decreases.

LIVING ORGANISMS

The biosphere comprieses of billions of microorganisms which improve scil productivity by decomposing rocks, minerals and organic matter. Of special interest in the influence of the biosphere on soil genesis, is the lichens which are symbiotic associations of algae and fungi. Algae, moss, and lichens grow on bare rock and respire to produce carbon dioxide, which reacts with water to form carbonic acid that dissolves primary minerals and releases the nutrient contained in them for plant growth. Algae use atmospheric nitrogen which is released in the soil upon the death of the algae. Thereafter higher plants grow in an environment richer in nutrients and water. Higher plants absorb-nutrients from the lower layers of the soil and leave them on the surface in the form of decaying leaves, which decompose, forming humus. Carbon dioxide produced from the decomposition of organic matter and respiration of microorganisms and plant roots continues to decompose primary minerals to form more clay and **convert** insoluble minerals to soluble ones.

$$CO_2 + H_2O \longrightarrow H_2CO_3$$

$$\underset{\text{Insoluble}}{CaCO_3} + H_2CO_3 \longrightarrow \underset{\text{Soluble}}{Ca(HCO_3)_2}$$

$$\underset{\text{Orthoclase of Microcline}}{4\,KAlSi_3O_8} + 2H_2CO_3 + 2H_2O \longrightarrow 2K_2CO_3 + \underset{\text{Keolinite}}{Si_4Al_4O_{10}(OH)_8} + 8SiO_2$$

When plants grow on soils the phosphorus and potassium content of soils increases but the calcium and magnesium content and the soil pH usually decreases.

"Root system of **grasses** are profusely developed and are uniformly distributed throughout the A horizon and the upper portion of the B horizon. These roots die and decay to form the humus which is incorporated with the soils of the A and the upper portion of the B horizon throughly enough to impart dark colour to them. Forest trees shed huge quantities of leaves on the forest floor. Hence forest land soils possesss the thick "O" horizon.

Usually grassland soils contain more organic matter and nitrogen and are heavier in texture and darker in colour, of higher cation exchange capacity and pH then the forest and soils. Pine trees which contain low amount of basic elements, grow on soils poor in basic elements and acidic in reaction.

When both grassland and forest land soils have developed from the same parent material under the same climatic conditions, the rate of removal of bases and the consequent development of soil acidity is more in forest land soil than in the grassland soil probably due to the three following reasons. Firstly the forest trees return fewer basic elements in the form of decayed leaves to the surface soil. Different species of trees return different amounts of bases to the soil. Secondly, rainwater leaches the basic elements and plant nutrients to a greater depth in before they are absorbed by roots in forest land soil then in grassland soil. Thirdly, water entering forest soils is more acidic in reaction because it contains organic acids formed from decomposition of forest vegetation. The organic acids react with the primary minerals of the forest soil to release the basic ions, which are washed down by high rainfall. So in forest land soil, the percentage of clay in the B horizon is usually more than that of the A horizon whereas the clay contents of the A and the B horizons of grassland soils are more or less equal. The decomposition of the organic matter in the forest land soils usually produces organic acids that humus and clay are washed down to B horizon. This horizon therefore possesses illuvial deposits of humus and iron and aluminium compounds in addition to the illuvial deposit of clay.

Soil animals like moles, earthworms, ants and termites bring the soil from B horizon to A horizon. Excessive ploughing of the land, burning of the natural vegetation and overgrazing of the land, decreases the soil productivity and retards soil development. The application of lime and fertilizers to the soil also retards the development of the soil profile but increases soil productivity. In the humid regions, this practice makes the soil environment more favourable for the growth of grasses then for trees. The soil profile also develops slowly when the soil is constantly mixed up by the animal and the people.

Living organism are influenced both by parent material and climate. Vegetation growing on soils developed from basic parent material are usually acidic parent material. Grasses are the dominent form of vegetation in semi arid areas whereas the forest are the dominanat vegetation in semi humid areas. Earthworms are usually found in soil richer in basic element and organic matter and alkaline in reaction in the humid region. Ants and termites usually occur in soils of temperatate regions and tropical and sub-tropical regions respectively.

TOPOGRAPHY

The relief of the land refers to the differences of elevation within it. The relief shown on the topographic map is called the topography of the land.

The parent material which is located on the fairly level to very gently sloping lands, quickly decomposes to form the soil profile because a major proportion of the rain water washed down some of the soluble salts and humus and fine clay down the soil. If the parent material has any hard pan to restrict the downward flow of water through it, the soil profile development will be slower. Water may be ponded on the surface of the level land where the organic matter may accumulate to form the peat and the muck soil. The rain water rapidly flows down the parent material which is located on the steep slope and carries away considerable amounts of the weathered materials. So less water passes down the parent material on steeper slopes. So less of parent material is decomposed to form less soil on steeper slopes. When the slope of the land is increased, more clay and basic ions move to the bottom land. Hence the bottom land soils are darker in colour heavier in texture, deeper and more alkaline in reaction than soils on steeper slopes, the soil located on steeper slopes are highly oxidized, red, acidic in reaction and of shallow depth. A group of soils which have developed from the same parent material in same climate but under different topographical condition is caled SOIL CATENA.

Topography influences vegetation. Grasses occupy the elevated areas and forest plants occupy the lower areas where the climate is neither too dry nor too wet.

TIME

Time has been regarded as one of the factors influencing formation soil because even chemical weathering of rocks to form soil requires some time to complete itself. The parent material passes through the following five stages of soil formation when it gradually decomposes to form soil:

(*i*) *Initial stage*: The parent material has not yet been weathered.

(*ii*) *Juvenile stage*: Weathering of parent material has just started.

(*iii*) *Verile stage*: Now the easily decomposable parent material has decomposed to form clay. So the percentage of clay in the soil has increased.

(*iv*) *Senile stage*: Slowly decomposable primary minerals have also decomposed.

(*v*) *Final stage*: Soil profile development is practically complete.

If conditions are favourable, the parent material may be transformed into immature soil or young soil in relatively less time in which organic matter has accumulated in the surface soil and a lesser quantity of parent material has decomposed to form the A horizon; very little of clay sesquioxide and humus has been eluviated from the A horizon. Only A and C horizons now exist. The soil inherits most of its character from the parent material. As time passes more parent material gradually decomposes to form the B horizon and some clay sesquioxide humus etc. have also been illuviated in the B horizon. All the horizons have been clearly developed and so the soil is mature. As time passes more primary minerals decompose to form the secondary minerals which are also gradually decomposed to form the hydrous oxides of iron and aluminium. When all the primary minerals have been converted to hydrous oxides of iron and aluminium, the soil has reached the old age stage.

3.6. SOIL FORMATION IN ARID REGIONS

Arid regions are commonly characterized by high temperature and low rainfall where chemical weathering of rocks and minerals predominates over their physical weathering. Little decomposition of primary minerals takes place under extremely low rainfall i.e. less than 125 mm per annum to release their constituents in the soluble form that remain in the extremely coarse textured surface soil. The desert soil thus formed is rich in calcium and magnesium and at some place also in sodium and is therefore alkaline in reaction.

A little increase in rainfall i.e. upto 375 mm/annum induces short grasses to come up which get withered during the dry season and rejuvenate with the advent of the short moist season of scanty rainfall. A little quantity of organic matter in the form of dead roots and shoots of grasses is added to the soil that decomposes to form the humus. The moisture and carbon dioxide generated from the respiration of grass roots and decomposition of dead root facilitates the decomposition of some meagre amount of primary minerals to form little amount of clay, Horizons are feebly developed. Whatever meagre amount of clay and humus have been formed, combine together to impart gray colour to the soil which is known as Sierozems which in Russian mean, gray earth. Its A and B horizons are of gray colour. White specks of lime and gypsum have also been seen in the B horizon. The soil is alkaline in reaction.

3.7. SOIL FORMATION IN SEMI ARID REGION

Temperature is less and annual rainfall (375 to 1250 mm) are a little more in the semi arid region than in the arid region. So chemical weathering of primary minerals takes place a little more readily, in the semi arid region than in the arid region. Increase in rainfall permits short grasses to come up. They add some organic matter in the form of their dead roots and shoots during the ensuing dry season moisture coupled with the carbon dioxide generated from the respiration of gross roots and decomposition of dead grasses facilitates the decomposition of primary minerals to form some clay (as shown in section 2.30). Humus combines with the clay to impart gray coloured surface soil. This is known as chestnut soil which is rich in calcium and magnesium and is therefore of alkaline in reaction. A lime layer occur near the surface. A gypsum layer may be present below the lime layer. Hence the soil forming processes is known as calcification.

A little increase in annual rainfall i.e. upto about 1250 mm, induces tall grasses to grow densely enough to profusely develop their underground sterms and roots. Some roots are regularly sloughed off and decomposed to form humus. The availability of relatively higher amount of moisture coupled with carbon dioxide generated from the respiration of the underground roots and stems of grass vegetation and also from the decomposition of dead grass roots and stems facilitates the chemical weathering of primary minerals to form the clay. Humus combines with clay to form the clay humus complex. Hence soils of A and B horizons are uniformly coloured dark gray. The soil is rich in calcium and magnesium and is therefore alkaline in reaction. Relatively higher rainfall is responsible for lowering the lime and gypsum layer. As the soil is rich in lime, so the soil forming process is known as calcification. However melanization which is the process of darkening of the soil mass by the incooperation of humus with the soil, is also responsible for the genesis of the soil which is the process of darkening of the soil mass by the incooperation of humus with the soil, is also responsible for the genesis of the soil which is known as chernozem.

3.8. SOIL FORMATION IN HUMID CLIMATE

Podzolisation is the chemical process of migration of iron and aluminium to the B horizon and precipitation of silica in the A horizon as described below:

The acidic parent material rich in silica and poor in basic elements, has developed from the weathering of acidic rocks like quartzite, that is rich in silica and poor in basic elements. The weathering of this acidic parent material results in the formation of acidic soils poor in basic elements in cold and moist climates. Some plants like Hemlock, pine, Heath etc. grow on these soils and deliver organic matter to the soil which is poor in basic elements and therefore decomposes to produce considerable amounts of organic acids, which decomposes clay to liberate silicon, aluminium, iron, and other mediums of the A_2 horizon which becomes as *h* grey in colour. Iron, aluminium ions, clay and humus are washed down to the B_2 horizon, where the positively charged iron and aluminium ions react with the negatively charged clay and humic micelle and precipitate there. In Russia, pod means under, zola means ash. They call the process podzolisation, and the soil, podzol because it possesses an ash agey A_2 horizon.

LATOSOLISATION (LATERISATION)

It is the chemical migration of silica from the solum and precipitation of iron and aluminium in the solum as described below.

Basic rocks like Basalt decompose when both rainfall and temperature are high, to form basic parent material, which in turn decomposes to form soil rich in basic the elements, on which some broad leaved forest plants grow and organic matter is added to the soil as dead leaves. The basic parent material is rich in ferromagnesian minerals. The organic matter is also rich in basic elements. The ferromagnesian minerals and organic matter decompose to release silicon, aluminium, iron and basic ions. The basic ions make the medium alkaline, which keeps the silicon ions in solution which is washed down by high rainfall. Iron is precipitated and later oxidized to ferric oxide which is red in colour. So the soil may be coloured red. Later, basic elements are gradually washed down when the soils become acidic in reaction, and kaolinite decomposes to hydrous oxides of iron and aluminium. This process is called latosolisation. Whenever drainage is restricted, a soft deposit of iron oxide occur at or near the water table. It is cut in the form of bricks which hadren on drying. In Latin later means bricks. So this process is known as Laterisation.

GLEIZATION

When soils remain saturated with water for long periods in the presence of organic matter, ferric compounds like ferric phosphate and ferric sulphide are reduced to ferrous compounds like ferrous phosphate or sulphide which are bluish grey in colur. So this sub-merged soil develops a bluish colour which change to brown when the soil is re-exposed to the atmosphere. This process of soil formation is known as gleization. Gleization usually takes place in low lying areas where water accumulate.

3.9. SALINIZATION, ALKALIZATION AND DEALKALIZATION

Soluble neutral salts of sodium, calcium and magnesium originate mainly from the decomposition of primary minerals in the soils of arid regions. When water evaporates from the surface of the soil, the water containing soluble salts moves from the deeper layers to the surface of the soil and deposits them at the surface of the fields, which are covered with a white crust of soluble salts in patches.

This process of soil development is known as salinization (or formation of solonchak).

If these soluble salts are removed to the lower layer by a limited amount of rainfall occurring in the arid regions, then calcium and other ions are replaced from the clay and humic micelle by sodium ions. Consequently the soil becomes sticky and plastic when wet and very hard when dry, black due to the dissolution of humus in the alkaline medium. This process of soil formation is known as alkalization (or formation of solonez).

If the rainfall increases a little, sodium ions are replaced from the clay and humic micelle by hydrogen ions, the silicate clay is decomposed to release silica which is deposited on the soil particles. Consequently an ash grey colour develops on the soil. The process a soil formation is known as dealkalization (or formation of Solod).

3.10. SOIL IS A NATURAL BODY

Soil is the collection of natural bodies occupying portions of the earths surface, possessing properties due the integrated effects of climate and living matter, action upon parent material as conditioned by relief over periods of time.

Soils are natural bodies that exhibit three dimensional sequences of characteristics. First, the properties gradually change downwards from the surface to the bed rock. The unconsolidated material lying above the bed rock is called the regolith. The upper portion of the regolith is different from the lower portion. Being nearer the atmosphere, this upper zone has been subjected to the weathering action of wind, water and heat. Plant roots are found in the zone. Organic matter delivered as fallen leaves on the surfaces and dead roots below the surface are decomposed to form humus. The humus and partially decomposed organic matter is mixed with the soil at the surface layer. Primary minerals are decomposed to form clay. This upper and biochemically weathered portion of the regolith is called the solum.

Scientists have considered the soil to be a natural body possessing both depth and surface area. The properties of soil change vertically because the intensity of weathering of primary minerals decreases from the surface downwards and also because the organic matter is incorporated into surface layer, and decomposed there. The characteristics of soils also changes in the horizontal direction due to change in topography and parent material e.g. soils on gently sloping upland is well drained and oxidized whereas the soil on the adjacent bottom land may be poorly drained and, consequently in a reduced condition. Soil developed from granite tends to be coarse textured, whereas soil developed from limestone tends to be fine textured.

Scientists have recognised the variation in characteristics of the soil in the horizontal direction. They have set up classification systems in which the soil is considred to be composed of a large number of individual soils, each with its own distinguishing characteristics.

An individual soil body is bounded laterally by other soil bodies. Adjacent soil bodies may be differentiated on the basis of the depth of solum e.g. a soil individual may have depth of 60 cm to 90 cm which means that it is bounded laterally by two other individual soils whose depths are less then 60 cm and more than 90 cm respectively. All these concepts are applied to find out the minimum size of a soil individual, which is called a Pedon. The lower limit of a pedon is the depth of the solum, its lateral dimensions are large enough to permit the study of the nature of any horizon present because a horizon may be variable in thickness, or may even be discontinuous. Its area range from 1 to 10 square metres, depending upon the variability in the horizon. Where horizons are intermitent or cyclic and recur at a linear interval of 2 to 7 metres (7 to 23ft), the pedon includes one-half of the cycle. Thus each pedon includes the range of horizon variability that occurs within these small areas. Where the cycle is 2 metres or all the horizons are continuous and of uniform thickness, the pedon has an area of 1 square metre. Again, under these limits, each pedon includes the ranges of horizon variability associated with that small area. The shape of the pedon is roughly hexagonal. One lateral dimension should not differ appreciably from any other.

3.11. SOIL AS A MEDIUM FOR PLANT GROWTH

Plant growth depends on the following six factors (*i*) Light (*ii*) Mechanical support (*iii*) Nutrient supply (*iv*) Water supply (*v*) Oxygen supply an (*vi*) Heat. Plant growth depends on the soil for all except the first factor. Roots go deeper in to the soil in search of nutrients and water, Roots anchored in the soil enable growing plants to remains erect. They cannot grow deep in poorly drained soil and absorb nutrients and water from the soil. Not less than sixteen elements are essential for the growth of crops. They get carbon and oxygen from the air, hydrogen from water and the remaining 13 elements from the soil which are called plant nutrients. Nitrogen, phosphorus, potassium, calcium, magnesium and sulphur which are required by crops in relatively larger amounts are called major elements or macronutrients and the remaining seven elements i.e. iron, manganese, zinc, copper, boron and molybdenum and chlorine, called minor elements or micronutrients are required in relatively smaller amounts. About 500 ppm or more of macronutrients are required whereas about 50 ppm or less of micronutrients are required by crops. Nitrogen, phosphorus and potassium which are added to the soil in the form of fertilizer, are called fertilizer elements. Calcium and magnesium, are added to the soil in the form of lime are called lime elements.

Primary minerals and organic matter are decomposed to release the nutrients to the soil water in the form of ions. This soil water containing nutrient ions is known as soil solution. Roots take up nutrients in the form of ions mentioned in Table 3.1.

TABLE 3.1 Ionic forms of Plant Nutrients

	Nutrients	*Ionic Form*
Crops absorb an enormous quantity of water,	Nitrogen	- NO_3^-–NH_4^+
most of which is most by transpiration. Only a	Phosphors	- $H_2PO_4^-$ HPO_4^{--}

small fraction of its becomes an integral part of the crop. Therefore the soil must retain sufficient water for absorption by plants. The soil should neither be acidic nor too alkaline for the growth of crops it should be well drained and aerated and free from pathogens.

Potassium - K^{++}
Calcium - Ca^{++}
Magnesium - Mg^{++}
Sulphur - SO_4
Iron - Fe^{++}
Manganese - Mn^{++}
Zinc - Zn^{++}
Copper - Cu^{++}
Boron - BO_3^{---}
Molybdenum - MoO_4^{---}
Chlorine - Cl^-

3.12. DIFFERENCE BETWEEN THE SOIL AND THE LAND

The soil is a dynamic natural body whose properties are due to the interaction of the five factors of soil formation while the characteristics of the land include the nature of the soil, slope, elevation, irrigation and drainage, location, and its suitability for specific use.

SELECTED READINGS

1. Arakeri, H.R. and Donahue, R.L. (1985), *Principles of Soil and Water Conservation and Water Management*, Oxford & IBH Publishing Co., 66 Janpath, New Delhi-1.
2. Biswas, T.D. and Mukherjee, S.K. (1987): *Text Book of Soil Science*, Tata-McGraw Hill Publishing Co. Ltd, Asaf Ali Road, New Delhi.
3. Brady, N.C. (1984), *Nature and Properties of Soil*, 9th Edn., Macmillan Publishing Co., New York.
4. Buol, S.W. *et. al.* (1973), *Soil Genesis and Classification*, Oxford & IBH Publishing Co., 66 Janpath, New Delhi.
5. Foth, H.D. (1984), *Fundamentals of Soil Science*, 7th Edn., John Wiley & Sons.
6. Jenny, H. (1941), *Factors of Soil Formation*, McGraw Hill Book Co., Inc. London & New York.
7. Joffe, J.S. (1949), *Pedology*, Pedology Publication, New Bruswich, New Jersy.
8. Robinson, G.W. (1949), *Soils, Their Origin, Constitution and Classification*, Thomas Murby & Co., London.
9. Russel, E.W. (1973), *Soil Conditions and Plant Growth.* 10th Edn., Eng. Lang. Book Society, London.
10. Stefferud, A. (1957), *Year Book of Agriculture Soils*, U.S.D.A.

4

SOME IMPORTANT PHYSICAL PROPERTIES OF SOILS

Physical properties of soils do not involve any chemical change in their manifestations.

Those properties which can be evaluated by observing and feeling the soil are known as physical properties of soils that can be measured against some scales like intensity or strength or size. For examples, size, density, temperature are some of the physical properties of soils. Physical properties of soils influence the air-water relationship of the soil, which in turn affects the development and proper functioning roots.

Mineral or inorganic soil consists of mineral or inorganic matter, organic matter, and air which are intimately mixed with one another. Inorganic soil should contain about 45 per cent inorganic matter, five per cent organic matter, 25 per cent water and 25 per cent air for the optimum growth of crops.

4.1. SPECIFIC SURFACES OF SOIL PARTICLES

Specific surface of soil particles is the amount of surface area possessed per unit volume of soil particle. It is expressed in square centimetre per cubic centimetre or square metre per gram. When the size of the soil particles are decreased, their specific surface increases. For examples, the specific surface of a one centimetre cube is 6 sq. cm. per cubic centimetre. It is subdivided into 1000 one millimetre cubes, then the specific surface of each of them is 0.06 × 1000 = 60 sq. cm. per cubic centimetre.

4.2. MECHANICAL COMPOSITION OF SOIL

The mechanical composition of the soil is the weight percentage of the mineral matters that occurs in each of the specified size fractions of the soil. Inorganic particles occurring in the soil have classified of the basis of their sizes as follows:

Stone is greater than 2 cm in diameter

Gravel is between 2 mm to 2 cm in diameter

Fine Earth is less than 2 mm in diameter

Fine Earth has been subdivided into the following groups called soil separates according to different systems.

(1) *International system*

Soil Separate	*Diameter in Millimetre*
Coarse sand	2—0.2 mm
Fine sand	0.2—0.02 mm
Silt	0.02—0.002 mm
Clay	less than 0.002 mm

(2) *United States Department of Agriculture (USDA) System*

Soil Separate	*Diameter in mm*
Very coarse sand	2—1.0 mm
Coarse sand	1.0—0.5 mm
Medium sand	0.5—0.25 mm
Fine sand	0.25—0.10 mm
Very fine sand	0.10—0.05 mm
Silt	0.05—0.002 mm
Clay	less than 0.002 mm

USDA System emphasises the sand fraction too much. Sand and silt fractions are equally important according to the British system mentioned below.

(3) *British System*

Soil Separate	*Diameter in mm*
Coarse sand	2—0.6 mm
Medium sand	0.6—0.2 mm
Fine sand	0.2—0.06 mm
Coarse silt	0.06—0.02 mm
Medium silt	0.02—0.006 mm
Fine silt	0.006—0.002 mm
Clay	less than 0.002 mm

(4) *The European*

System equally emphasises sand. Silt and clay fractions as mentioned below.

Soil Separate	*Diameter in mm*
Coarse sand	2—0.6 mm
Medium sand	0.6—0.2 mm
Fine sand	0.2—0.06 mm
Coarse silt	0.6—0.02 mm
Medium silt	0.02—0.006 mm
Fine silt	0.006—0.002 mm
Coarse clay	0.002—0.0006 mm
Medium clay	0.0006—0.0002 mm
Fine clay	less than 0.0002 mm

4.3. PHYSICAL NATURE OF SOIL SEPARATES

Sand particles may be round or irregular in shape, when sand particles are not coated with silt and clay particles, they are not sticky and plastic when wet. The empty between them are too large to retain water by capillary action. Sandy soils are well drained and aerated but cannot retain enough water for crop growth. The average specific surface of fine sand is 0.1 sq. metre per gm.

Silt particles are micro sand particles. They are very irregular in shape. Usually some clay particles adhere with the silt particles. Hence silt particles possess some plasticity (the capacity of being moulded to any shape), stickiness etc. The specific surface of slit particles is about 1 square metre per gm.

Clay particles are plate like in shape and therefore possess an enormous surface area. The specific surface of clay particles usually ranges from 10 to 1000 sq. metre per gm. Clay is very plastic and sticky when wet and very hard when dry. When clay is dried it absorbs a considerable amount of heat, which is evolved again when it is rewetted. This heat is called the heat of wetting. Since the specific surface of clay is very high it can absorb a large amount of water, ions and gases. Clayey soils are usually poorly drained and aerated but can retain enough water for crop growth.

4.4. MINERALOGICAL NATURE OF SOIL SEPARATES

The coarsest sand particles are often fragments of rocks as well as mineral. Quartz commonly dominates the finer grades of sand. In addition to quartz, small quantities of other primary minerals like mica feldspar or calcite may be present. Other secondary minerals like haematite or limonite may be present as coatings around the sand grains when the wand is of red or yellow colour. Quartz is also the chief constituent of silt, which, however, contains a little more other secondary minerals like hydrous oxide of iron and aluminium than the sand and a little secondary silicate minerals or clay minerals as coalings on their surfaces.

Clay mainly consists of secondary silicate minerals like metamorillonite, vermiatite, illite, kaolinite chlorite, allophane and small quantities of other secondary minerals like hydrous oxides of iron and aluminium.

4.5. CHEMICAL NATURE OF SOIL SEPARATES

Quartz is the chief constituent of sand. Primary minerals occurring in sand are too insoluble to release the nutrients contained in them for the growth of crops. The main function of sand seems to be the maintenance of good physical condition for proper growth and functioning of root systems of plants. Hence sand is considered to be the skeleton of the soil.

Some other secondary minerals and in little clay usually adhere to the silt particles. Some potash bearing primary minerals may also be present. Therefore silt can absorb and retain some nutrients for crop growth.

Clays are complex aluminium silicates which absorb nutrient irons and water molecules and release them by a phenomenon called base exchange or cation exchange for crop growth. They are the most active portion of the soil.

4.6. DEFLOCCULATION AND FLOCCULATION OF THE CLAY

Clay particles consist of negatively charged particle called MICELLE (*See* Sec. 7.7 Acid Clay) and numerous positively charged cations which form a diffuse double layer around the micelle. Positively charged cations adsorb water molecules when they increase in size as shown below:

Size in Angstrom (Å)

Ions	*Hydrated*	*Unhydrated*
Lithium	10.03 Å	0.98Å
Sodium	7.19Å	0.98Å

1cm.-= 10^8 Å

When highly hydrated sodium or lithium ions replace other cations from the diffuse double layer surrounding the micelle, the thickness of the diffuse double layer is increased, because they cannot approach the negative inner layer very closely, due to their bigger size. Hence a residual negative charge develops on clay particles which continue to repel each other in a soil-water suspension and do not settle down to the bottom. This phenomenon is called the deflocculation or dispersion of clay.

Whenever a neutral salt like sodium chloride is added to a declocculated clay-water suspension, the large number of sodium and chloride ions intermingle with the cations in the diffuse double layer around the negative clay micelle, to make it less diffuse. So its thickness is reduced. Consequently the residual negative charge on the clay micelle is reduced. Therefore the clay particles do not repel each other but unite to form units called floccules. This phenomenon is called flocculation of clay.

4.7. STOKES LAW

The relationship between the size (radius) of particles and their rate of fall down a liquid medium was first determined by G.G. Stoke in 1851. He stated that: "The resistance offered by a liquid medium against the fall particles down the same liquid medium, varies with the radius of the particles" according to the formula

$$v = \frac{2(dp-d)gr^2}{9\eta}$$

Since $v = \frac{h}{t}$

$$\therefore \quad \frac{h}{t} = \frac{2(dp-d)gr^2}{9\eta} = \frac{2(dp-d)gD^2}{9\eta \times 4}$$

$$= \frac{(dp-d)gD^2}{18\eta}$$

$$\therefore \quad t = \frac{18\eta h}{(dp-d)gD^2}$$

v = Velocity of fall of particles in cms/sec
r = Radius of particles in cms

D = Diameter of particles in cm
dp = Density of particles in gm/cc
d = Density of the medium in gm/cc
η is pronounced as Etah in Greek
η = Viscosity of the medium in poise
g = acceleration due to gravity
= 980 cm/per sec/per sec
h = height of fall of particles in cms/per second

The value of 't' in the above equation increases when the value of D decreases. Therefore as the size of particles are decreased the time taken by them to fall down the same depth increases. Stokes law is used to calculate the time or settling of soil particles in mechanical analysis of soils.

4.8. MECHANICAL ANALYSIS OF SOILS

Mechanical analysis is the analytical procedure by which the percentages of sand, silt and clay in soil are determined in the laboratory.

Principles of Mechanical Analysis

It is based on Stokes' law, according to which when the size of the soil particles is decreased, their time of settling below a constant of about 10 cm increases. So the time t secs of setting of particles larger than a given size say 0.05 mm which is the upper limit of the size of silt particle, is calculated from Stokes' law. The soil-water suspension is shaken and left undisturbed for t seconds, after which only silt clay is left at a 10 cm depth. A sample of a known volume of soil-water suspension (e.g. 25 ml) is withdrawn from this 10 cm depth of the soil water suspension with a pipette. In a similar way, only clay is present at a 10 cm depth after t_1 second. When another sample of known volume, *i.e.* 25 ml, is withdrawn from this depth. Silt + clay present in the first sample and only clay is present in the second sample. Therefore their difference gives the amount of silt present in the first sample.

Pipette Method of Mechanical Analysis

Take 10 gm of soil in a beaker and add 20 ml of 30 per cent Hydrogen Peroxide to it; boil to destroy the soil organic matter. Cool and add a little dilute hydro-chloric acid to destroy calcium carbonate present in the soil and to replace the divalent cations from the clay micelle by hydrogen ions as shown in the following equation.

$$CaCO_3 + 2HCl \longrightarrow CaCl_2 + CO_2\uparrow + H_2O$$

$$K^+ \underset{Ca^{++}\ Mg^{++}\ Ca^{++}}{\overset{Ca^{++}\ Mg^{++}\ Ca^{++}}{\boxed{\text{MICELLE}}}} Na^+ + 14HCl \rightleftharpoons H^+ \underset{H^+ H^+ H^+ H^+ H^+ H^+}{\overset{H^+ H^+ H^+ H^+ H^+ H^{+}+}{\boxed{\text{MICELLE}}}} H^+ + 4CaCl_2 + 2MgCl_2 + NaCl + KCl$$

Filter the soil and transfer it on the filter paper. Wash it with distilled water till it is free from chloride *i.e.* when no white precipitate is obtained with silver nitrate solution. Than transfer the soil back to the original beaker, with distilled water and add a little dilute sodium hydroxide to deflocculate the clay and stir it. Hydrogen ions are replaced from the clay micelle by sodium ions as shown in the following equation.

$$\mathrm{H^+}\ \boxed{\overset{H^+H^+H^+H^+H^+H^+}{\underset{H^+H^+H^+H^+H^+H^+}{\text{MICELLE}}}}\ \mathrm{H^2\ H^+ + 14NaOH} \rightleftharpoons \mathrm{Na^+}\ \boxed{\overset{Na^+Na^+Na^+Na^+Na^+Na^+}{\underset{Na^+Na^+Na^+Na^+Na^+Na^+}{\text{MICELLE}}}}\ \mathrm{Na^+ + 14H.OH}$$

Transfer it to a tall 1000 ml cylinder and make up the volume to 1000 ml, shake and leave the cylinder undisturbed for the time t_1 seconds, taken from the silt column of the mechanical analysis chart. After the time t_1 sec is over, withdraw 25 ml of soil-water suspension from the requisite depth, which is usually 10 cm, 22 cm or 28 cm depending on the temperature. Dry it in an oven and weigh it as silt+clay. Similarly, shake it and leave it undisturbed for time t_2 taken from the clay column of the Mechanical Analysis chart. After this time is over, withdraw 25 ml of the soil water suspension oven dry it and weigh it as clay.

Calculation

Suppose wt. of soil taken = 10 gms

Total vol. of soil water suspension = 1000 ml

Vol. of soil water suspension withdrawn at 1st and 2nd pipetting = 25 ml.

25 ml or 1st pipetting sample contains 0.18 gms of silt + clay

25 ml or 2nd pipetting sample contains 0.12 gms of clay

$\therefore$ 25 ml or the soil water suspension contains 0.18 – 0.12 = 0.06 gms of silt and 0.12 gms. of clay

$$\therefore \text{ 1000 ml of the soil water suspension contains } \frac{0.06 \times 1000}{25} = 2.4 \text{ gms}$$

$$\text{of silt and } \frac{0.12 \times 1000}{25} = 4.8 \text{ gms of clay}$$

$\therefore$ 10 gms of the soil contains 2.4 gms of silt and 4.8 gms of clay

$\therefore$ 100 gms of the soil contains 24 gms of silt and 48 gms of clay

Pour away most of the supernatant liquid from the tall cylinder and transfer the soil particles to a beaker. Add water and stir, leave it undisturbed for time taken from the sand column of the mechanical analysis chart. After this, slowly pour away the water. Repeat this process a number of times till it is absolutely clear. Then dry it in an oven and weigh as sand.

Suppose the weight of this sand is 2.8 gms

Then 10 gms of soil contains 28 gms of sand.

Then the soil contain 28 per cent sand, 24 per cent silt and 48% clay.

4.9. SOIL TEXTURE

Definition of Soil Texture

Texture means size of the particles. Soil texture is the average size of the soil particle which depends on the relative proportion of sand, silt and clay in the soil. If the proportion of sand in the soil is increased, the average size or the soil particles increases and the resultant soil becomes coarser in texture. On the other hand if the proportion of clay in the soil is increased, the average size of the soil particles decreases and the resultant soil becomes finer in texture.

Determination of Soil Texture in the Laboratory

Soil texture is determined with the help of the triangular texture diagram as shown in Fig 4.1, drawn as follows:

First draw an equilateral triangle of sides 10 cm each, divide them into 10 equal parts and mark them 10, 20, 30, 40, ...100. The left, right and bottom of the triangles are named clay, silt and sand side respectively. Locate the point 55 on the clay line and the point 45 on the sand line, and draw lines through them but do not complete as shown in Fig. 4.1. Locate the point 60 on the silt side and the point 40 on the clay side and draw a line through them till they intersect the first drawn 55-45 line. Similarly locate the point 73 on the silt side and the point 27 on the clay side and draw a line through them till they intersect the first draw 55-45 line at the point as shown in Fig. 4.1. Again locate the point 28 on the silt side and the point 72 on the sand side and draw a line through them which will pass through the point (Fig. 4.1). Draw another line through the point 20 on the clay side and the point 80 on the silt side. It intersects the previously drawn 28-72 line at *b* (Fig. 4.1) Again locate the point 48 on the clay side and the point 52 on the sand side and draw a line through them which passes through the point *b*. Again draw another line through the point 7 on the clay side and the point 93 on the silt side. It intersects the 48-52 line at C as shown in (Fig 4.1). Draw a

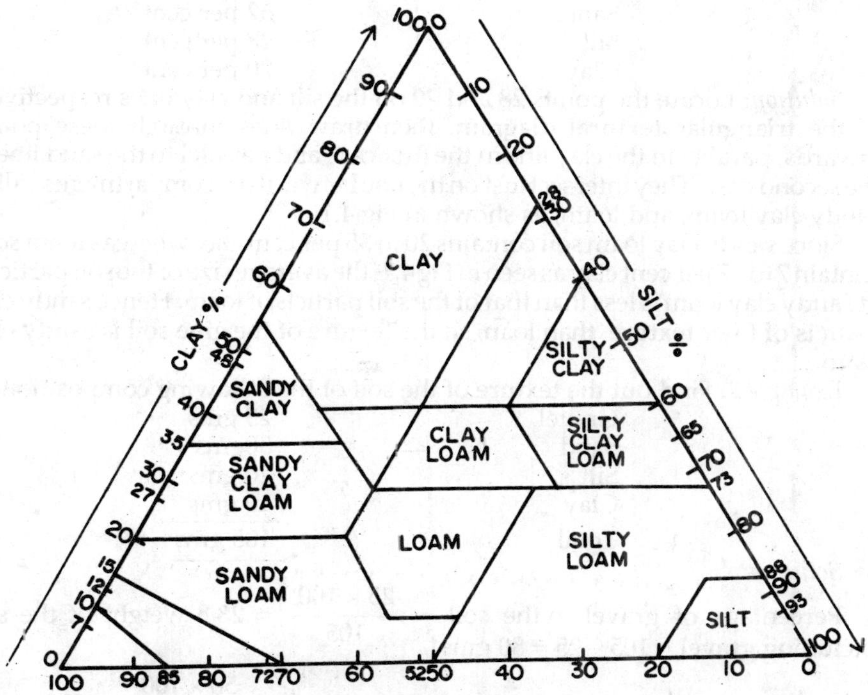

Fig 4.1. Triangular textural diagram on the basis of USDA system for determining soil texture (Used to solve Ex. 1)

line parallel to the clay side, through the point 50 on the sand side till it intersects the previously drawn 73-27 line (Fig. 4.1). Draw a line parallel to the clay side through the point 40 on the silt side till it intersects the previously drawn 60-40 line at D as shown in Fig 4.1, through which draw another line is drawn parallel to the silt side till it intersects the previously drawn 73-27 line at *e* (Fig 4.1.) Draw a line parallel to clay side through the point 20 on the sand side and another line parallel to the sand side through the point 88 on the silt side till they intersect one another. Join the points 70 on the sand side and 15 on the clay side. Join 85 on the sand side and 10 on the clay side. Erase the unnecessary portions of different lines. Then the diagram looks like the Fig. 4.1. Name the different compartments as shown in Fig. 4.1.

Use of the Triangular Textural Diagram

Locate the percentages of silt and clay on the silt and clay side lines respectively of the triangular textural diagram. Then draw lines through these points, inwards, parallel to the clay side in the first case and to the sand side in the second case. The textural name of the compartment in which they intersect is the textural name of the soil concerned. If they intersect just on the line between the two compartments, then the textural name of the finer compartment is to be taken.

Example 1: Find out the texture of the soil of the following composition

Sand	52 per cent
Silt	28 per cent
Clay	20 per cent

Solution: Locate the points 28 and 20 on the silt and clay lines respectively of the triangular textural diagram, then draw lines through these points inwards, parallel to the clay line in the first case and parallel to the sand line in the second case. They intersect just on the line between the compartments called sandy clay loam, and loam, as shown in Fig 4.1.

Since sandy clay loam soil contains 20 to 35 per cent clay whereas loam soils contain 7 to 27 per cent clay as seen in Fig 4.1, the average size of the soil particles of sandy clay loam is less than that of the soil particls of loam. Hence sandy clay loam is of finer texture than loan, so the texture of the give soil is sandy clay loam.

Example 2: Find out the texture of the soil of the following composition

Gravel	25 gms
Sand	50gms
Silt	20 gms
Clay	10 gms
Total	105 gms

Solution:

Percentage of gravel in the soil $= \dfrac{25 \times 100}{105} = 23.8$ weight of the soil excluding gravel = 105 – 25 = 80 gms

Percentage of sand in the soil excluding gravel $= \dfrac{50 \times 100}{80} = 62.5$

Percentage of silt in the soil excluding gravel $= \dfrac{20 \times 100}{80} = 25.0$

$$\text{Percentage of clay in the soil excluding gravel} = \frac{10 \times 100}{80} = 12.5$$

Locate the point 25 on the silt line and the point 12.5 on the clay line. Then draw lines through 25 and 12.5 inwards, parallel to the clay line in the first case and parallel to the sand-line in the second case (Fig. 4.2). They intersect in the sandy loam compartment. As the soil contains more than 20 per cent gravel, the word "Gravelly" is to be written before the determined textural class-sandy loam. Hence the texture of the given soil is gravelly sandy loam. The word "Gravelly" is not to be written before the determined textural class if the soil contains 20 per cent or less gravel.

Textural Classification of Soils

Textural classification of soil means classification of soils on the basis of their texture into different groups or classes called, clay, silty, clay sandy clay, clay loam, sandy clay loam, loam, silty loam, silt, sandy loam, loamy sand and sand as shown in the triangular textural diagram.

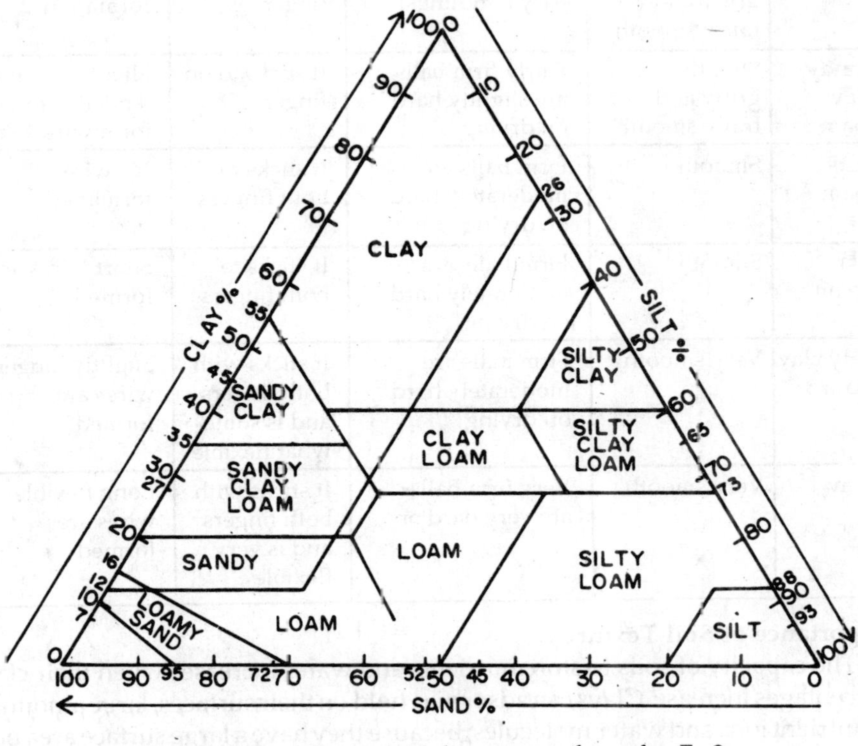

Fig 4.2. Trinagular textural diagram used to solve Ex.2.

Determination of Soil Texture in the Field

Sand particles are gritty, which means that if you rub them between your fingers, you feel that you are rubbing some minute coarse grains between your fingers. This feeling of coarseness is called grittiness. Similarly, silt particles have the feel of talcum powder. They are extremely smooth. Moist clay can be rolled between the hands to form a long thin wire. A ball can be made from moist clayey soil. The ball becomes firmer when the percentage of clay in the soil increases. Grittiness decreases and ball formation, stickiness and wire formation increase as the soil texture becomes finer as shown in Table 4.1.

TABLE 4.1: Textural classification of soils by feeling them

Soil Text. Class	*Feel of Moist soil*	*Ball formation form slightly moist soil*	*Stickiness of we soil*	*Wire formation from we soil*
Loamy Sand	Very gritty	Very weak balls that break down very easily	It dirties finger slightly	No wire formation
Sandy Loam	Moderately gritty	Balls can bear careful handling	It dirties finger slightly	No wire formation
Loam	Slightly gritty and fairly smooth	Balls can bear easy handlings	It dirties the finger	No wire formation
Sandy clay loam	Slightly gritty and fraily smooth	Fairly firm balls are slightly hard on drying	It sticks to one finger	Slight tendency to form wire
Silty loam	Smooth	Firm balls are moderately hard on drying	It sticks to both fingers	No wire formation
Clay Loam	Smooth	Firm balls are moderately hard on drying	It sticks to both fingers	Short wires are formed
Silty clay loam	Very smooth	Firm balls are moderately hard on drying	It sticks with both fingers and is somewhat flexible	Slightly longer wires are formed
Clay	Very smooth	Very firm balls are very hard on	It sticks with both fingers and is very flexible	Long flexible wires are formed

Importance of Soil Texture

The capacity of soils to store nutrients and water increases when their clay percentages increase. Clays can adsorb i.e. hold on their surfaces, large amounts of nutrient ions and water molecules, because they have a large surface area per unit volume, are negatively charged and therefore have positive ions and dipolar water molecules, which can be absorbed by the roots whenever required. Hence fertilizer and water may be added to clayey soils at longer intervals than to the sandy soil.

Wet clayey soil sticks to implements. Ploughing a clayey land soil produces very hard clods. Therefore a clayey land soil should be ploughed to an intermediate soil moisture content, when it is dry enough to loss its stickness, yet moist enough not to produce hard clods. More power is required to plough a clayey land soil than a sandy land soil. So clayey soils are called heavy soils, and sandy soils are called light soils.

Clayey soils are usually poorly drained and aerated, whereas sandy soils are excessively drained and aerated, but cannot retain enough water and nutrients for the growth of crops. Loam soils are therefore usually best for growing most crops because they are reasonably well drained and acrated and can retain enough water for growth of must crops.

4.10. SOIL STRUCTURE

Primary soil particles sand, silt and clay are united together to form secondary soil particles, which are called soil aggregates. A naturally occurring soil aggregate is called a 'ped', while a clod is formed artificially by ploughing a dry clayey land soil etc. A broken piece of a clod is called a fragment. The arrangement of peds with respect to each other, determines soil structure, which is categorised as (*i*) type of soil structure; (*ii*) class of soil structure; and (*iii*) grades of soil structure.

Types of Soil Structure

The type of soil structure refers to the general shape of peds.

The general shape of the peds is as described below:

(*a*) *Platy Structure:* The horizontal azes of peds are much larger than their vertical axis. This means that the peds are in the form of thin plates.

(*b*) *Prismatic Structure:* The vertical axis of peds are larger than their horizontal azes. Tall peds are bounded by plane faces intersecting each other at sharp angles. Prismatic structure has been subdivided into two sub types:

- (*i*) *Prismatic:* The top of the ped is flat.
- (*ii*) *Columner:* The top of the ped is rounded.

(*c*) *Blocky Structure:* Horizontal and vertical axes of peds are roughly equal, which means that the peds are roughly in the form of cubes. Blocky structure has been subdivided into two types:

- (*i*) *Angular blocky:* Peds are bounded by plane faces intersecting each other at sharp angles.
- (*ii*) *Sub-angular blocky:* Peds are bounded by both plane faces intersecting each other at sharp angles and round faces.

  In the case of platy, prismatic and blocky structure peds are in close contact with each other so that there is less air space between them.

(*d*) *Spheroidal structure:* The horizontal and vertical axes of peds are more or less equal; peds are not in close contact with each other, so there is plenty of air space between them because they are irregular in shape. Spheroidal structure has been subdivided into the following two types:

- (*i*) Granules are relatively less porous.
- (*ii*) Crumbs are very porous.

Classes of Soil Structure

This refers to the average size of ped as shown in Table 4.2 (Diameter in millimetres).

TABLE 4.2: Class of soil structure

Class Structure of Soil Structure	*Platy*	*Prismatic Columber Block*	*Angular and Sub-angular*	*Spheroidal*	
				Granule	*Crumbs*
1. Very fine or very thin in case of platy	Less than 1 mm	Less than 10 mm	Less than 5 mm	Less than 1 mm	Less than 1 mm
2. Fine of thin case of platy	1—2 mm	10—20 mm	5—10 mm	1—2 mm	1—2 mm
3. Medium	2—5 mm	20—50 mm	10—20 mm	2—5 mm	2—5 mm
4. Coarse or thick in case of platy	5—10 mm	50—100 mm	20—50 mm	5—10 mm	—
5. Very coarse over very thick in case of platy	more than 10 mm	more than 100 mm	more than 50 mm	more than 10 mm	

Grades of soil Structure

Grades of soil structure refers to the durability of peds and distinctness with which they can be observed in undisturbed soil, as described below:

(*i*) *Structureless:* No peds are observed in undisturbed soil.

(*ii*) *Weak*: Poorly formed indistinct peds are found in undisturbed soil. When the soil structure is disturbed, it breaks down into a mixture of a few entire peds, many broken peds and a large amount of unaggregated material.

(*iii*) *Moderate:* Fairly well formed peds are found in undisturbed, soil that are not very distinct in undisturbed soil. When the soil structure is disturbed it breaks down into a mixture of many entire peds, some broken peds and a little unaggregated material.

Strong: Well formed distinct peds are bounded by planes of weakness when the soil structure is disturbed, then it breaks down into almost entire peds, and a little broken peds and no unaggregated material.

Formation of Soil Aggregates Under Natural Conditions

Water molecules are dipolar. One side of a water molecule is positively charged, while the opposite side is negatively charged. The negative end of one water molecule attracts the positive end of a second water molecule, the negative end of which attracts the positive and of third water molecule, and so on. In this way, long chains of water molecules are formed in nature. Positive cations are hydrated. They attract the negative ends of a chain of dipolar water molecules, the positive end of which are attracted by a clay particle. Hence clay particles are held by bridges of oriented water molecules and positive cations as shown in Fig. 4.3.

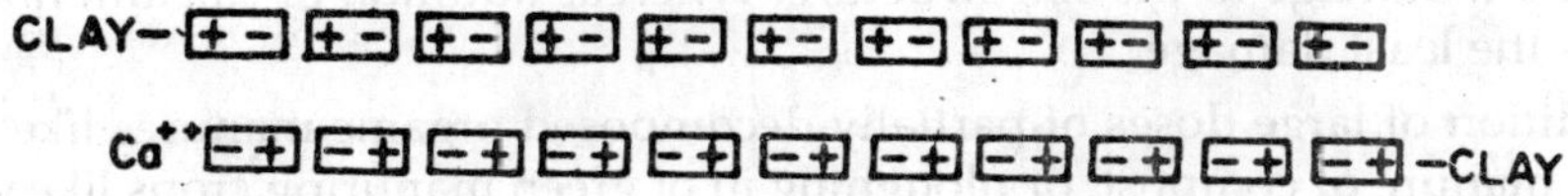

Fig. 4.3 : Two negatively charged clay particles have been joined together by a bridges oriented water molecules and a divalent cation *i.e.* calcium.

When the soil dries up, the chain of water molecule shortens more and more, so the clay particles are ultimately united. They are united together by some organic compounds formed during the decomposition of soil organic matter. Actually several clay particles are united in this way to form compound particles which, in turn, are bound together by cementing agents like clay, iron and aluminium oxides to form the soil aggregates. Soil micro-organism decompose soil organic matter to form dark, gummy substances which bind the clay and sand and silt particles to form the soil aggregate. Root hairs and fungal hyphae also bind the compound particles of the soil as well as sand and silt particles to form soil aggregates.

Destruction of Soil Structure Under Natural Conditions

Wet soil aggregate break down when the negative clay micelle is saturated with highly hydrated cations like sodium ions. Clay is deflocculted. This may be corrected by adding gypsum when calcium ions replace sodium ions.

Soil aggregates are rapidly wetted by heavy rains. The outer layer of the soil aggregates are detached from their main body. This process is repeated several times and the soil structure ultimately breaks down to mud. The fast falling raindrops strike the soil aggregates with great force and break them up. Therefore the surface of the soil should be kept covered with organic matter. Ploughing a wet clayey soil also destroys soil structure. Therefore, clayey soils should be ploughed to an intermediate soil moisture content, when they are dry enough to lose their stickiness yet moist enough not to form hard clods.

Improvement of Soil Structure

Sandy soils should be worked with a little pressure when the soil pores are just filled with water. Consequently, smaller peds are united with each other to form bigger peds. Large clods of sandy soils are easily broken down to smaller ones. The large clods of clayey soils are broken down when they are slowly wetted by a prolonged drizzle. Larger clods of clayey soils can also be broken down to smaller ones when water slowly freezes to form ice. Within the pore space and exert pressure on the moist soil. Large clods of clayey soils are broken down by repeated freezing and thawing.

Addition of lime to acidic soils improves their structure because lime stimulates the growth of micro-organisms which are probably responsible for improvementy of soil structure. Addition of fertilizers to bare uncultivated soil reduces the stability of soil aggregates. Addition of sodium nitrate cause maximum damage to the soil structure. Whereas addition of calcium nitrate causes the least damage.

Addition of large doses of partially decomposed organic manures like farmyard manure or compost, or ploughing in of green manuring crops like sunhemp, (crotalaria juncea) or Dhaincha, (Sesbania bispinosa) greatly improves soil structure in about three weeks. Lime increases the speed of decomposition of organic matter by micro-organisms, thus improving the soil structure.

The surface of the land should be kept convered with waste organic materials like grasses, leaves, straw etc., which are known as mulches. They keep the surface soil moist and cool, where micro-organisms multiply rapidly and improve soil structure. Mulches also protect the surface from the beating action of the raindrops.

Roots of grasses bind the soil particles to form the soil aggregates. It legumes like horsegram Kudzu etc. are grown with grasses the soil structure is further improved, because the micro-organisms get the additional nitrogen which has been fixed by the symbiotic bacteria living in the root nodules of the legumes.

Importance of Soil Structure

A good soil structure is essential for the circulation of a proper proportion of air and water in the soil which, in turn, is essential for the growth and functioning and root system. A good soil structure is especially essential for certain crops like potato, sugarbeet maize etc. which need better soil aeration than other crops.

4.11. SOIL FABRIC AND SOIL PLASMA

The soil fabric refers to the arrangement of the primary and secondary paticles and the space between them in the soil is the "soil material" capable of being or having been moved, reorganized and/or concentrated by the process of soil formation (Buol et al., 1973), *i.e.* the small sacle soil structure is known as the soil fabric. The soil plasma is an amorphous combination of humus, clay and iron oxide/lime etc. It has been produced by the secondary weathering processes and also by the incorporation of organic matter with the inorganic matter. Inorganic and organic components of the soil plasma are virtually inseparable from each other.

4.12. PORE SPACE

The pore space in a soil represents that part of the soil volume which has been occupied by air or water. Pore space in a soil may be of the following three type:

(*i*) Macro pore space occur between peds. They are more than 75 micron in diameter. Water and air rapidly pass through them. They are necessary for rapid drainage of the soil.

(*ii*) Mesopores are 75 to 30 microns in diameter. Relatively larger mesopores occur between the peds while relatively smaller ones occur within the peds. Mesopores are required for storage of water for crop growth.

(*iii*) Micropores are less than 30 microns in diameter. Water is held within the micropores tightly enough to be unabailable for plant growth.

Factors Affecting Pore Space Per cent of Soils

(i) Texture: Total and Micropore spaces of soils increase when their clay percentage increases *i.e.* their texture becomes finer. The micropore spaces of clayey soils are much more than those of the sandy soils, because clay particles unite to form soil aggregate within which micropore space occur.

(ii) Organic matter: Organic matter is decomposed by the soil micro organisms to form humus, which usually binds the soil particles to form aggregates. Besides this, humus also stimulates the growth of soil micro organisms, which mechanically bind the soil particles to form aggregates within which micropore spaces occur. Hence the micro and total pore spaces increase.

(iii) Nature of crops and cultivation: Excessive cultivation of soils increases the oxidation of humus. Consequently, total and micropore space of soils are decreased if soils are more and more cultivated. Less cultivation is required for growing grasses, the roots of which bind the soil particles to form soil aggregates. Crops like potato, maize etc. require the maximum amount of soil cultivation for soil aeration which increases the oxidation of humus. Hence grasses increase the percentage of pore space of the soil, whereas aerable crops like maize, potato etc. decreases it.

(iv) Soil depth: Since plant roots and organic matter occure more in surface soil then in sub-soil, so pore space per cent of the surface soil is usually much more than that of the sub-soil.

Importance of Pore Space

Micropore space must contain enough oxygen for respiration, aerobic micro-organism and plant roots. Excess water drains down the macropore space. Micropore space stores water for crop growth.

4.13. PARTICLE DENSITY OF THE SOIL

The density of soil solids is known as the particle density of the soil, which may be defined as the dry weight of soil solids per unit volume of dry soil solids. For exemple if one cubic centimetre of the dry soil solid weighs 2.65 gms, then the particle density of the concerned soil is 2.65 gms. per cubic centimetre. The particle density of the soil can be determined by adding the densities of the soil constituents as shown in the following formula.

Particle Density = $D_1P_1+D_2P_2$......

Where D and P are the respective density and proportion of soil canstituents 1, 2......

Example: Find out the particle density of a soil of the following composition:

Soil Constituent	*Percentage*	*Average density in gm/cc*
Quartz	35	2.65
Kaolinite	60	2.40
Hematite	4.5	5.10
Humus	0.5	1.40

Solution:

Particle Density = $D_1P_1+D_2P_2D_3P_3+D_4P_4+D_2P_2+D_3P_3+D_4P_4$

Particle Density of the soil is

$$= 2.65 \times \frac{35}{100} + 2.40 \times \frac{60}{100} + 5.10 \times \frac{4.50}{100} + 1.40 \times \frac{0.5}{100}$$

$$= 0.93 + 1.44 + 0.23 + 0.07 = 2.67 \text{ gms/cc}$$

4.14. SPECIFIC GRAVITY OF THE SOIL

The specific gravity of the soil is the ratio of the weight of a given volume of soil solids to the weight of a equal volume of water. Since one cubic centimeter of water weighs one gram, specific gravity and particle density of the soil are numerically equal. Since the specific gravity is a ratio it is dimensionless e.g. one cc of soil weighs 2.67 gms. in the above example, and one cc. of water weighs 1 gm. So the specific gravity of the soil is

$$\frac{2.67}{1} = 2.67$$

The particle density and specific gravity of soils depends upon the densities and relative proportions of different soil constituents. They vary from 2.60 to 2.70, with an average value of 2.65 gms/cc in the case of particle density and dimensionless in the case of specific gravity, of average mineral or inorganic soil.

4.15. BULK DENSITY OF SOILS

The density of the soil is called the 'bulk' density of the soil. Bulk density of soils is the weight of dry soil solid per unit volume of soil. This soil volume is the sum of the volume occupied by soil solids and volume occupied by air.

$$\text{So bulk density of the soil (BD)} = \frac{\text{Weight of soil solids}}{\text{Vol. of soil solids + Vol. of air}}$$

$$\text{But particle density of the soil} = \frac{\text{Wt. of soil solids}}{\text{Vol. of soil solids}}$$

Wt. of soil solids in the second case is PD × Vol of Soil Solids

Wt. of soil solids in the first case is BD × (Vol soil solids + Vol of air)

PD (Vol of soil solids) = BD (Vol. of soil + Vol of air)

$$\text{or} \quad \frac{\text{Vol. of soil solids}}{\text{Vol. of soil solids + Vol. of air}} = \frac{\text{BD}}{\text{PD}}$$

As, Vol. of soil solid percentage

$$= \frac{\text{Vol. of soil solids}}{\text{Vol. of soil solids + Vol. of soil air}} \times 100$$

So Vol. of soil pore space percentage

$$= 100 - \frac{\text{Vol. of soil solids}}{\text{Vol. of soil solids + Vol. of soil air}} \times 100$$

$$= 100\left(1 - \frac{\text{Vol. of soil solids}}{\text{Vol. of soil + Vol. of soil air}}\right)$$

or Pore space per cent of soil $= \left(1 - \frac{\text{BD}}{\text{PD}}\right) \times 100$

Factors Affecting the Bulk Density of Inorganic Soils

(1) *Texture*: Dry soils of a finer texture have more spaces filled with air around the mineral particles than those of a coarser texture. So the weight per unit volume of the soil is reduced. Thus a clayey soil in which soil particles are united to form soil aggregates have a lower Bulk Density than a sandy soil in which soil particles lie close to each other.

(2) *Humus*: When organic matter decomposes to form humus in the soil, some organic compounds are formed, which bind the primary soil particles sand, silt and clay to form the soil aggregate. Pore space occurs within and between soil aggregates. So the weight of unit volume of soil decreases. Hence the bulk density of soils decreases when their percentage of humus increases.

(3) *Soil depth*: Bulk density of surface soil is usually lower than that of the sub soil because surface soil contains more humus than sub soil. Plant roots present in the surface soil bind the soil particles to from soil aggregate, and therefore reduce the bulk density of the surface soil.

(4) *Nature of crops*: The bulk density of soils decreases if grasses are grown, because the grass roots bind the soil particles to form soil aggregate. In addition, grasses add organic matter to the soil in the form of dead roots which decompose to form humus that binds the soil particles to forms more soil aggregates. On the other hand, excessive amounts of cultivation required for growing aerable crops like potato oxidizes soil humus and therefore destroys soil aggregates, which increase the bulk density of the soil. Hence bulk density of grassland soils is less than that of the aerable land soils where crops like potato, maize etc. are grown. The bulk density of sandy soils and clayey soils should be less than 1.4 gms/cc and 1.2 gms./cc respectively for optimum crop growth, soils having a bulk density of about 1.1 gms/cc contain sufficient pore space to retain enough water and air for plant growth.

The bulk density of undisturbed soil is found by inserting a small metallic cylinder of convenient internal diameter and height, into the soil. The cylinder containing the soil core is then taken out, oven dried and weighed.

$$\text{Bulk density} = \frac{\text{Wt. of soil core}}{\text{Vol. of soil core}}$$

$$\text{Void ratio} = \frac{\text{Volume of voids (Pores)}}{\text{Volume of soil solids}}$$

Example: A metallic cylinder was introduced into the soil. Then it was taken out and the soil contained in it was oven dried and weighed.

Determine the bulk density, pore space percentage, and void ratio of soils from the following data:

Height of the cylinder = 8 cms
Internal Diameter of the cylinder = 6 cms
Weight of Dry soil = 271.2 gms
Particle Density of soils = 2.65 gm/cc

Solution: Volume of the soil core = Internal volume of the cylinder

$\pi r^2 h = 3.14 \times 6/2 \times 6/2 \times 8 = 226.08$ cc

$$\therefore \text{Bulk Density of the soil} = \frac{271.2}{226.08} = 1.20 \text{ gms/cc.}$$

$$\text{Pore space \%} = \left(1 - \frac{BD}{PD}\right) \times 100 = \left(1 - \frac{1.20}{2.65}\right) \times 100 = 54.7 \%$$

Vol. of pore is 54.7 cc. Hence the volume of soil solids is 100 – 54.7 = 45.3 cc

$$\text{Void ratio} = \frac{54.7}{45.3} 1.18$$

The soil seems to be finer texture because its percentage of pore space is more than 50 per cent.

4.16. VOLUME WEIGHT OF SOIL OR BULK SPECIFIC GRAVITY OF SOILS

It is defined as the ratio of the weight of a given volume of soil (This Vol. of soil = Vol. of soil solid + Vol. of pore space) to the weight of an equal volume of water. Since the weight of one gram of water is one cc, the bulk density of soils and the volume weight of soils are numerically equal.

4.17. WEIGHT OF SOIL

We know that the density of water is one, which means that one cubic centimetre of water weight one gram.

Convert cubic centimetre to cubic feet and gram to pounds (lb)
Since 1 cc = 0.0000353 cubic fit.
and 1 gm = 0.002204365 pound (lb)
0.0000353 cu.ft. of water weighs 0.0022043656 lbs

or 1 cubic foot of water weighs $= \frac{0.002204365}{0.0000353} = 62.44$ lbs

Hence Density of water is 62.44 lbs per Cubic foot.

In a similar way, if the bulk density of a soil is 1.2gm/cc, this means that 1 cc of the soil weighs 1.2 gms. Convert both cc and gms to cu.ft. and lbs respectivbely. Then 0.0000353 cu.ft. of soil weighs 1.2 × 0.002204365 lbs. or

1 cubic foot of soil weighs $= \frac{1.2 \times 0.002204365}{0.0000353} = 1.2 \times 62.44 = 74.928$ lbs.

Hence the weight of one cubic foot of a dry soil is determined multiplying its bulk density, 1.2 gms/cc, by 62.44 which is the weight of one cubic foot water is pounds.

Again in this example, if we convert cubic centimetre to cubic metre by dividing cubic centimetre by 1000000 and gram to kilogram by dividing gram by 1000.

Then $\frac{1}{1000000}$ cubic metre of soil weighs $\frac{1.2}{1000}$ kg.

or 1 cubic metre of soil weighs $\frac{1.2}{1000} \times 10,00,000 = 1200$ kg.

"Therefore the weight of one cubic metre of a soil can be determined simply by multiplying the soil bulk density in gm/cc by one thousand.

Weights of One Acre Foot of Soils and One Acre Furrow Slice of Soils

One Acre foot of soil is the bvolume of the soil of one Acre in Area and one foot in depth.

One Acre Furrow slice of soil is the volume of the soil of one acre in area and six inches in depth.

Volume of one Acre Foot of soil = 4840 × 9 × 1 = 43560 cu.ft.

Volume of one Acre Furrow slice of soil = 4840 × 9 × ½ = 21780 cu.ft.

If the average Bulk Density of the surface soil of one foot deep is 1.2 gms/cc

Then wt. of 1 cu.ft. soil = 1.2 × 62.44 = 74.928 cu.ft.

Wt. of one Acre Foot of soil = 43560 × 74.928 = 3263863.68 lbs

And wt. of one Acre Furrow slice of soil = 21780 × 74.928 = 1631931.84 lbs.

Weight of one acre furrow slice of soil is 2 million pounds in the case of average mineral soils.

Weight of One Hectare Furrow Slice of Soils

One hectare furrow slice of a soil is the volume of soil of one hectare in area and 15 cms in depth. 1 hectare = 10000 square metres.

Volume of one Hectare Furrow slice of soil of 15 cm depth is

$$10000 \times \frac{15}{100} = 1500 \text{ cubic metres}$$

If soil bulk density is 1.2gm/cc the weight of one cubic metre of soil is 1.2 × 1000 = 1200 kg.

The weight of one hectare furrow slice of soil is
1200 × 1500 = 1800000 kg of soil = 1.8 million kg;

Usually the weight of one hectare furrow slice of soil is 22 million kg in most cases.

Example: The soil contains 0.003 per cent N. The bulk Density of the soil 1.2 gms/cc. The wheat crop requires 80 kg of nitrogen per nitrogen per hectare. How much nitrogen is to be added to the soil per hectare.

Solution:

Volume of one hectare furrow (15 cm deep) = $10000 \times \frac{15}{100}$ = 1500 cu. metres

Wt. of one cubic meter of soil is $\frac{1.2}{100} \times 1000000 = 1200$ kg

Wt. of one hectare furrow slice of soil = 1500 × 1200 = 1800000 kg

1800000 kg of soil contain $\frac{0.003 \times 1800000}{100} = 54$ kg

So the soil contain 54 kg of nitrogen per hectare. Then 80 – 54 = 26 kg of nitrogen is to be added to the hectare wheat field.

4.18. SOIL CONSISTENCE

Soil consistence is term used to describe the action of the physical forces of Cohesion (attraction between soil particle or between water molecules) and Adhesion (attraction between soil particles and water molecules) in soils different moisture contents, determined by the resistance of the soil against deformation by the applied force, the tendency of the soil mass to stick to the implement or finger, and the feel of the said fingers the observers when rubbed against the soil.

Soil consistence varies with the soil moisture content, the texture, the organic matter content and the nature of the soil colloid.

Soils flow at a high moisture content when they are viscous. As the soil moisture content is gradually decreased, the soil becomes sticky the plastic. When the soil moisture content is reduced further it becomes friable when soft. Finally it becomes very hard and coherent when it has been completely dried.

Soil consistence is described at three soil moisture levels, wet, moist and dry. Soil consistency is described of wet soil when the soil moisture content is at or slightly above the field capacity (Soil moisture content after excess gravitational water has drained away) by stickiness and plasticity. Stickiness is the property of the soil to stick or adhere to the finger or implement as described below:

(*i*) *Non-sticky*: Soil mass does not stick to the finger.
(*ii*) *Slightly sticky*: Soil mass sticks to one finger only.
(*iii*) *Sticky*: Soil mass sticks to both fingers.
(*iv*) *Very sticky*: Soil mass sticks with both fingers and is very flexible.

Plasticity of the soil is its property to change its shape under the influence of the applied force, and to retain its new shape even when the force stops being exerted on it. For field description of plasticity, roll a wet soil mass between your hands and observe whether a then wise can be formed as shown below.

(*i*) *Non plastic*: No wire is formed
(*ii*) *Slightly plastic*: A wire less than 1 cm long is formed.
(*iii*) *Plastic*: A wire more than 1 cm long is formed
(*iv*) *Very Plastic*: A wire more than 1 cm is formed and much pressure is required to deform the soil mass.

Soil consistence is described of moist soil at a soil moisture content intermediate between the soil at field capacity and the air dry soil, by the resistance of the soil against crushing between the thumb and the force finger as shown below:

(*i*) *Loose*: Soil particles do not adhere to each other then exist as individual particles.
(*ii*) *Very friable*: This soil mass can be very easily crushed.
(*iii*) *Friable*: The soil mass can be easily crushed.
(*iv*) *Firm*: The soil mass can be crushed with moderate pressure.
(*v*) *Very firm*: The soil mass can be crushed with moderate strength.
(*vi*) *Extremely firm*: The soil mass cannot be crushed.

Friability characterizes the case of crumbling of soil mass. The moisture range in which soils are friable, is also the range in which conditions are optimum for tillage.

Soil consistence is described as of dry soil by the resistance of the soil against breaking by the applied force as shown below:

(*i*) *Loose*: Soil particles are not united.
(ii) *Soft*: Soil mass can be broken very easily between the thumb and fore fingers.
(*iii*) *Slightly hard*: Soil mass can be easily broken between thumb and fore fingers.
(*iv*) *Hard*: Soil mass can be easily broken by hand.
(*v*) *Very hard*: Soil mass can be broken by hand with great difficulty.
(*vi*) *Extremely hard*: Soil mass cannot be broken by hand.

Cementation of soil mass means a brittle, hard consistency caused by some cementing substance like lime, oxides of iron and aluminium and silica as described below:

(*i*) *Weakly cemented*: Cemented mass can be broken in the hand easily.
(*ii*) *Strongly cemented*: Cemented mass can be broken by hammer easily.
(*iii*) *Indurated*: Cemented mass can be broken only with a strong blow of hammer that produces a metallic sound.

4.19. DIFFERENCE BETWEEN SOIL STRUCTURE AND SOIL CONSISTENCE

Soil structure describes the shape, size and durabilities of peds which varies with the change in the physical forces of cohesion and adhesion with the soil while the soil consistence describes the nature of such forces themselves.

4.20. TILLAGE

Soil are mechanically manipulated (for example, ploughed in order to make them favourable for the growth of more crops. The kinds of mechanical manipulations performed on soil to make them most favourable for the growth of crops are called Tillage of soils.

Soil Till

The physical condition of the soil in relation to plant growth is called soil till. This means that soils which are in good till, allow rain water to enter in them rapidly. They are well drained and aerated, yet retain enough moisture for the growth of crops. They are mellow and friable and can be ploughed easily.

Ploughing

The plough is the most important tillage implement for crop production. It cuts loose, granulates and inverts a certain volume of soil called furrow slice and also turns under of residues, weeds and green manures. The most important reason for ploughing the soil is loosening it so that seeds can germinate and crops grow. The curved mould board is used for turning under surface residues.

The Seed Bed

The seed bed is the plough layer of soils where seeds can readily germinate and roots can develop and grow. The lower part of the seedbed should contain the finest soil granules and its upper part should contain the coarse soil granules. The immediate surface layer of the soil of 6 to 9 inches depth should be well aerated and should retain enough moisture for the germination of seeds and growth of crops.

Cultivation

Cultivation is the tillage of soils after sowing the seeds or the germination of seeds. The fundamental reason for cultivation is to maintain the seedbed in good tillh during the growth of crops. Soils are cultivated in order to control weeds and to make them well aerated. Poorer the soil structure, the greater will be the need for frequent loosening of the soil surface, and more rapid will be eventual deterioration of the soil structure.

4.21. SOIL COLOUR

Causes of Soil Colour

The Colour of soils is due to the colours of their constituents. For example, humus is black in colour. So a black coloured soil may be rich in humus. Ferric oxide is red in colour. So red soils are rich in it. Quartz is white, or lighter in colour and clay is darker. So lighter coloured soils are richer in sand than the darker soils. Hence darker coloured soils are richer in clay and therefore more fertile than the lighter coloured soils.

Importance of Soil Colour

Soil colour tells us about soil conditions. For example, red soils are rich in ferric oxide, highly weathered and of poor fertility. Black coloured soils are expected to be rich in humus and therefore fertile. Darker coloured soils are richer in clay and therefore more fertile than the lighter coloured soils, because they can retain enough water and nutrients for plant growth.

Description of Soil Colour

Soil colour is described by the following three variables:

(*i*) Hue is the dominant spectral colour.

(*ii*) Value is the relative lightness of the soil colour.

(*iii*) Chroma is the relative strength of purity of spectral colour. It increases with decreasing greyness.

All the three variables have been systematically arranged in a loose leaf note book called the Munsell colour chart. All the colour chips shown in each page of the Munsell colour chart are of same hue which is written at the top right hand corner of the page as shown in Fig. 4.4. Vertically their value increases, and horizontally their chrome increases which means that vertically they become lighter and horizontally they become purer and less-grey. Opposite each page of the Munsell colour chips is a page where the corresponding English name of the concerned soil colour is written.

In order to describe the soil colour, it is compared with the colour chips of the Munsell colour chart. Its colour is described by the Munsell Notation of that colour chip with which it matches best.

4.22. SOIL AERATION

Definition

Soil aeration is phenomenon of rapid exchange of oxygen and carbon dioxide between the soil pore space and the atmosphere, in order to prevent the deficiency of oxygen and/or toxicity of carbon dioxide in the soil air. The well aerated soil contains enough oxygen for respiration of roots and aerobic microbes and for oxidation reaction to proceed at optimum rate.

Causes of Poor Aeration

Compact soils of finer textures suffer from poor aeration. Cultivation (working the soil when crops are growing) of soil prevents it. Water logging is another important cause of poor aeration especially in the case of soils of finer texture. The gaseous exchange between the soil air and the atmosphere may not be rapid enough to remove carbon dioxide from the soil air and to supply oxygen to the growing roots. This may happen if an excessive amount of readily decomposible organic matter has been added to the soil.

Mechanism of Gaseous Exchange

The exchange of gases between the soil air and the atmosphere takes place mainly by the following two mechanisms:

(*i*) *Mass flow:* Gases may move in a mass in the soil or out of it. The soil temperature is higher than the atmospheric temperature at midday when the soil gases expand and move out of the soil pore space into the atmosphere. The soil is cooler than the atmosphere during right when the atmosphere. The soil is cooler than the atmosphere during night when the atmospheric gases enter the soil.

When the atmospheric pressure is increased, volumes of gases present in the

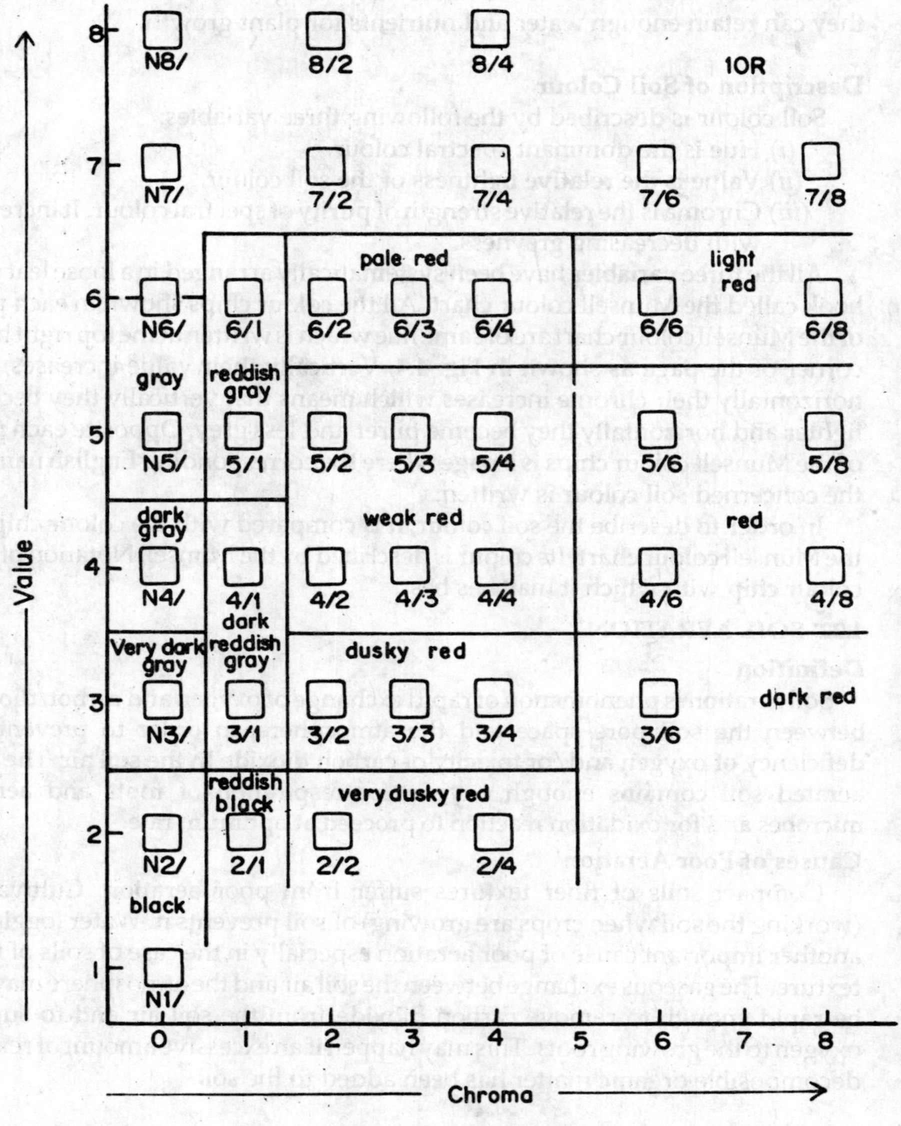

Fig. 4.4. Page of munsell colour chart shown of several combination of value for hue of 10R

soil are decreased and therefore atmospheric gases enter the soil. Rain water displaces soil gases in the pore space and also carries gases dissolved in it to the soil. Rain fall usually account for 1/12 to 1/16th of the normal soil aeration. Variations in temperature and pressure between the soil and at the atmosphere play an insignificant role in soil aeration.

(*ii*) *Diffusion:* Most of the gaseous interchange between the soil and the atmosphere takes place by diffusion. Duffusion is the process by which each gas tends to move in the space occupied by another as determined by the partial pressure of each gas. The partial pressure of a gas is the pressure which the gas would exert if it were present alone in the volume which has been occupied by the mixture of gases. The atmosphere contains higher amounts of oxygen than the pore spaces of soils which contain more carbon dioxide than the atmosphere. So the partial pressure of oxygen is higher in the atmosphere than in the soil pore space and the partial pressure of carbon dioxide is higher in soil pore spaces than in the atmosphere even though the total pressure in the atmosphere and the soil pore spaces may be the same. So oxygen moves in the soil and carbon dioxide moves out of the soil.

Aeration Status of Soils

It can be determined in the three ways *i.e.* (*i*) Percentage oxygen and carbon dioxide content of the soil, (*ii*) Oxygen diffusion rate, and (*iii*) The oxidation reduction potential (Redox potential).

(*i*) *Percentage composition of soil air*. The average inorganic soil contains about eight times more carbon dioxide and a little less oxygen than the atmosphere as shown below:

TABLE 4.3 : Average composition of soil air an atmosphere

(*% by Vol*)

	Atmosphere	*Soil*
Nitrogen	79.00	79.15
Oxygen	20.97	20.60
Carbon dioxide	0.03	0.25

Well aggregated soils contain enough macropores to keep the soil aerated for proper growth and functioning of roots and micro-organisms. After a heavy rian, the macropores are filled up with water but the soil may still contain some quantity of air dissolved in water. So micro-organisms can grow for a short time only, after which the soil must be drained so that the macropores are re-filled with air.

(*ii*) *Oxygen diffusion rate* (ODR): It determines the rate at which the oxygen should be supplied to the soil when it is being continuously used for the respiration of roots and soil micro-organisms. The growth of roots of most crops ceases when the oxygen diffusion rate decreases to about 20×10^{-8} gm/sq. cm/min. It should be above 40×10^{-8} gm/sq.cm/min. for good growth of most crops.

(*iii*) *Oxidation-reduction potential of soil*. The oxidation potential of chemical systems including soil is a measure of the tendency of the oxidation reaction to occur in that system, including soils. This means that the system is in a reduced condition. Highly reduced soils have a high oxidation potential of +0.50 volts.

The reduction potential of a chemical system including soil is a measure of the tendency of the reduction reaction to occur in that system. This means that the system is an oxidised condition.

Reduction Potential or Redox Potential (Eh) is the opposite in sign to the oxidation potential. So a highly reduced soil which has a tendency to be oxidised, has a Reduction Potential or Redox Potential Eh. of– 0.50 volts. Well drained and aerated soils which are highly oxidized, usually have a redox potential of +0.50 volts. The value of the Redox Potential increases when the oxygen content of soils decreases.

Factors Affecting Soil Aeration

(*i*) *Soil organic matter:* When organic matter is added to the soil, it is readily decomposed by the soil micro-organisms to liberate the carbon dioxide content of the soil air.

(*ii*) Since the top soil contains much more macropore space than the sub-soil, the opportunity for gaseous exchange is more in top soil than in the sub soil. Hence the oxygen content of the top soil is greater than that of the sub soil.

(*iii*) *Soil moisture*: The macropores are filled up with water immediately after heavy rain when the oxygen content falls to near zero. When the soil is artificially drained again, the macropores are filled up with air and the oxygen content of the soil increases.

4.23. IMPORTANCE OF SOIL AERATION

Soil aeration affects the availability of some nutrients elements to plant roots. Manganese and iron occurs in the well aerated soil in their higher valent forms (Mn^{++++}, Mn^{+++}, Fe^{+++}) and in poorly aerated soils in their lower valent forms (Mn^{++}, Fe^{++}). They are available to plants only in their lower valent forms. Crops suffer from manganese toxicity if excessive amounts of manganese occurs in the soil in the soluble form. Manganese toxicity to plant roots, under this circumstance, may be corrected either by making the soil more aerated by tilling the soil and improving drainage of the soil or by increasing the soil pH by applying lime to the soil. When iron and manganese are in short supply to the soil, then the soil may be subjected to anaerobic condition by applying readily decomposable organic matter to it. Carbon dioxide would be produced from the decomposition of organic matter to make the soil relatively more anaerobic when manganese and iron will be reduced from their respective higher volent (Mn^{++++}, Mn^{+++}, Fe^{+++}) forms to their respective lower valent form (Mn^{++}, Fe^{++}) *i.e.* divolent forms and would be available to plant roots. Ferric phosphate would be reduced to ferrous phosphate. Carbon dioxide produced from the decomposition of organic matter reacts with water to form carbonic acid which slowly dissolves insoluble phosphate. So the availability of phosphates would be increased to the plant roots.

Sulphur occurs as sulphate in well aerated soil. Plant roots assimilate sulphate. Sulphate is reduced to sulphide in poorly aerated (water logged) soils. Hydrogen sulphide is toxic to plant roots which suffers from it in water logged soil.

Organic matter is decomposed by aerobic bacteria in well aerated soil when complex organic nitrogen and phosphorus compounds are decomposed to their respective simple inorganic compounds which plant roots readily assimilate symbiotic and non-symbiotic nitrogen fixation takes place only in well aerated soils. Nitrates are reduced to oxides of nitrogen and nitrogen gases in poorly aerated soils. These gases escape to the atmosphere, long light coloured roots develop in well aerated soils. Root hairs develop best under well aerated condition. Roots gets thicker, shorter and darker in anaerobic soils that also retard the development of root hairs. Poor aeration causes abnormal development of roots, e.g. abnormal shaped sugar beet and carrot roots have been found in poorly aerated soils.

Nutrient absorption is an energy consuming process. Energy is available from respiration is expended in absorbing nutrient ions from the soil. Hence nutrient absorption is retarded in poorly aerated soils.

If excessive amounts of readily decomposable organic matter has been added to the soil, then it would decompose to evolve high amounts of carbon dioxide to the soil. Consequently root growth and germination of seeds would be adversely affected.

Some crops become infested with pathogens in poorly aerated soils. The incidence of will disease caused by the fungus (Fusarium sp) has been attributed to poor aeration. Citrus and suffers from die-back in poorly aerated soils (Daji 1970) Rai (1981) have also reviewed the works of some investigators who have observed that poorly aerated soils (waterlogged soils) has an effect on the pathogenecity of root infesting fungi.

4.24. SOIL TEMPERATURE

Importance of Soil Temperature

Temperature begins to affect soil characteristics from the stage of weathering of its parent rock. Temperature fluctuates frequently enough to crack and exfoliote hard rocks which are finally disintegrated enough to facilitate chemical weathering to take place. Increase in temerpature accelerate chemical weathering of primary minerals in presence of adequate amount of moisture.

Soil micro-organisms are able to function within a wide range of soil temperature i.e. from 25°C to 30°C. They decompose organic matter. The higher the temperature, more rapid is the decomposition of organic matter to release the nutrient especially nitrogen contained in it in the soluble form which are either absorbed by plant roots or lost by teaching. Hence soil organic matter content depend on mean annual soil temperature. So temperate region soils contain more organic matter than tropical region soils. The intensity of other microbial process taking place in the soil depend on soil temperature. For example nitrification takes place best at a temperature varying from about 27°C to 32°C.

The soil must be warm and moist enough for the seeds to germinate. Seeds of different crops germinate at different temperature e.g. maize begins to germinate at a temperature varying from 7° to 10°C. The soil temperature should be 10° to 29°C and 4°C to 10°C for maximum germination and growth of maize and wheat and peas respectively. Potato tubers sprout best and potato crop grows best at 16°C to 21°C. The optimum temperature for emergence of cabbage and pinach seeds varies from 8° to 11°C, that for emergence of beets and cauliflower seeds varies from 11° to 18°C.

If the temperature of the soil is very low then seeds either do not germinate or germinate very feebly. Seeds may be injured in very hot soil. Temperature affects the rate of entry of water in the seed. It therefore controls the germination of seeds when it affects the development of plumule and the redicle when plant grows, photosynthesis and respiration are influenced by temperature of the soil which also influence the growth of roots and absorption of nutrients by them. Most of the warm season *i.e.* kharif crops grow best at a temperature of about 25°C.

Cold soils tend to retard the absorption of phosphate by plant roots from it. This may be remedied by draining them. Drainage tends to warm these soils up.

Alternate freezing and thawing of soils results in the lifting up of roots of plant. This phenomenon is known as having of soils. Roots get broken by this phenomenon. However alternate freezing and thawing of soil improves the structure of cloddy soil if it contains moderate amount of moisture but destroy the soil structure if the soil contains excessive moisture.

Anhydrous ammonia should be applied at a depth of 10 cm when the soil is at 10°C or less at 10 cm depth because ammonium ions are very slowly converted to nitrate below 10°C. So leaching loss of nitrate is less.

4.25. FACTORS AFFECTING SOIL TEMPERATURE

(*i*) *Nature of the soil — Inorganic vs. Organic*: As the average specific heat of dry inorganic soil is about 0.22 cal/gm and that of humus is about 0.44 cal/gm, so dry inorganic soil would warm up quicker and also would cool down quicker than the dry organic soil. This means that organic soils, once they have been warmed up, will remain warm for longer period, than inorganic soils. Hence organic soils remain warmer in winter and cooler in summer than the inorganic soils because the former warms up slowly in summer and cools down slowly in winter in comparison to the latter.

(*ii*) *Soil moisture content*. As the specific heat of water is 1 cal/gm and the average specific heat of dry inorganic soil is 0.22 cal/gm, so the specific heat of the soil increases with the increase in its moisture content as shown below:

Suppose the soil contains 20 per cent moisture on weight basis.

This means that 100 gms of a dry mineral soil has absorbed 20 gms of water.

Heat required to raise the temperature of 100 gms of dry inorganic soil by 1°C is $0.22 \times 100 = 22$ calories.

Heat required to raise the temperature of 20 gms of water by 1°C is 20 calories.

Therefore the heat required to raise the temperature of $100 + 20 = 120$ gms of moist soil by 1°C is $22 + 20 = 42$ calories

Therefore the heat required to raise the temperature of 1 gm of moist soil will be $42/120 = 0.35$ calories.

Therefore the specific heat of the inorganic soil has increased from 0.22 cal/gm to 0.35 cal/gm when the soil moisture content has increased from 0 to 20 per cent.

Hence moist soil remains cool in summer and warm in winter because it warms up slowly in summer and cools down slowly in winter. As water evaporates especially in summer from moist soil, so it looses heat and cools down. Hence moist soil remains cooler than dry soil in summer.

(*iii*) *Soil texture:* The specific heat of pure quartz is 0.17 cal/gm. and that of pure kaolinite is 0.24 gms/cal (As quoted by Daji 1970) clayey soils contain more humus and more moisture, so the specific heat of average clayey soil is more than that of the sandy soil. Therefore sandy soils would warm up quicker in summer and cool down quicker in winter than clayey soils. In other words, clayey soils remain cooler in summer and warmer in winter than sandy soils because the former warms up more slowly in summer and cools down more slowly in winter than the latter.

(*iv*) *Slope of the land*: The angle at which the sun rays meet the land surface, affect the amount of solar radiation reaching the unit area of the land surface. If the same solar radiation reaches the plane surface of the land, and the land surface inclined at an angle against the horizon line, then it would meet more area of the land in the latter case than in the former case as illustrated in Fig. 4.5. So the amount of solar radiation received by the unit area of the land surface would be more on plane land surface than on the inclined land surface. Hence the soil would warm up quicker in the former case than in the latter case.

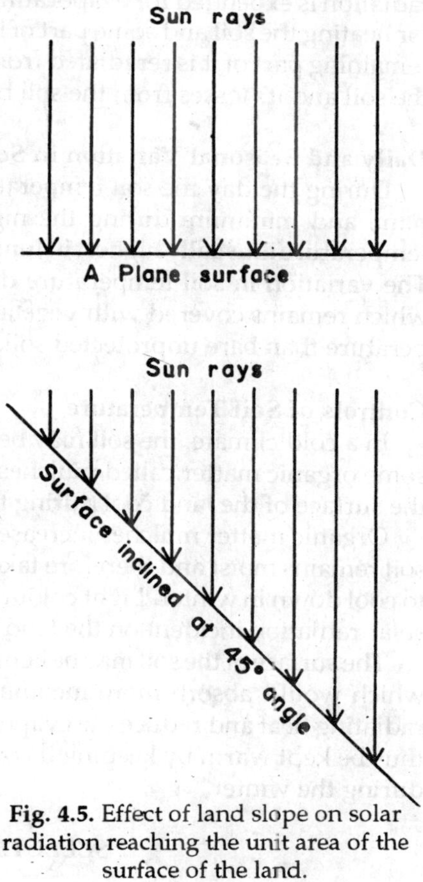

Fig. 4.5. Effect of land slope on solar radiation reaching the unit area of the surface of the land.

(*v*) *Vegetative cover of the land surface*: Bare soil absorbs heat and becomes very hot during summer very quickly and becomes very cold during the winter. This will not happen, if the surface of the land is kept covered with organic matter which would insulate the soil. So the soil will neither become too hot nor too cold.

(*vi*) *Soil depth:* Heat is conducted very slowly from the surface to the subsurfae and back. So the temperature of the subsurface soil remains lower in summer and higher in winter than the surface soil.

(*vii*) *Soil colour:* Suppose a black coloured soil and a white coloured soil are located side by side and both the soils are absolutely dry and do not contain any humus. As black coloured substance absorbs heat and a white coloured soil reflects heat incident on them, so under this circumstances only, the black coloured soil would be warmer than the white coloured soil. However usually a black/darker coloured soil contain more clay and therefore more humus and moisture than a white/light coloured soil, so the former may not be warmer than the latter.

4.26. HEAT BALANCE OF SOILS

The heat balance of a soil refers to the gains and losses of heat energy. Some portion of the incoming radiation is reflected back to the atmosphere while some portion of it is absorbed by the soil surface. A dark coloured soil may absorb 80 per cent of the incoming solar radiation while a light coloured sand my absorb 30 per cent of it. The albedo is the amount of reflected solar radiation. Albedo of soil and water are less than 10 per cent and about 20 per cent respectively. More than half of the incoming solar radiation is reflected from the snow whose albedo is more than 50 per cent. Some part of the solar radiation is expended for evaporating water, some other part of it is expended for heating the soil and some part of it is expended for heating the air while the remaining part of it is reradiated from the surface of the earth. Gains of heat to the soil and its losses from the soil balances each other in the long run.

Daily and Seasonal Variation in Soil Temperature

During the day the soil temperature is maximum between noon and 2.00 p.m., and minimum during the night and morning. The average monthly temperature is usually highest in June and minimum in December and January. The variation in soil temperature decreases as the soil depth increases. Soil which remains covered with vegetation has undergone less variation is temperature than bare unprotected soil.

Controls of Soil Temperature

In a cold climate, the soil may be kept warmer by keeping it covered with some organic matter called mulches and draining the soil. Mulches also keep the surface of the land cool during the hot summer season.

Organic matter mulches increase the infiltration of water in the soil; so the soil remains moist and therefore takes longer time to warm up in summer and to cool down in winter. Light coloured organic matter mulch also reflect some solar radiation incident on the land surface.

The surface of the soil may be kept covered by dark coloured plastic mulches which would absorb more incoming solar radiation, prevent the soil from radiating heat and reduce the evaporation of water from the soil. The soil may thus be kept warm by keeping it covered with dark coloured plastic mulches during the winter.

SELECTED READINGS

1. Baver, L.D., Gardener, W.H. and Fardner, W.R. (1955), *Soil Physics*, 4th Ed. Wiley Eastern Ltd., New Delhi 110002.

2. Biswas, T.D. and Mukherjee, S.K. (1987): *Text-Book of Soil Science* Tata-McGraw Hill Publishing Co. Ltd. Asaf Ali Road, New Delhi
3. Brady, N.C. (1984), *Nature and Properties of Soil* 9th Edn., Macmillan Publishing Co., New York.
4. Buol, S.W., Hole, F.D. and Mc Cracken, R.J. (1973): Soil Genesis, and Classification. Iowa State University Press, Ames, Iowa 50010.
5. Daji, J.A. (1968), *A Textbook of Soil Science,* Media Promoters & Publishing, Bombay.
6. Danahue, R.L. (1964), *Soils: An Introduction to Soils and Plant Growth,* Prentice Hall, Englewood Cliff, New Jersy 07632.
7. Foth, H.D. (1984), *Fundamentals of Soil Science,* 7th Ed. John Wiley & Sons.
8. Knonke, H.K. (1968), *Soil Physics,* Tata McGraw-Hill Publishing Co. Ltd.
9. Rai, M.M. (1981), *Principle of Soil Science,* 2nd Ed. Macmillan India Ltd.
10. Stefferud, A. (Ed.) (1957), *Soils,* Year Book of Agriculture, U.S.D.A.
11. Tamhane, R.V., Motiramani, D.P. Bali, Y.P. and Donahue, R.L. (1964), *Soils: Their Chemistry and Fertility in Tropical Asia.* Prentice Hall of India Ltd. New Delhi.

5
SOIL WATER

No living organism can survive without water. Water is the chief constituent of plants. It is essential for certain vital processes like photosynthesis and is also responsible for maintaining the turgidity of plant cells. Nutrients are dissolved in the soil water and absorbed by the crop. The soil water containing nutrient ions is known as the soil solution. Water affects weathering of rocks and formation of soil. Soil water regulates the physical, chemical and biological properties of the soil. Although plants absorb some water directly from rain and dew, the major portion of water utilized by the crop comes from the soil.

5.1. SOURCE OF SOIL WATER

The original source of soil water is rainfall, which occurs during the months of June to September and December-January. All parts of India do not receive equal rainfall. The moisture bearing south-west monsoon wind blows over to the Arabian sea and the Bay of Bengal and strikes the Western Ghat mountains on the west coast of India and the Himalayan mountain ranges at Assam. Thus it rains heavily on the western side of the Western Ghat mountain ranges and Assam beginning from the last week of May. A broad and rather irregular belt of low rainfall exists on the eastern side of the Western Ghat mountains, which stretches from the interior of Tamil Nadu to the Deccan Plateau and Madhya Pradesh. The south-west monsoon wind is deflected by the Himalayan mountain ranges towards upper India, during June to September when it rains heavily in West Bengal, Bihar, Uttar Pradesh and northern Pubjab. Since the Aravalli mountain ranges lies parallel to the direction of the south-west monsoon wind, so it passes over Gujarat, Rajasthan and Haryana where it rains little during June to September. The pressure increases in northern India during the latter part of the monsoon season. This obstructs and pushes back the monsoon wind towards southern India. So heavy rainfall occurs on the east coast of Tamil Nadu during October-November.

The original source of irrigation water is the river system. The six major river systems of India have been assessed to supply the following quantities of water.

River system	Estimated water in million cubic metres
(*i*) The Indus River System	49300
(*ii*) The Ganges River System	185000
(*iii*) The Brahamputra River System	12300
(*iv*) The Mahanadi and other following rivers	123400
(*v*) The Narmada, Tapti, Godavari, Krishna Cauvery and other southern rivers	246700
(*vi*) The west flowing rivers excluding Narmada and Tapti.	49300
	666000

Irrigation systems are being developed in order to utilize the above water resources of the river system. Well water and pond water are also used for irrigation at many places of India.

5.2. STRUCTURE AND RELEATED PROPERTIES OF WATER MOLECULES

One oxygen atom combines with two hydrogen atoms by sharing a pair of electrons with each of them. But the electrons are not shared equally. They are nearer to the oxygen than to the hydrogens. Consequently, a weak negative charge develops on the oxygen side of the water molecules and a weak positive charge develops on the opposite hydrogen side of it. Hence water molecules are dipolar which means that they possess positive and negative poles. Two hydrogens are connected with one oxygen at an angle of 105° as shown in the Fig. 5.1.

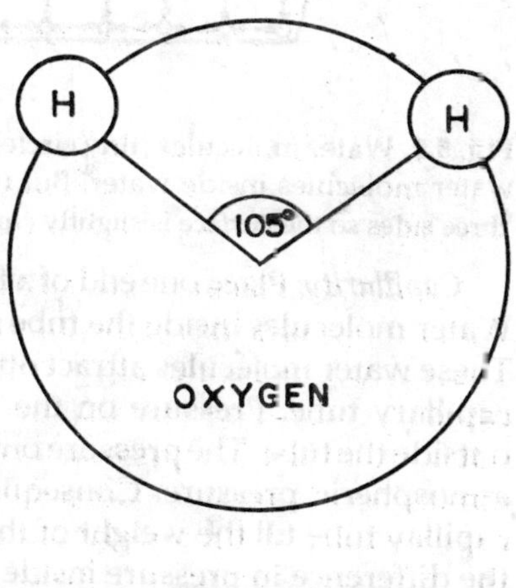

Fig. 5.1. Two dimensional representation of a water molecule in which two hydrogens are attached with one oxygen at 105° angle.

The negative poles of the first water molecule attract the positive pole of the second water molecule, the negative pole of which attracts the positive pole of the third water molecule and so on. In this way a long chain of water molecules are formed in nature. This phenomenon by which hydrogen atoms act as connecting links between water molecules is called Hydrogen Bonding.

Surface Tension is the force in dynes per centimetre acting upon an imaginary line of one centimetre length on the surface of water, which tends to compress the surface area. Water molecules (tiny circles in Fig. 5.2) within water are attracted in all directions by the surrounding water molecules.

In Fig. 5.2 water molecules (tiny circles) are attracted from all sides by the surface of water are pulled in all direction except vertically upwards. Hence, they are pulled in the vertically downward direction as shown in Fig. 5.2. The surface of water is under some tension, which is known as surface tension. In a similar way the solid surface also attracts water molecules, which is known as adhesion. The attraction between water molecule is known as cohesion.

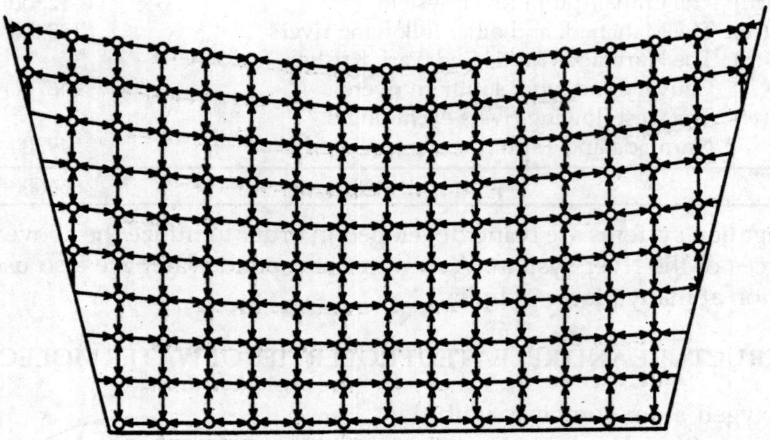

Fig. 5.2. Water molecules (tiny circles) are attracted from all sides by the neighbouring water molecules inside water. But water molecules at the surface are attracted from three sides so the surface is slightly curved in the downward direction.

Capillarity: Place one end of a fine glass tube called a capillary tube in water. Water molecules inside the tube are attracted by the inner surface of the tube. These water molecules attract other water molecules. Hence water rises in the capillary tube. Pressure on the water surface is lower inside the tube than outside the tube. The pressure on the surface of the water outside the tube is the atmospheric pressure. Consequently water is slowly pushed up inside the capillay tube till the weight of the column of water inside the tube is equal to the difference in pressure inside and outside the tube.

5.3. FORMS OF SOIL WATER

Soil water is commonly of the following three forms:

(*i*) Gravitational or free water than moves under the influence of gravity, down the macropores. It is held by the soil with a force of less than 0.1 bars. It is harmful to crops because it removes air and nutrients from the soil. Hence crops suffer from lack of enough oxygen and nutrient. Moreover, they cannot use this kind of water.

(*ii*) Capillary water is held on the surface of the soil particles and within the micropores by surface tension forces (0.1 to 31 bars). Plants can use it.

(*iii*) Hygroscopic water is held with a force greater than 31 bars, mostly in non-liquid or vapour form. Higher plants cannot absorb it.

5.4. INFILTRATION

Infiltration means the downward entry of water into the soil. A loose porous well aggregated surface soil allows more water to infiltrate it than a compact soil. If the land surface is covered with vegetation, it remains loose and porous. Hence, more water will infiltrate it than a bare unprotected soil. More water infiltrates to clayey soils when they dry. So more water initially infiltrates clayey soil. But later, infiltration decreases, when the clayey soils swell and cracks are closed. Infiltration rates are classified in Table 5.1.

TABLE 5.1

Infiltration class	*Rate inches/hour*	*Nature of soil*
Very slow	Less than 0.1	Soils that contain high amount of clay.
Slow	0.1 to 0.5	Soils that contain high amount of clay or low amount of organic matter. Shallow soils.
Moderate	05. to 1.0	Sandy loams and Silt loams.
Rapid	More than 1.0	Deep sand and well aggregated silt and loam.

5.5. PERMEABILITY AND PERCOLATION

Soil characteristics that determine the rate of movement of water and air in the soil is called permeability of the soil. The soil characteristics that determine the rate of movement of water down into the soil is called the hydraulic conductivity of the soil. Permeability and hydraulic conductivity of soils depends mainly on the amount of macropores (non-capillary pores) in the soil. Permeability and hydraulic conductivity of the surface soil is more than that of the sub soil because the surface soil contains more macropores than the sub soil. If the soil possesses an impermeable layer, then the permeability of the whole soil is reduced. These impermeable layers must be broken by sub-soiling to improve the permeability of the whole soil. Permeability of soils is reduced by continuous tillage and improved by growing deep rooted crops. Permeability rates of soils are shown in Table 5.2.

Percolation means the downward movement of water down the column of the soil. An excess of rain or irrigation water that percolates down the column of soil also carries away considerable amounts of plant nutrients and recharges the ground water in the humid region. Percolation is negligible in the arid regions. More water percolates down the sandy soil than the clayey soils. If crops are grown on the field, they will utilize the water and less water will percolate down the soil. Permeability rates are classified as has been shown in Table 5.2.

TABLE 5.2

Permeability class	*Rate in inches per hour*
1. Very slow	Less than 0.051
2. Slow	0.05 to 0.20
3. Moderately slow	0.20 to 0.80
4. Moderate	0.80 to 2.50
5. Moderately rapid	2.50 to 5.00
6. Rapid	5.0 to 10.0
7. Very rapid	more than 10.0

5.6. ENERGY CONCEPT OF SOIL WATER

Energy is involved in the retention of water by the soil, its movement, its absorption by crops and loss to the atmosphere. Potential energy is mainly involved. Potential energy = mgh where m is the mass of water, h is the height or distance the water has been moved and g is the acceleration due to gravity. When work is done on soil water, for example it is raised to some higher elevation, then its potential energy is increased. But when the soil water itself does work, for example it is absorbed by the clay particle or exchangeable cations, then its potential energy is decreased. Water has a tendency to move from a state of higher potential energy to a state of lower potential energy.

FREE ENERGY OF SOIL WATER

Free energy of soil water is a measure of its tendency to change, react or escape. Water has a tendency to move from state of higher free energy to a state of lower free energy. Free energy is the same as potential energy because like potential energy, free energy of soil water is increased if work is done on it. If, for example, it is raised to a higher elevation and free energy of soil water is decreased, if it itself does work, for example, if it is absorbed by the clay particles or exchangeable cations.

Factors Affecting/Free Potential Energy of Soil Water

(*i*) Gravity tends to pull water down. Work is done to raise water to higher elevation when the applied kinetic energy is converted to potential energy. Hence potential or free energy of soil water is increased. Therefore, the potential or free energy of soil water located at a higher elevation is always higher than the same of the soil water located at some reference point at the lower elevation.

(*ii*) Water molecules are attracted by the clay particles and exchangeable cations. Water molecules, while moving towards them, do work. So the potential or free energy is decreased. Hence potential or free energy of soil water that has been held by clay particles and exchangeable cations, is always less than the pure free water located at the same elevation as the soil water.

POTENTIAL OF SOIL WATER

Potential energy possessed by a unit quantity of water is called potential of soil water. It is very difficult to measure the absolute level of potential of soil water. So the difference of potential or free energy of soil water and that of pure free water at some reference point, usually at the lower edge of the soil profile is usually measured. Potential or free energy of water at this reference point is always considered zero. If the potential or free energy of water at a particular point is more than the pure free water at the reference point, it is considered a positive, if it is less, it is considered negative.

The total potential of soil water is the sum of the potential of soil water due to gravity and due to the attraction between it and clay and humic particles, and exchangeable cations.

The total potential of soil water = Gravitational potential + Matric potential + Osmotic potential + Gravitational potential. It is the amount of work that must be done per unit quantity of pure water in order to transport isothermally (at the same temperature) and reversibly (which can also be put in a backward direction) an infinestimal quantity (extremely small quantity) of water from the reference point to the concerned point in soil. Gravitational potential is usually positive because work has been done on the water to raise it from the lower edge of the soil profile, to the surface from where it tends to flow down, under the influence of gravity.

Matric potential may be defined as the amount of work that must be done per unit quantity of pure water in order to transport isothermally and reversibly an infinestimal quantity of water from the reference point, to the concerned point located at the same elevation as the reference point in the soil. It is due to the attraction of water by clay and humic particles which decreases the potential of water at the concerned point. Hence matric potential is always negative.

Osmotic potential is the amount of work done per unit quantity of pure water in order to transport isothermally and reversibly and infinestimal quantity of water from the reference point to the concerned point located at the same elevation as the reference point in the soil. It is due to the attraction of water by exchangeable cations, and soluble salts which decreases the potential of soil water. Hence the Osmotic potential is always negative.

SUCTION AND TENSION OF SOIL WATER

Both the matric and esmotic attractive forces reduce the potential or free energy of water with respect to potential of pure water at the reference point, considered zero. Thus these are negative and are also called suction or tension of soil water.

METHOD OF EXPRESSING THE NEGATIVE POTENTIAL OR SUCTION OR TENSION OF SOIL WATER

Most commonly, the negative potential is expressed in terms of the height in centimetres of a column of water of one square centimetre cross-sectional area, whose weight is just equal to the negative potential or tension of soil water.

P^F is the logarithm* to the base 10 of the height in cms of a water column of

*Logarithm of a number with respect to any base other than one is the index to which the base must be raised in order to get the number. For example we know that $2 \times 2 \times 2 = 2^3 = 8$. Here 8 is the concerned number, 2 is called the base and 3 is called the index. Hence logarithm of 8 with respect to the base 2 is 3. So mathematically we can write : $\text{Log } 8_2 = 3$.

Similarly we know that $10^3 = 1000$, So log 1000 = 3

and $10^4 = 10000$, So log 10000 = 4 and so on.

1 sq. cm. cross-sectional area whose weight is just equal to the negative potential or tension of soil water.

Another method is to express it in terms of Bars. One Bar is the pressure exerted by a column of water of length 1023 cm and of cross-sectional area of one sq. cm. One bar is approximately equal to one atmosphere, which is equal to a pressure of 14.7 pounds/sq inch or 760 mm. of mercury at the mean sea level.

5.7. ENERGY CONCEPT OF RETENTION OF WATER BY THE SOIL

Dry soil absorbs water whenever it is added either in the form of irrigation or rain, when water loses potential or free energy which reappears in the form of heat, called heat of wetting. The first layer of water molecules adsorbed on the clay and the humus loses the maximum amount of potential or free energy and therefore possesses the highest negative potential or lowest free energy. The second layer of water molecules absorbed over the first layer of water molecules by the clay and humus loses a little less potential or free energy because they have travelled a little less distance than the first layer of water molecules. So they possess little less negative potential. In a similar way, the negative potential of the third layer of water molecules absorbed over the second layer is still less than that of the second layer. Therefore the negative potential of the succeeding layer of water molecules gradually decreases as their distance from the clay and the humus increases *i.e.*, the thickness of the moisture film around the clay and humus increases till it becomes first zero and then acquires low positive value. This stage is reached immediately after a heavy rain or irrigation when even the macropores are full of water. Now gravity begins to pull the layer of water molecules continuously downwards untill the force of attraction between clay, humus and water is just equal to the gravitational force when the downward movement of water stops and the soil is said to be at its field capacity. Crops growing on the soil absorb water and the negative potential of the outermost layer of water molecules gradually increases till it is equal to the force with which roots extract water from the soil. At this stage, roots cannot absorb any more water from the soil and so the crop wilts permanently. Soil moisture at this stage represents the permanent wilting point or coefficient. Water continues to be lost from the soil by evaporation and the negative potential of the outermost layer of water molecules gradually increases until water vapourises. The soil moisture at this stage represents the hygroscopic coefficient.

5.8. SOIL MOISTURE CONSTANTS

(i) Field capacity is the amount of moisture held by the soil after the excess gravitational water has drained away. Negative potential or free energy of soil water is 0.1 to 0.3 bar. Approximate p^F value is 2.54.

(ii) Moisture equivalent is the amount of water held by the soil against a centrifugal force 1000 times that of gravity. The field capacity of sand is higher than its moisture equivalent, but the field capacity of loams is approximately

equal to their moisture equivalent.

(*iii*) The wilting coefficient represents the soil moisture content which cannot be taken up by roots and so the crop wilts. The negative potential or free energy of soil water is approximagely 15 bar.

Approximate p^F value is 4.18.

Field capacity = 0.86 Moisture equivalent + 2.6 Wilting coefficient

$$\text{Wilting coefficient} = \frac{\text{Moisture equivalent}}{1.84}$$

(*iv*) Hygroscopic coefficient represents that soil moisture content held by the soil in the vapour form. Negative potential or free energy is 31 bars. Approximate p^F value is 4.5.

5.9. AVAILABLE MOISTURE

Crops can take up moisture held between a negative potential of 0.1 to 15 bars *i.e.* between the field capacity and wilting coefficient. Therefore the available moisture of soils is the moisture retained by them between the field capacity and wilting coefficient.

The available water holding capacity of soils depend upon the following factors:

(*i*) *Soil texture:* The pore space volume and the surface area of soils affect their capacity to hold water. Soils of finer texture possess the maximum total water-holding capacity whereas the soils of medium texture *i.e.* loam to light clay loam, possess the maximum available water holding capacity. Wilting coefficient of soils increases with the increase in the fineness of their texture but their field capacity increases with the increase in the fineness of their texture up to clay loam only. Research, as quoted by Foth (1984) has shown that the available water content of many soils is closely correlated with their silt and fine sand contents. As the soil texture become finer, a smaller percentage of water held by the soils at their field capacity, is available for plant growth.

(*ii*) *Nature of the clay:* Available water holding capacity of black cotton soils is higher than those of red soils. The most important clay minerals in black cotton soil and red soil are montmorillonite and kaolinite respectively. As the cation exchange capacity of Montmorillonite (80 to 150 milliequivalent per 100 gms) is much higher than that of kaolinite (3 to 15 milliequivalent per 100 gms), a montmorillonitic soil can retain much more moisture than a kaolinitic soil.

(*iii*) *Soil structure:* As soil moisture at the field capacity is mainly present in the micropores and on the surface of the soil particle, the available water holding capacity of soils increases when their structure is improved because the micropore space and specific surface of soil particles increases.

(*iv*) *Soil humus:* Both the field capacity and wilting coefficient of humus are very high. Hence humus improves the available water holding capacity of soils mainly by improving their structure.

(*v*) *Soluble salt* decreases the available water holding capacity of soils because they increase the osmotic potential of soil water which in turn increases the wilting coefficient of soils. The available water holding capacity which is the difference between the field capacity and the wilting coefficient, decreases.

(*vi*) *Depth of soil:* All other factors being favourable, the available water holding capacity of the deeper soil is usually considered higher than that of shallow soil because the volume of soils from which roots extract moisture is higher in the deeper soils than shallower soils. The presence of hard pan restricts root development and therefore reduces the soil volume from which the roots extract moisture. Hence in such cases, the total available water in that soil is decreased.

(*vii*) Rooting habits and drought.

Other factors like the rooting habit of crops and their capacity to tolerate dry period, their stage and rate of growth, usually influences the rate of absorption of moisture by crops.

5.10. MECHANISM OF ABSORPTION OF MOISTURE BY THE ROOT

When root hairs begin to absorb water from any point in the soil the soil moisture content at that point or zone decreases where the negative potential of the remaining water increases. This means that water molecules are held by the soil particles with a greater force than before, at that point in the root zone. Therefore, water from the surrounding, relatively wetter zone where the negative potential of the soil water is much less than in the root zone, move to the root zone. The magnitude of the movement of the soil water depends upon the difference of the negative potential of the soil water at the root zone and that of the soil water at the surrounding zones, and the ease with which the soil pores allow the water to move through them. This kind of movement of the soil water is relatively more rapid in sandy soils than in the clayey soils. The soil water usually does not move by more than a few millimetres by this mechanism.

Roots also rapidly elongate towards the surrounding moist zones and absorb moisture. Well aggregated soils are usually well permeated with roots and root hairs where the soil-water movement of a few millimetres is quite sufficient meet the water demands of the crop.

5.11. MEASUREMENT OF NEGATIVE POTENTIAL OR TENSION OF SOIL WATER

Negative potential or tension of soil water is measured with a tensiometer, a simple form of which is shown in Fig. 5.3. One porous clay bottle has been attached with a long glass tube provided with a narrow side tube of one square centimetre cross-sectional area that has been bent twice at right angles, near its upper end as shown in the Fig. 5.3. The tensiometer is filled with water and is placed in the soil. The lower end of the bent side tube is placed in a dish containing mercury. Water moves out of the clay bottle and therefore the mercury rises in the manometer till the force of attraction between clay and humus and water is satisfied. The mercury column becomes stationery at this stage. The weight of this mercury column per square centimetre represents the force with which water is held by the soil. The product of the height in centimetres of this mercury column, and the density of mercury gives the height of an equivalent column of water in centimetres, the logarithm to the base ten of which gives the p^F value of the soil moisture. For example, suppose the height of the mercury column of unit cross-sectional area is 25 cm.

Then equivalent height of the water column of unit cross-sectional area is 25 × 13.6 = 340 cms. Here density of mercury is 13.6 gm/cc.

$\therefore p^F$ of the soil water = $\log_{10} 340 = 2.54$

The tensiometer works only up to a tension of 0.85 atmospheres, beyond which air enters into it, after which it no longer operate. So tensiometers are more suitable for measuring the tension of soil moisture in sandy soils. They may also be used for crops that require a lot of water, e.g. potato. Tensiometers actually used in the field are provided with a vacuum gauge, which measures the tension or negative potential of soil water directly in bars or atmospheres.

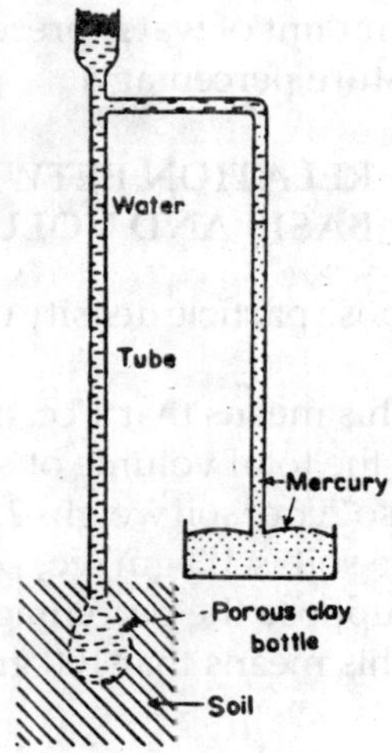

Fig. 5.3. Tensiometer

5.12. MEASUREMENT OF SOIL MOISTURE

Gravemetric Method

The moist soil is weighed. Then it is oven dried and re-weighted.

$$\text{Soil moisture percentage} = \frac{\text{Wt. of moist soil} - \text{Wt. of oven dried soil}}{\text{Wt. of oven dried soil}} \times 100$$

Electrical Resistance (Gypsum Block Method)

Two electrodes are embedded in a gypsum block, which is buried in the soil permanently. When the soil moisture is to be measure, these two electrodes are connected with the two terminals of a battery operated moisture meter. The gypsum block absorb moisture till its moisture content is equal to the soil moisture content. This increases the flow of electricity through the gypsum block, which is recorded by the moisture meter that has already been calibrated to read zero at wilting coefficient and 100 at field capacity. This reading of the moisture meter thus expresses the amount of available water as the percentage of the available water holding capacity of the soil.

Neutron Scattering Method

A neutron emitting material, for example radium, and a detector of radio activity is lowered in a hole dug in the soil. The high velocity neutrons emitted by the radio active material collide with the hydrogen nuclei of the water molecule, and are slowed down and deflected. A certain percentage of these slowed neutrons ultimately reach the detector. This is proportional to the number hydrogen nuclei of water molecules. The reading of the detector gives the amount of water present in the soil as it has been calibrated against the soil moisture percentage.

5.13. RELATION BETWEEN SOIL MOISTURE CONTENT ON WEIGHT BASIS AND VOLUME BASIS

Suppose particle density of the soil is 2.6 gms/cc, and its percentage pore space is 50.

This means that 1 cc. of soil solids weigh 2.6 gms and if pore spæe % is 50, then the total volume of soil = vol. of soil solids + vol. of pore space = 1 + 1 = 2 cc. so 2cc of soil weighs 2.6 gms and 1 cc. of soil weighs 1.3 gms. So bulk density of the soil is 1.3 gm/cc.

Suppose the percentage of soil moisture on the basis of weight is 20.

This means that 100 gms of soil has absorbed 20 gms of water

$$\text{and } 1 \text{ gm of soil has absorbed } \frac{20}{100} = 0.20$$

$$\text{As } \frac{\text{Mass}}{\text{Volume}} = \text{Density, Therefore volume} = \frac{\text{Mass}}{\text{Density}}$$

$$\text{volume of this 1 gm of soils is } \frac{\text{its wt.}}{\text{its bulk density}} = \frac{1}{1:3} \text{ cc.}$$

And volume of this 0.20 gm of water is 0.20 cc.

$$\text{So } \frac{1}{1:3} \text{ cc. of the soil has absorbed 0.20 cc. of water.}$$

$\therefore$ 100 cc of the soil has absorbed 0.20 × 1.3 × 100 = 26 cc. of water,

So the volumetric soil water percentage is 26. The same value is obtained if the gravemetric soil water per cent is multiplied by the bulk density of the soil.

$\therefore$ Volumetric soil water per cent = 20 × 1.3 = 26

Hence volumetric soil water percentage = Gravemetric soil water percentage × bulk density of the soil

Again Vol. soil water per cent =

$$\frac{\text{Volume of water absorbed}}{\text{Vol. of soil that has absorbed the water}} \times 100$$

$$\text{or } \frac{\text{Depth of water} \times \text{Area of Water}}{\text{Depth of soil} \times \text{Area of soil}} \times 100 = \frac{\text{Depth of Water}}{\text{Depth of soil}} \times 100$$

Since area of the water and the soil are the same in the field *i.e.* one acre, for example

$$\therefore \frac{\text{Depth of Water}}{\text{Depth of soil}} \times 100 = \text{Vol. soil water \%} = \text{Gravemetric Soil Water per cent} \times \text{Bulk Density of the soil.}$$

$$\text{or Depth of water} = \frac{\text{Gravementric soil water per cent}}{100} \times \text{Bulk density of the soil.} \times \text{Depth of the soil.}$$

or inches of water per foot depth of the soil

$$= \frac{\text{Gravemetric soil water \%}}{100} \times \text{Bulk Density of the soil} \times 12$$

and cm of water per metre depth of soil

$$= \frac{\text{Gravemetric soil water \%}}{100} \times \text{Bulk density of the soil} \times 100$$

Since 12 inches = 1 foot and 100 cm = 1 metre.

So inches of water per foot depth of the soil actually means acre inches of water per acre foot of soil. Centimetres of water per metre depth of soil means hectare cm of water per hectare metre depth of soil.

Acre inches of water means volume of water of one acre in area and one inch deep.

Hectare cm of water means volume of water of one hectare in area and 1 cm deep.

Example 1: Determine the soil moisture percentage of a field from which moist soil sample were collected and placed in the soil moisture box; oven dried and weighed. The measurements were:

Weight of the empty moisture box = 45.2 gms
Weight of moisture box + moist soil = 125.8 gm
Weight of moisture box + oven dried soil = 116.3 gms

Solution:

Weight of moist soil = 125.8–45.2 = 80.6 gms
Weight of oven dry soil = 116.3–45.2 = 71.1 gms
Weight of moisture = 80.6–71.1 = 9.5 gm

$$\text{Soil moisture per cent} = \frac{\text{Wt. of moisture}}{\text{Weight of oven dry soil}} \times 100$$

$$= \frac{9.5}{71.1} \times 100 = 13.5$$

Example 2: Find out the volumetric soil moisture characteristic in Inches of water per foot depth of soil and cms of water per metre depth of soil from the following gravemetric data.

Maximum water holding capacity = 50%
Field capacity = 30%
Wilting coefficiency = 10%
Bulki density of soil = 1.5 gms/cc

Solution: Inches of water per foot depth of soil

At maximum water holding capacity, the soil contains

$$= \frac{15}{100} \times 1.5 \times 12 = 9.0 \text{ inches of water per foot depth of soil.}$$

At field capacity, the soil contains $\frac{30}{100} \times 1.5 \times 12 = 5.40$ inches of water per foot depth of soil.

At wilting coefficent the soil contains $\frac{10}{100} \times 1.5 \times 12 = 1.8$ inches of water per foot depth of soil.

The volume of water that drains away 9.0–5.4 = 3.6 inches of water per foot depth of soil. Volume of water available for crop growth = Field capacity – wilting coefficient = 5.4–1.8 = 3.6 inches of water per foot depth of soil.

In cm. of water per metre depth of Soil.

At Max. W.H.C., the soil contains $\frac{50 \times 1.5 \times 100}{100}$ = 75 cm of water/metre depth of soil.

At field capacity, the soil contains $\frac{30}{100} \times 1.5 \times 100 = 45$ cm of water/metre depth of soil.

∴ The water that drains away = 75 – 45 = 30 cm/metre depth of soil.

At a willing coefficient, the soil contains $\frac{10}{100} \times 1.5 \times 100 = 100 = 15$ cm. of water/ metre depth of soil.

Water available for plant growth = 45 – 15 = 30 cm of water/metre depth of soil.

Example 3:

Soil moisture content of at the field capacity is	24%
Soils moisture content at the wilting coeff. is	16%
Bulk density of the soil is	1.20 gms/cc
Present soil moisture content is	15%

Find out

(1) the inches of water per foot depth of the soil at the present movement.

(2) the depth of soil to be wetted by applying 1.2 inch of irrigation water.

(3) the available water holding capacity of the soil.

Solution :

(1) At present, the soil contains $\frac{15}{100} \times 1.2 \times 12 = 2.16$ inches of water per foot depth of soil.

(2) Let d inches of soil be wetted by applying 1.2 inches of irrigation water.

$$\text{Then } 1.2 = \frac{24-15}{100} \times 1.2 \times d = \frac{9}{100} \times 1.2 \times d = \frac{10.8}{100} \times d = 0.108\, d$$

$$\text{or } d = \frac{1.2}{0.108} = 11.11 \text{ inches}$$

so 11.11 inches depth of soil is to be wetted by applying 1.2 inches of water

(3) Available water holding capacity of the soil is

$\frac{24-16}{100} \times 1.2 \times 12 = \frac{8}{100} \times 1.2 \times 12 = 1.152$ inches of water per foot depth of the soil.

5.14. MOVEMENT OF WATER IN THE SOIL

A Movement of Liquid Water

Liquid water moves in the soil under saturated and unsaturated conditions.

Saturated flow of water: Soil is saturated with water which means that even the macropores are full of water, and water at the surface of the soil possesses a high positive potential with respect to the water located at the lower level of the soil profile. As water moves from the state of higher positive potential to the state of lower the adjacent drier soil. But water can only flow through the pore space which are very irregular or even discontinuous in shape.

Saturated flow of water depends on the difference of potential which is the force causing the flow of water between different points the soil and the case with which the soil allows the ready movement of water. This latter factor is called the Hydraulic Conductivity of the soil which depends on the macropores and the viscosity of water.

Unsaturated flow of water: The soil is unsaturated with water which means that the macropores are filled up with air and the micropores are filled up with water. Water flows through micropores and the surface of the soil particles from a zone of lower negative potential to another zone of a higher negative potential. Clay and humus attract water molecules. Water moves from moist soil where it is under a lower negative potential to dry soil. The water absorbed by dry clay possesses a very high negative potential.

Movement of Water Vapour

The movement of water vapour is of two kinds:

(*i*) ***Mass flow:*** Water vapour flow in a mass along with the movement of other soil gases. When the atmospheric pressure decreases soil gases including water vapour expand and move out the soil. When the atmospheric pressure is increased, soil gases and water vapour contracts and therefore gases and water vapour enter the soil.

(*ii*) ***Diffusion:*** Diffusion is the property by which two gases readily mix with each other when they are brought in contact with each other. Vapour pressure of water is a measure of its tendency to evaporate. Water evaporates from the moist soil and the vapour diffuses into the adjacent dry soil. Water also evaporates from warm soil and the water vapour diffuses into the adjacent cool soil. Diffusion of water vapour due to the difference in relative humidity may balance the diffusion of water vaopur due to the difference in temperature so that the net movement of water vapour may be zero as shown in Fig. 5.4A.

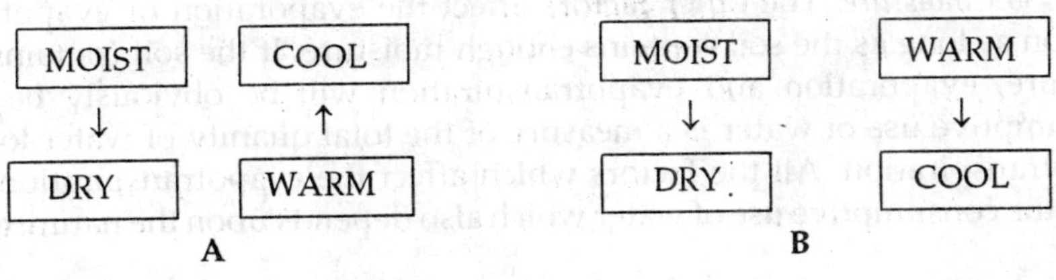

Fig. 5.4. Movement of water vapour by diffusion.

Water vapour moves from one zone to another zone only when both of them work in one direction only as shown in the Fig. 5.4B.

5.15. LOSS OF SOIL MOISTURE

Moisture is lost from the soil in the vapour form and the liquid form.

A Loss of Moisture in the Vapour Form

About 5 to 20 per cent of the annual rainfall is intercepted by the foliage of green plants and evaporated back to the atmosphere even before it reaches the surface of the field. More moisture is lost in this way if the rain is received as light showers.

Water vapour is lost from the soil by evaporation and transpiration. Evaporation is the process by which the liquid water is gradually changed at the surface of the soil or water to the vapour.

Transpiration is the process by which water vapour leaves the living foliage of green plants and enters the atmosphere.

The combined loss of moisture by evaporation and transpiration is called evapotranspiration. Water vapour exerts a certain amount of pressure which is called the vapour pressure. The difference of vapour presure at the soil surface or plant surface and the atmosphere is called the vapour pressure gradient. Evapotranspiration depends mainly on the vapour pressure gradient which depends on the following factors.

(*i*) *Radiant energy:* More radient energy is received from the sun in the arid regions than in the humid regions and also during cloudless days than during cloudy days in any region. This radiant energy evaporates water from the surface of the soil and the plant.

(*ii*) *Atmospheric vapour pressure:* If the vapour pressure of the atmosphere is much lower than the vapour pressure at the surface of the green plant of soil, more water is lost by evapotranspiration. If it is high, much less water is lost by evapotranspiration.

(*iii*) *Temperature:* When temperature increases, then vapour pressure at the surface of the green plant of soil increases, but the vapour pressure of the atmosphere is not increased. So evapotranspiration increase.

(*iv*) *Wind velocity:* Wind blowing at a high velocity continuously removes moist air near the surface of the plant or soil. Hence more water evaporates from the surface of the plant or atmosphere.

(*v*) *Soil moisture:* The other factors affect the evaporation or evapotranspiration as long as the soil contains enough moisture. If the soil contains less moisture, evaporation and evapotranspiration will be obviously be less. Consumptive use of water is a measure of the total quanity of water lost by evapotranspiration. All the factors which affect the evapotranspiration also affect the consumptive use of water which also depends upon the nature of the crop.

Efficiency of Water Use

A certain number of crops can be grown if a given amount of water is suplied to the soil. The efficiency of water use may be expressed in terms of consumptive use, which is the total quantity of water required for evapotranspiration (evaporation + transpiration) and metabolic activities of the crop in any specified time. It is the kilogram of water consumed to produce each pound of plant tissue or may also be expressed in terms of the transpiration ratio which is the amount of water transpired by the crop for producing unit quantity of dry matter. Like consumptive use, Transpiration ratio also depends on the nature of the crop. e.g. this ratio of pea, wheat and maize have been found to be 563, 544 and 337 respectively at Pusa in Bihar (India).

Reducing Loss of Moisture by Evaporation

The surface of the land should be kept covered with waste organic material in order to reduce evaporation. This material left on the surface of the land in order to reduce evaporation is called the mulch. In USA plastic sheets are spread between the rows of crops, to reduce evaporation and weed growth. Evaporation may be conveniently reduced if the refuse of the previous crop, the stubble mulch, is left on surface. The seed bed is prepare and seeds are sown.

Sometimes no crop is sown on certain portion of the field. This practice is called fallowing. This is done when the soil is not moist enough to produce crops in that season and the water can, be stored in the soil for the crop which will be grown in the next season.

Loss of Moisture in the Liquid Form

Water moves down the soil profile. This phenomenon is called percolation, and depends upon rainfall characteristics, soil characteristics and crop characteristics. When the water which then is more than the field capacity of the soil, enters the field and the excess water perfolates down the profile. Percolation loss increases when rainfall increases. The presence of crops of the field reduces percolation loss, because they absorb considerable amount of water.

Considerable amount of plant nutrients are lost from the soil along with the percolation water. More important among them are usually nitrate, potassium and calcium. This loss of nutrients is usually greater in sandy soils than in clayey soils because the sandy soils cannot retain them in sufficient quantities. Therefore, fertilizers should be applied to sandy soils in split doses at the root zone at the time when the crop requires them.

5.16. IRRIGATION

Irrigation means the application of water to the field in order to moisten the root zone of the crop to the field capacity level.

Necessity for Irrigation

Indian agriculture is severely handicapped by the erratic distribution of rainfall in different seasons. It not only varies from one region to another but, also from year to year in the same region. If irrigation water is available during the period of rapid growth of crops and their flowering and grain formation stages, the yields of crops and the resultant money returns from them could can be maximised by efficient utilisation of all the agricultural resources. Vegetable farming is not possible without irrigation. The tremendous increase in agricultural production in India can only be attributed to the judicious use of irrigation water, fertilisers and high yielding varieties during the last two decades.

Problems of Irrigation Farming

Some well waters especially in Uttar Pradesh, Haryana and Rajasthan—are saline, and their application to land has made lands saline and have dissolved the salts in the superficial layer of the soil profile, which have moved upwards along with the upward movement of water in the dry season to accumulate on the surface of the lands. Water seeping out of the irrigation canals have elevated the ground water at many places in Uttar Pradesh and Haryana.

Units of Measurement of Water

The rate of flow of water one cubic foot per second is called one cusec of water. The weight of one cubic foot of water is 62.4 pounds. So when one cusec of water flows for one hour, $62.4 \times 60 \times 60 = 224640$ lbs = 100.28 tons of water is delivered to the field.

One acre inch of water is the volume of water which would cover one acre field one inch deep. One acre inch of water weighs approximately 101 tons.

The duty of water means the number of acres of field where only one crop is growing, that have been irrigated by one cusec of water flowing through the growing season of that crop. Hence the duty of water varies with the nature of the crop, the soil and the season.

Principles of Irrigation

The amount of irrigation water and its frequency of application depends on compensating the water deficit in the root zone just before the crop wilts with sufficient excess water to wash down the accumulated salt if any into the drains.

Wilting coefficient of soils varies from about three per cent in coarse sandy soils to about twenty per cent in fine clayey soils, whereas the field capacity ranges from about eight per cent in coarse sandy soils to about forty per cent in fine clayey soils. Only that much water should be applied to fields which in necessary to raise the soil moisture content at root zone of the concerned crop to the fied capacity level. Water should not be wasted by applying too much to the field as this will unnecessarily leach down the plant nutrients from the root zone. Usually about 0.5, 1.0, 2.0, 2.5 and 3.0 acre inches of water are required to wet sandy soils, sandy loam soils, loam soils clayey loam soils and clayey soils respectively to a depth of one foot.

Since the moisture retention capacity of soils increases when their texture become finer, so the soils of finer textures and those of coarser textures may be irrigated at an approximate interval of seven days and ten days respectively under average conditions. A handful of the soil of the concerned field is manipulated in the hand in an attempt to make a ball. The field should be irrigated if the soil exhibits the following characteristics.

(*i*) The sandy loam soil will not form the ball. It will become lighter in colour when it is dried.

(*ii*) The loamy soil will form the ball which will break down when it is tossed upwards again caught in the hand.

(*iii*) The clayey soil will form the ball which will crack open when it is squeezed.

The appearance of the standing crop may also indicate whether or not it needs irrigation. The field should be irrigated when the largest leaves of the crop wilt in the mid afternoon, or it turns unusually dark green in colour. The concerned field should continue to be irrigated till a blunt stick can easily be pushed to a depth of one foot and thereafter it should be stopped.

Most of the crops grow best and produce maximum yield if the soil moisture content is maintained at the field capacity level at least during the period of rapid vegetative growth, flowering and grain and fruit formation stages. A high moisture level helps to develop succulence and softness in vegetable and fodder crops. Young plants should be more frequently irrigated than at later stages because their root system have not developed deep enough to utilize the moisture from the deeper layer of the soil.

Method of Irrigating Fields

Methods of irrigation may be grouped as follows:

Method of irrigation
- Surface irrigation
  - Flood irrigation
  - Bed irrigation
  - Basin irrigation
  - Furrow irrigation
- Sub-surface irrigation
- Aerial/overhead/ Sprinkler irrigation

Flood Irrigation

Irrigation water is let in the field usually at one side of the field, so that it can flow freely in all directions. Some portions of the field may be over irrigated while other portions may be under irrigated. Some portion of the field may be water-logged.

Bed Irrigation

Small beds are made on the levelled field. The beds are surrounded by low 6 to 12 inches high earthen embankments or bunds. Two rows of beds are irrigated from one sub-channel that runs between the rows. Bed irrigation is usually practised for some vegetables growing on any soil except sandy soils.

Basin Irrigation

Small circular basins are made around fruit trees. A small circular area around the stump of the fruit free is kept dry be erecting small bunds of 6 to 9 inches height. Water is suplied to each basin and the sub channels that connect the basins to the main irrigation channel. Basin irrigation is suitable for soils of any texture. It makes the most efficient use of irrigation water.

Furrow Irrigation

Crops are grown on ridges that have been made on levelled land and water is applied in furrows on either side of the row of crop on the ridge. One sub-channel is made at one side of the field for supplying water to the furrows.

Sub-surface Irrigation

If any impervious layer exists below the root zone, then ditches are dug up to the impervious layer. Water is applied to the ditches when it moves laterally and vertically to wet the root zone. If no impervious layer occurs below the root zone then porous or perforated pipes are laid there for supplying water to the root zone of the crop.

Sprinkler Irrigation

The water is sprayed to the crop from overhead pipes which comprise of the main line and the laterals. The main pipes are usually made up or iron or steel. Aluminium lateral pipes are usually portable, and are fitted with riser pipes, which in turn are fitted with sprinkler nozzles each of which applies water to a circular area of a diameter that depends upon the size of the nozzle and the pressure of the water. The rate of water application varies from one fourth to three-fourth inches per hour. This is controlled by using nozzles of a proper size and water pressure in the laterals and the discharge of individual sprinklers are controlled by valves that have been fitted with the pipe lines.

5.17. DRAINAGE OF AGRICULTURAL FIELDS

Irrigation and drainage cannot be separated from each other. Irrigation is the process of supplying an adequate amount of water to the crop growing in a particular field, whereas drainage is the process of removal of the excess water from the root zone of the crop as quickly as possible. Both irrigation and drainage are equally important for improved agriculture, so that the crop gets the right amount of water.

Importance of Drainage

Excess water is removed from the macropores which are then filled up with air for respiration of roots and aerobic micro-organisms. When the field is drained, roots grow deeper and extract nutrients and water from the deeper layers of the soil. Therefore the crops can tolerate the dry season better than crops that are growing in poorly drained soil. Micro-organisms also carry on favourable activities like decomposition of organic matter and formation of humus, conversion of unavailable organic forms of nutrients to available in organic nutrients; this process is known as the mineralisation of nutrients.

When the excess water is drained away, the specific heat of the soil decreases. Hence less solar energy is required to warm up the field during the winter season. At the same time the loss of heat due to the cooling of soil, caused by the evaporation of water within, it is also reduced. Hence drained soils warm up quicker during winter than the undrained soil.

Soil water freezes in cold countries and in this process roots are lifted up from the soil and broken. This phenomenon is known as heaving. Drainage of the land prevents heaving.

Red, yellow, gray spots blue and black colour of the soil indicates the drainage requirements of the soil.

Types of Drainage

(1) **Surface drainage**: Drains may be made on the surface and excess water is removed through them before water enters the soil. Surface drains usually remove the excess water quickly, but they occupy some land where crops cannot be grown. It is used on impermeable soils and in areas of heavy rainfall.

(2) **Underground drainage**: It is needed where water accumulates within a depth of two to four feet.

(*i*) *Mole drainage:* A mole drain is an unwalled, cylindrical channel of about 10 cm diameter located at depth of about 75 cm from the surface of the soil. This channel is made by means of a special plough equipped with a sharp blade. A pointed cylindrical metal plug is attached with a sharp blade. When this plug is pulled through the soil, a compressed wall channel is formed. This can be practised on clayey soils.

(*ii*) *Tile drainage:* Burnt clay pipes of about 30 cm length called tiles, are laid end to end at the bottom of a trench of about 75 cm to 1 metre deep and above the impermeable layer. They are then covered with straw, and surface soil. The trench is then filled up with the earth. Water enters the tiles through joints, mostly from the sides and bottom. So the top of the joints should be covered with cloth. The spacing between drains and their depth on the soil texture is shown below.

Suggested spacing between tiles drains and their depth

Soil texture	*Permeability*	*Spacing metre*	*Depth cm centimetre*
Clay	Very slow	9 to 15	75 to 90
Sily clay loam, Clay loam	Slow	12 to 21	90 to 100
Sandy clayloam	Moderately slow	20 to 27	100 to 115
Loam	Moderate	27 to 35	115 to 125
Sandy loam	Moderately rapid	35 to 50	125 to 135
Loamy sand & sand	Rapid	50 to 90	135 to 145

Drains should be properly graded, so that they carry water away at non-erosive velocities. The grade of drains should not exceed 0.2 per cent for sandy loam soils and 1 per cent for clayey soils. In solving any drainage problem, the root cause of accumulation of water should be investigated and remedial measures undertaken.

SELECTED READINGS

1. Baver, L.D. (1956), *Soil Physics*. 3rd ed. John Wiley and Sons Inc., New York.
2. Baver, L.D., Gardner, W.H. and Gardner, W.R., 4th ed. Wiley Eastern Ltd. New Delhi 110002.

3. Biswas, T.D. and Mukherjee, Mukherjee, S.K. (1987): Textbook of Soil Science. Tata McGraw Hill Publishing Co. Ltd. Asaf Ali Road, New Delhi.
4. Brady, N.C. (1984), *Nature and Properties of Soils*, 9th ed. Macmillan Publishing Co., New York.
5. Buckman, H.O. and Brady, N.C. (1964), *Nature and Properties of Soils*. Eurasia Publishing House, Ram Nagar, New Delhi-110001.
6. Khonke, H.K. (1965), *Soil Physics*. Tata McGraw Hill Publishing Co. Ltd.
7. Kanitkar, N.V. *et al* (1960), *Dry Farming in India*. I.C.A.R. New Delhi.
8. Michael, A.M. and Ojha, T.P. (1965), *Principles of Agriculatural Engineering* Vol. II. Jain Brother, 873 East Park Road, New Delhi-110055.
9. Michael, A.M. (1978), *Irrigation. Theory and Practice*. Vikas Publishing House Pvt. Ltd. 5, Ansari Road, New Delhi-110002.
10. Stefferud. A Editor, (1955), *Water Year Book of Agriculture*. 1955 U.S.D.A. Oxford & IBH Publishing Co., 66, Janpath, New Delhi-110001.
11. Thomson, L.N. and Froch F.R. (1975), *Soils and Soil Fertility*. McGraw Hill Publication-Ltd.

6

SOIL EROSION AND CONSERVATION

Soil erosion is the phenomenon of wearing away of the land surface by the action of natural agencies such as water and wind.

6.1. TYPES OF SOIL EROSION

Basically soil erosion is of two kinds (*i*) Normal or geologic erosion, and (*ii*) Accelerated erosion. Normal or geologic erosion is an extremely slow natural process by which the surface of the land tends level off to ultimately form a plain land called pene-plains. This process is so slow that the removal of surface soil is balanced by the formation of soil from the weathering of rocks and minerals contained in them. Accelerated erosion means the quick removal of soil from the land surface due to mismanagement of the land, which includes excessive grazing and deforestation of lands and cultivation of soil, exposing crops without adopting proper soil conservation measures. Accelerated erosion includes wind erosion and water erosion.

6.2. WIND EROSION

The wind is responsible for the removal of soil particles by the following three kinds of erosion is semi-arid and arid region:

(*i*) *Saltation:* The soil particles of size 0.1 to 0.5 mm are pushed along the surface of the land by the wind. They collide with other soil particles and leap vertically upwards. This phenomenon is called saltation. Some soil particles of a 0.5 mm size rise only a little distance upwards. Others may move 30 cm upwards.

(*ii*) *Suspension:* Very fine soil particles, of less than 0.1 mm are kicked upwards by the action of soil particles in saltation to form suspensions of very fine soil particles in air which are carried away.

(*iii*) *Surface creep:* Soil particles of size betwen 0.5 to 1 mm are pushed along the surface by the impact of soil particles in saltation.

About 90 per cent of the soil particles move within 30 cm and about 50 per cent of the soil particles move within 5 cms from the ground surface, by the action of the wind. The control of wind erosion is based on reduced or elimination of particles in saltation.

6.3. TYPES OF SOIL EROSION CAUSED BY WATER

Soil erosion caused by water may be of the following kinds:

(*i*) *Splash erosion:* The raindrops which fall at an approximate speed of 75 cms per second on ground strike it with a force of about 14 times its own weight. The impact of this great force causes the soil particles to detach each other to form flowing mud. This mud is then splashed as much as 30 cms high and 12 cms down the land slope. Fine sandy soil and silty soils are dislodged easily. Such a removal of the soil by the impact of the fast falling raindrops is known as splash erosion.

(*ii*) *Sheet erosion:* Thin uniform layers of soils are removed from the surface of the land by rain water flowing over the land during the initial stage of soil erosion called wheet erosion when muddy water flows over the land.

(*iii*) *Rill erosion:* If sheet erosion is not controlled immediately, the downward flow of muddy rain water concentrates to form minute finger shaped channels called rills over the entire field during the later stage of soil erosion called rill erosion.

(*iv*) *Gully erosion:* If rill erosion is not controlled in time, then rills are deepened by the downward flow of water. Ultimately rills are converted to deep channels called gullies, which may be "U" or "V" shaped "U" shaped gullies are formed if the subsoil has a coarse texture whereas "V" gullies are formed if the sub-soil has a fine texture. Gullies become deeper and wider during heavy rains. Large fields are gradually fragmented and ultimately destroyed.

(*v*) *Slip erosion:* Large masses of rocks and soil materials are detached from the sides of mountains during the rainy season and are carried down by the action of gravity. This phenomenon is knows as slip erosion.

(*vi*) *Stream bank erosion:* Rivers e.g. Kosi in Bihar frequently changes its course when one bank is cut and the soil materials are carried down stream. The finer soil particles are deposited along the other bank. This phenomenon is known as stream bank erosion.

6.4. FACTORS AFFECTING SOIL EROSION CAUSED BY WATER

Rainfall

Intensity of rainfall is the volume of rainfall recorded per unit time. Whenever the intensity of rainfall exceeds the infiltration capacity of the soil, the excess rain water runs down the slope of the land and carries away considerable amounts of soil with it. It has been observed on black soil. That rain water ran off the slope of the land when it rained, more than two inches per day at Manjri and Sholapur in Maharashtra and Bijapur in Karnataka. Rainfalls of between one to two inches per day and between half to one inch per day caused a run off in 80 and 40 per cent of the cases respectively.

Land Slope

According to the law of falling bodies, if the vertical drop is increased four times, the velocity of run off water is doubled and the erosive power of the run off water is increased by four times, the amount of soil particles of a particular size that is carried away by run off water, in increased by 32 times and the size of soil particles that are carried away by the run off water increases 64 times. It has been observed at the Deochanda Experiment station of the Damodar Valley Corporation that in red and yellow soils, the percentage of rainfall that runs off increases from 6 to 29 when the land slope is increased from 2 to 5 per cent. The corresponding soil loss in tonnes per hectare has increased from 3.3 to 23.6. More plant nutrients are lost from lands when the soil loss is increased due to the increase in the degree of the slope of the land. For example, it has been observed at Kanpur that when the slope of lands are increased from 0.5 to 3.0 per cent, the loss of organic matter, nitrogen, phosphoric acid, potash, lime and magnesia from the alluvial soils increases from 87, 9, 11, 43, 53 and 41 kilogram per hectare to 180, 11, 23, 118, 203 and 212 kilogram per hectare respectively.

If the land slope length is increased, the velocity of run off water gradually increases as the run off water flows down the long slope. This means that the erosive power of the run off water gradually increases. Hence more soil is lost from the lower portion of the land than the upper portion of the land.

Nature of Soil

Water infilterates into the soil and percolates down the soil through the micropore spaces, which are more in the sandy soil than in clayey soils. Hence less rain water would run off the sandy land soil than the clayey land soil. Red loam, red and yellow soils are much more resistant to soil erosion than black soils because kaolinite and hydrous oxides of iron and aluminium dominate the clay mineralogy of the former soils whereas the smectite group of clay minerals dominates the clay mineralogy of the latter soils which expand and contract with the changed in soil moisture content. Consequently soil masses are easily dislodged and carried away by rain water.

Crop Management Practices

The fast falling raindrops strike the bare surface of soil with great force to make a soil water suspension, which clogs the pore spaces of the surface soil and reduces the infiltration capacity of the surface soil. Hence more rain water runs off the sloping land if soil exposing crops like maize, are grown than if the soil covering crops like groundnut are grown. For example, it has been observed that more soil is lost from the red and yellow soil of the Deochanda experiment station of the Damodar Valley Corporation if maize is cultivated (soil loss is 24 tonnes per hectare) than if groundnut is cultivated (soil loss is 14 tonnes per hectare).

6.5. WATER EROSION SOIL LOSS EQUATION

The universal soil loss equation has been developed by Wischameir and Smith on the basis of numerous run off and soil loss studies on small field plots and small water shed in order to predict soil loss from agricultural lands under different management practices. The soil loss equation is:

$A = RKLSCP$

where A = Computed soil loss in tonnes per hectare.

R = Rainfall and runoff erosivity index by geographical location.

K = Soil erodibility factor which is the average soil loss in tonnes per hectare per unit of erosion index for a particular soil in a cultivated continuous fallow field with a slope length of 22 metres and slope steepness of 9 per cent.

$L = \left(\frac{1}{22}\right)^x$ where x is a constant

Land slope %	*Value of X*
less than 3%	0.3
3 to 4%	0.4
more than 4%	0.5

l = slope length is metres

$$S = \frac{0.43 + 0.30 + 0.043\,s^2}{6.574}$$

s = slope percentage

C = Crop management factor which is the ration of soil loss for a given condition to the same from continuous fallow.

C value has been calculated for different crops under different slopes P = Soil conservation factor which is the ratio of the soil loss for a given practice to the same from up and down the slope. For example, P is 0.10, 0.12 and 0.16 for land slope of 2.1–7 per cent, 7.1–12 per cent and 12–.118 per cent respectively.

6.6. SOIL CONSERVATION MEASURES

Soil conservation measures besically aim at increasing the infiltration of rain water in soils of fields and decreasing the run off and soil loss from fields. All agricultural lands must be used only for that purpose for which they are best suited, in order to get the maximum possible return from them without any physical damage or decrease in their productivity. Agronomic soil conservation measures may be adopted to control soil erosion up to land capability class II beyond which Agronomic soil conservatioin measures must be supplemented with mechanical soil conservatioin measures.

6.6.1. AGRONOMIC SOIL CONSERVATION MEASURES

Agronomic soil conservation measures basically aim at keeping the land under vegetative cover at least during the period or season when the land most susceptible to soil erosion. The vegetative cover of the land protects it against the beating action of fast falling raindrops and provides organic matter to the soil that decomposes, and improves the physical condition of the soil. The agronomic soil conservation measures include (*i*) Crop rotation, (*ii*) Mulching, (*iii*) Contour farming, (*iv*) Strip cropping, and (*v*) Growing grasses and forest trees.

(*i*) *Crop rotation:* It is the growing of a set of crops in a regular succession on the same field within a certain period. A good crop rotation comprises of a soil exposing crop like maize, a densely grown small grain like upland Paddy

(Dular Paddy, Gora Paddy in Bihar), a spreading legumes like groundnut and a deep rooted crop like pigeon pea. In these ways, nutrients are removed by different crops from different depths and the land is kept covered.

(*ii*) *Mulches:* Organic materials like leaves, straw etc. which have been left on the surface of the land to cover it are called mulches. They protect the surface of the land against the beating action of rain drops. Later they decay to form humus.

(*iii*) *Contour farming:* All the points on a contour line are on the same elevation. Contour Farming means growing crops in lines parallel to contour lines. It reduces the speed of the rain water flowing down the soil and consequent soil loss.

(*iv*) *Strip cropping:* It means the growing of soil exposing crops like maize, and soil covering crops like groundnut in alternate patches or strips across the slope of the land. It is suitable for gently sloping land. It may be of the following four kinds:

(*a*) *Contour strip cropping:* Strips of crops are parallel to the contour lines.

(*b*) *Field strip cropping:* Strips of crops are parallel to the general slope of the land.

(*c*) *Buffer strip cropping:* Severally eroded portions of land are permanently kept under grasses and contour strip cropping is practised in the rest of the area.

(*d*) *Wind strip corpping:* Strips of crops are across the direction of the wind.

Width of strips of crops in strip cropping

Land slope per cent	*Width of erosion permiting crop*	*Width of erosion resistant crop*
Less than 1 per cent	45 metres	9 metres
between 1–2 per cent	24 metres	6 metres
between 2–3 per cent	12 metres	4 metres

(*v*) *Growing of grasses and forest trees:* Grasses completely cover the surface of the land. Their root system binds the soil particles to form soil aggregates. Hence the physical condition of the soil is improved, so more rain water infilterates, into the soil and less rain water flows the slope. Hence the rate of soil erosion is decreased. Grasses should be planted on the grasses waterway and terraces in order to stabilize them. Perennial grasses like Pennisetum pedicellatum and Pennisetum polystachyon and annual grasses like Panicum antidotale, Pennisetum purpurium. Chloris gayanan, Dicantheum nodosum, and Cenchrus ciliaris may be grown permanently on class V to class VI land, to reduce soil erosion from them.

6.6.2. MECHANICAL SOIL CONSERVATION MEASURES

Mechanical control measures aim at dividing a long slope of the land into a series of shorter ones in order to reduce the velocity of run off water. Water is impounded over the land for long periods. Hence more water infiltrates in the soil and less water flows down the slope of the land at non erosive velocity. The important mechanical soil conservation measures are as follows:

(i) *Basin listing:* The land is ploughed to make furrows parallel to the contour lines. Than these furrows are closed at short intervals to make a series of shallow basins on the contour line. This process is called Basin Listing. It is effective on gently slopping land in low rainfall areas.

(ii) *Sub-soiling:* The impermeable layers at a depth of 30 to 60 cms are broken at an interval of 90 to 180 cms apart with a sub-soiler across the slope of the land in order to increase the permeability of the land.

(iii) *Terracing:* Low earthen embankments which have been constructed across the slope of the land are called terraces. These are of the following kinds:

(a) *Channel terrace:* Shallow channels are made across the slope of the land either exactly on the contour line or with a slight grade (0.1 to 0.2%). The excavated soil is placed on the lower side of the channel.

(b) *Broad based terrace:* Wide based, very low earthen embankments are made on the contour lines by excavating the soil from both the sides of terraces. This is practised in areas with a relatively low rainfall.

(c) *Narrow based terraces:* Narrow based low earthen embankments are made across the slope of the land at suitable intervals in high rainfall areas. The spacing between Terraces may be calculated as follows:

Verticle Interval = 0.3 $(S+2)$ Where S = Slope percentage. A water way must be provided for terreced field so that the excess rain water in the field may drain at a non-erosive velocity. It should be designed by using following formulae: $Q = CIA$ where C = Run off coefficient. I = Intensity of Rainfall. A= Catment area. C is the ratio of the rates of run off and rainfall.

If the soil texture becomes finer or any impermeable layer is present in the soil profile, then the proportion of rainwater percolating down the soil profile decreases and therefore the value of the run off coefficient C increases. For example, the value of C is 0.29, 0.40 and 0.5 for sandy soils, loamy soils without any impermeable layer, and clayey soils with clay pan respectively, in the case of the cultivated land soil. More rain water percolates down the profile of the cutivated land soils because excessive cultivation of the soil destroys the soils structure. Hence the value of C is less in grass land soil and in the forest land soils than the cultivated land soils. For example its value for sandy soils, loamy soils without any impermeable layer and clayey soils or soils with clay pan is 0.15, 0.35 and 0.45 respectively in the case of grassland soils, and 0.10, 0.30 and 0.40 respectively in the case of the forest land soils.

The value of the intensity of rainfall "I" varies in different geographical locations.

$$I = \frac{KT^a}{(t+b)^n}$$ where T = return period of rain in years, usually ten and t = duration of rain in hours, i.e. time of concentration of the rain in the watershed.

The values of a, b, m and k have been calculated for the eastern (Orissa, Bihar, West Bengal and Assam), southern (Andhra Pradesh, Karnataka, Tamil Nadu and Kerala), western (Maharashtra and Gujarat), northern (Uttar Pradesh, Haryana, Punjab, Rajasthan, Kashmir and Himachal Pradesh) and central (Madhya Pradesh) zones of India as mentioned below. The value of "*K*" is 0.6933, 6.311, 3.974, 3.914 and 7.4645 for eastern, southern, western, northern and central zones of India respectively. The value of "*a*" is 0.1353, 0.1523, 0.1647, 0.1623, 0.1712 for eastern, southern, western, northern and central zones of India respectively. The value of "*b*" is 0.50 for eastern, southern, and northern zones of India and 0.15 and 0.75 for western and central zones of India respectively. The value of "*n*" is 0.8801, 0.9465, 0.7327, 1.0127 and 0.9599 for eastern, southern, western, northern and central zones of India "respectively".

The value of *t*, the duration of rain in hours may also be calculated as : $t = 0.01947\ k^{0.77}$ and $k = \sqrt{L^2/H}$

where L = maximum length of travel in metres and H is the difference in elevation between the most remote point and the outlet in metres.

The value of *t* (in seconds) may also be calculated as shown below:

$$\underset{\text{(in seconds)}}{t} = \frac{\text{Distance most remote point the outlet}}{\text{Average velocity in metre per sec.}}$$

The average velocity may be taken as shown below:

Av. slope % of the channel from the remotest point of the watershed to the outlet	Av. velocity in metres per second
1 to 2 per cent	0.6
2 to 4 per cent	0.9
4 to 6 per cent	1.2
6 to 10 per cent	1.5

The area of cross-section of the channel *a* is calculated by using the formula $Q = a\ V$. Where V is the velocity in metres per second of the run-off water. It usually varies from 0.3 to 0.6 metres per second depending on the availability of the grass sod. e.g. 0.6 metres/sec if good grass is available and 0.3 metres/sec if the land is bare, calculating *a*, the channel is tentatively designed by choosing the arbitrary top width, bottom width and depth., after calculating "*a*".

For example, if $a = 3$ square feet, then top width, bottom width and depth of the trapezoidal channel would be 8, 4 and 1/2 foot respectively, because

$$a = \frac{(8+4)}{2} \times \frac{1}{2} = 3 \text{ sq. ft.}$$

Then these calculations are verified by calculating V by using the Mannings formula $V = \frac{1.486}{n} R^{2/3} S^{1/2}$

Where
V = Velocity in ft. per second
n = roughness coefficient (0.04)
R = Hydraulic radius = a/p
a = Cross–section of the channel
P = Wetted perimeter ofthe channel, approximately equal to the top width of the channel
S = Rate of fall of channel in foot per foot

If the value of V thus calculated by using the Manning's Formula, closely agrees with the value of V choosen for designing the channel, then the channel has been correctly designed, otherwise another value of V is to be choosen and the channel i to be resigned.

(d) Bench terraces: They consist of a series of platforms with a suitable vertical drop along the contour lines. The vertical drop may vary from vary 2 to 6 ft. The soil materials that have been excavated from the top portion have been used to fill the lower portion. The depth of the soil must be at least 15 cms deeper than the vertical interval to ensure sufficient soil depth on the inner side after levelling. The vertical interval for the bench terrace is calculated by using the following formula:

$VI = \frac{WS}{100}$ where W = width of the bench in metres and S = slope percentage of the land.

The boundary of benches should be kept parallel to the contour line, to reduce the earthwork to the minimum. The earthwork in cubic metres per hectare is 12.50 WS for level bench terrace. It is 12.50 W (S-s) for outwardly sloping benches and 12.5 ($S + s$) for inwardly sloping benches where s is the inward or outward slope of finished benches.

(iv) Contour trenching: A series of Trenches two feet wide and one foot deep are excavated on the contour lines. The excavated soil is kept on the lower edge of the trenches where forest trees are planted.

6.7 FACTORS AFFECTING WIND EROSION

Factors affecting wind erosion are included in the following wind erosion equation:

$E = I\,K\,C\,L\,V$ where E = Soil loss in tonnes per acre
I = Soil Erodibility
K = Soil roughness
L = Length of Field
V = Vegetative cover

The soil erodibility factor depends upon the soil texture. Finer texture soils are more aggregated and therefore more resistant to soil erosion than the coarser textured soils.

The wind velocity is decreased near the surface of the ground if it is roughened by ploughing across in the direction of the prevailing wind. The soil particles moved by the wind are trapped on the rough ground surface.

The climate affects the soil moisture and wind velocity. Wet soils do not blow. The soil moisture content is reduced by hot dry winds to a permanent wilting point before wind erosion takes place.

Wind erosion is zero near the edge of the field. It increases as the distance from the field edge is increased. Organic mulches are left on the surface of the land, as mulches control wind erosion. Wind erosion is also controlled if crops are planted across the direction of wind blowing.

6.8. CONTROL OF WIND EROSION

If the sub soil has a finer texture than the surface soil, the land should be ploughed deep in order to bring the finer textured soil to the surface, because finer textured soils are aggregated and hence more resistant to soil erosion than the courser textured soil.

The surface of the land must be roughened by disc ploughing. The land ridges may be constructed at certain intervals across the direction of the prevailing wind.

Soil exposing crops like cotton, and soil covering crops like guar (Cyamppsis tetragonolba) should be planted in alternate strips across the direction of the prevailing wind. The soil surface should be kept covered by waste organic material called mulches.

SELECTED READINGS

1. Araker; H.R. and Donahye, R.L. (1985), *Principles of Soil and Conservation and Water Management*. Oxford and IBH Publishing Co., 66 Janpath, New Delhi.
2. Bennet, H.H. (1955), *Element of Soil Consvervation*, McGraw-Hill Book Co., Inc., New York.
3. Datta, S.K. (1986), *Soil Conservation and Land Management*. Internation Book Distributors 9/3 Rajpur Road, Dehra Dun-248001.
4. Rama Rao, M.S.V. (1962), *Soil Conservation if India*. Indian Council of Agricultural Research, New Delhi.
5. Schwab, G.O., Frevert, R.K., Edminster, T.W. and Barnes, K.K. (1981): *Soil Conservation Engineering*, John Wiley & Sons, New York.
6. Tamhane, R.V., Motiramari, D.P., Bali, Y.P., and Donahen, R.L. (1964): *Soils: Their Chemistry and Fertility in Tropical Asia*. Prentice Hall of India (Private) Ltd., New Delhi.
7. Tejwami, K.G. (1976), *Soil and Water Conservation*. Chapter in Hand Book of Agriculture, Indian Council of Agricultural Research, New Delhi.
8. U.S.D.A. (1954), *A Manual on Soil & Water Conservation*. U.S.D.A., Soil Conservation Service Hand Book No. 61.

7
CHEMICAL PROPERTIES OF SOILS

The chemical properties of soils refers to the nature of the chemical changes taking place in them. These chemical changes depend upon their chemical compositions and the nature of the inorganic and organic materials contained in them, which have originated from the gradual decomposition of the sial and organic materials, mainly of plant origin.

7.1. CHEMICAL COMPOSITION OF SOILS

Since the sial is mainly made up of oxygen (46.7%), silicon, (27%), aluminium (8.1%) and iron (5.0%), the inorganic component of soils are mainly made up of them. Plant nutrients like calcium, magnesium, potassium, sodium, phosphorus and sulphur are present in the minerals and the soil solution. The elemental composition of different soils very due to the differences in the rates of chemical weathering and leaching. Oxygen, silicon and aluminium occur as constitutents of minerals and as oxides. Iron occurs mainly in the form of oxides and ferromagnesium minerals. The proportion of titenium in soils increases when the intensity of chemical weathering increases, beause the titenium containing minerals, rutile and anatase are resistant to weathering. Calcium occurs mainly in calcite, gypsum, apatite, dolomite, while magnesium is present mainly in dolomite, hornblende, augite etc. Potassium occurs mainly in orthoclase, mcrocline, mica and illite, some of which slowly weather to release potassium in the soil soultion. These potash bearing minerals are present in the fine sand and silt fraction of most soils. Aerable soils usually contain insignificant amounts of sodium, which, however, is present in high quantities in saline and alkaline soils such as chloride, sulphate and carbonate. The major portion of inorganic phosphorus occurs as iron and aluminium phosphate in the acidic soils and calcium phosphate in the alkaline soils. Phosphorus also ocurs in the organic form, e.g. phospholipids, inositol choline etc., while the soil nitrogen occurs mainly in the organic form, e.g. proteins, amino acids etc. All micronutrients *i.e.* iron, manganese, zinc, copper, boron and molybdenum mainly occur in the inorganic form.

7.2. MINERALOGICAL ORGANISATION OF SECONDARY MINERALS SILICATE CLAYS

Silicate clays are classified on the basis of the number and arrangement of the tetrahedral and octahedral layers (Sec. 2.2 Fig. 2.1) as:

(*i*) 1 : 1 type of clay minerals which consist of one silica tetrahedral layer and one alumina octahedral layer.

(*ii*). 2 : 1 type of clay minerals which consist of two silica tetrahedral layers and one aluminium octahedral layer.

(*iii*) 2 : 1 : 1 type of clay mineral, which consists of two layers of silica tetrahedra, one layer of aluminium octahedron and one layer of magnesium octahedron called brucite.

7.2.1 TYPES OF CLAY MINERALS

Kaolinite: The most prominent member of the 1 : 1 type of clay minerals is kaolinite, in which one silica tetrahedral layer is joined with one aluminium octahedral layer when the top oxygen of the silica tetrahedral layer occupies the position of the oxygen of the aluminium octahedral layer and is common to the tetrahedral and octahedral layer as shown in Fig. 7.1. Practically no isomorphous substitution of ions taken places. A weak negative charge develops on the surface of the basal oxygen of the silica tetrahedral layer and a weak positive charge develops on the surface of the hydrogen of the hydroxyl of the aluminium octahedral layer. The positive charge on the hydroxyl surface attracts the negative charge on the oxygen surface. Hence the kaolinite units are tightly bound by hydrogen bonds. Two-thirds of the octahedral positions are occupied by aluminium ions and one-third is empty as shown in Fig. 7.1. The cation exchange capacity is 3 to 15 millequivalent per 100 gm. The total surface area is 37 to 45 sq. metres/gm. The external surface area is 30 to 35 sq. metre/gm. Internal surface area is 7 to 10 sq. metres/gm.

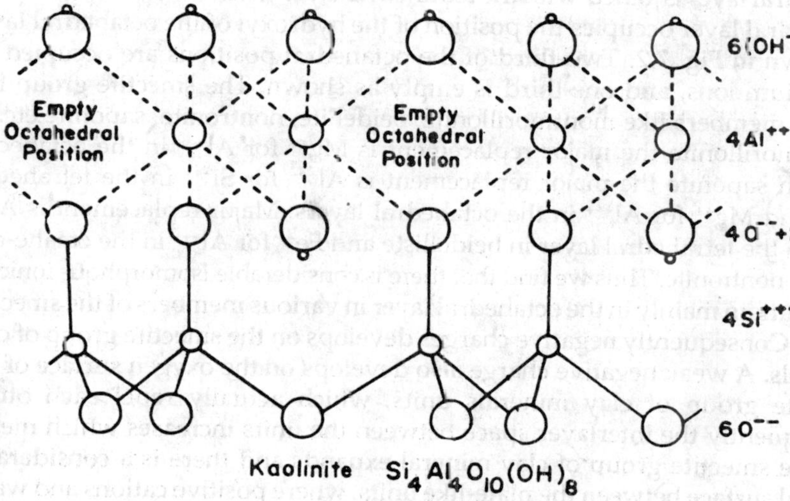

Fig. 7.1. Structure of Kaolinite.

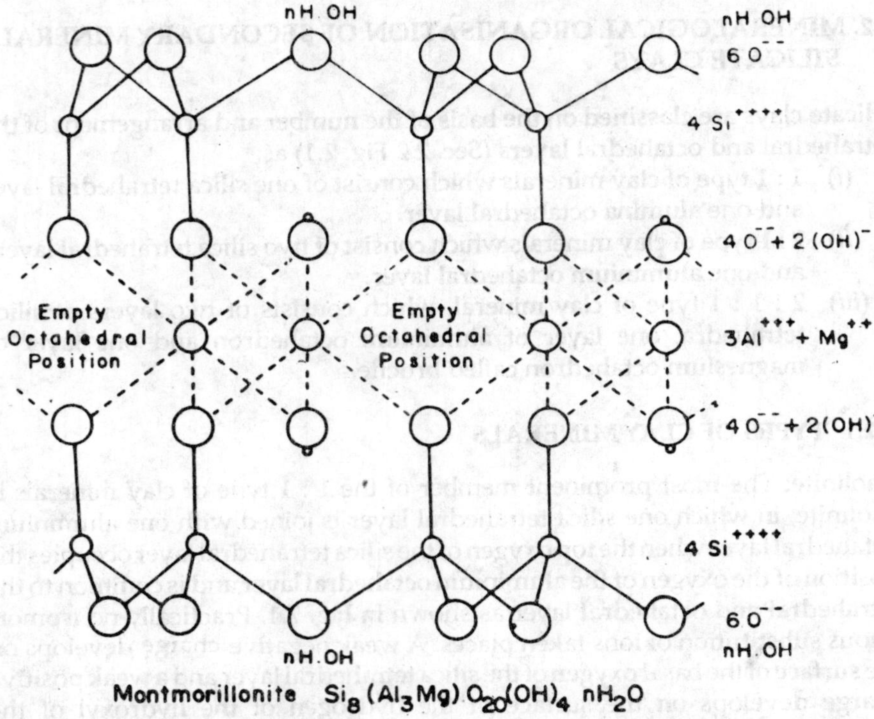

Fig. 7.2. Structure of montmorillonite. (Note that one Mg^{++} ion has isomorphously substitues one Al^{+++} ion to make the montmorillonite structure negatively charged).

7.2.2 TYPE OF CLAY MINERALS

Smectite group: One central layer of aluminium octahedron is attached with two layers of a silical tetrahedra, one at the top and another at the bottom. The octahedral layer is joined with the tetrahedral layer when the top oxygen of the tetrahedral layer occupies the position of the hydroxyl of the octahedral layer as shown in Fig. 7.2. Two-third of the octahedral positions are occupied by aluminium ions, and one-third is empty as shown. The smectite group has several members like montmorillonite, beidellite, nontronite, saponite etc. In montomorillonite the major replacement is Mg^{++} for Al^{+++} in the octahedral layer. In saponite the major replacement is Al^{+++} for Si^{+++} in the tetrahedral layer and Mg^{++} for Al^{+++} in the octahedral layers. Major replacement is Al^{+++} Si^{++++} in the tetrahedral layer in beidelliste and Fe^{++} for Al^{+++} in the octahe-dral layer in nontronite. Thus we find that there is considerable isomorphous ionic substitutioin mainly in the octahedral layer in various members of the smectite group. Consequently negative charges develops on the smectite group of clay minerals. A weak negative charge also develops on the oxygen surface of the smectite group of clay minerals units, which actually repel each other. Consequently the interlayer space between the units increases which means that the smectite group of clay mineral expands and there is a considerable internal surface between the plate-like units, where positive cations and water

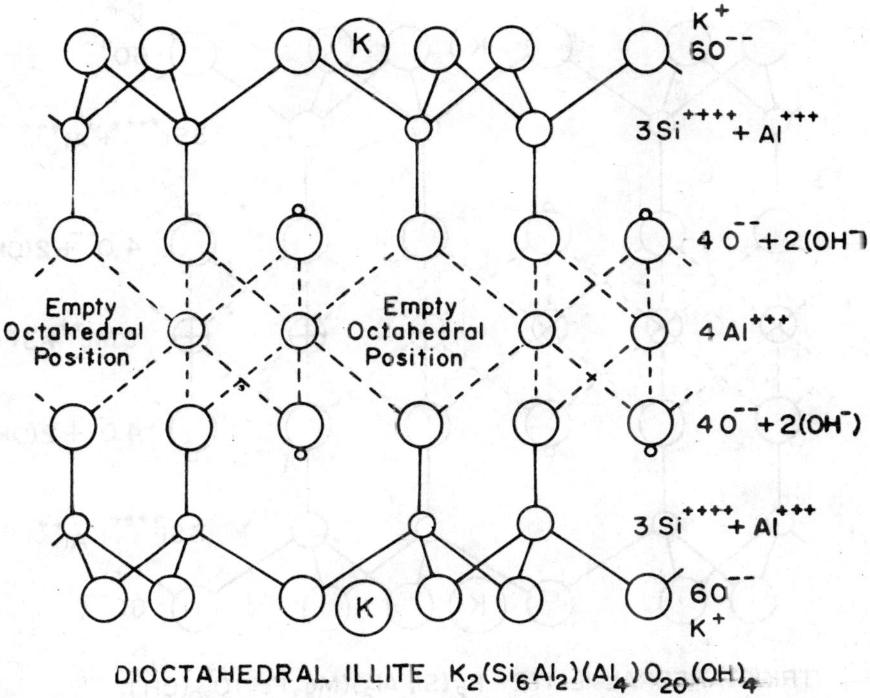

Fig. 7.3. Structure of dioctahedral illite.

molecules are attracted. So they enter there the cation exchange capacity of montmorillonite is 80 to 150 milliequivalent per 100 gm. The total surface area is 580 to 750 square metre per 100 gm. The external surface area is 80 to 150 sq. m./gm. The internal surface area is 500 to 600 sq. metres/gm.

Illite group: The structure of illite is the same as that of mica (Figs. 2.5 A and B). One octahedral layer occurs between the two tetrahedral layers of silica, one at the top and the other at the bottom. The octahedral layer is joined with the tetrahedral layer when the top oxygen of the tetrahedral layer occupies the position of the hydroxyl of the octahedral layer as shown in Fig. 7.3 and Fig. 7.4. One-sixth of the silicon ions Si^{+++} has been replaced from tetrahedral positions by aluminium ions, Al^{++++}. Hence the potassium ions K^+ occur in the hexagonal cavity betwen the two units of illite and bind the illite units tightly. Illites for this are non-expanding. When two-thirds of the octahedral positions are occupied by aluminium ions and one-third is empty, then it is called dioctahedral illite (Fig. 7.3). When all the octahedral positions are occupied by magnesium Mg^{++} and ferrous, Fe^{++} ions, then it is called trioctahedral illite (Fig. 7.4). The cation exchange capacity of illite is 15 to 40 milliequivalents per 100 gm. Total, external and internal surface areas are 120 to 170, 50 to 70 and 70 to 100 square metre per gm respectively.

Vermiculite: Vermiculite consists of two silica tetrahedral layers and one octahedral layer, joined together when the top oxygen of the silica tetrahedral layer occupies the position of hydroxyl ion of the octahedral layer. Some silicon

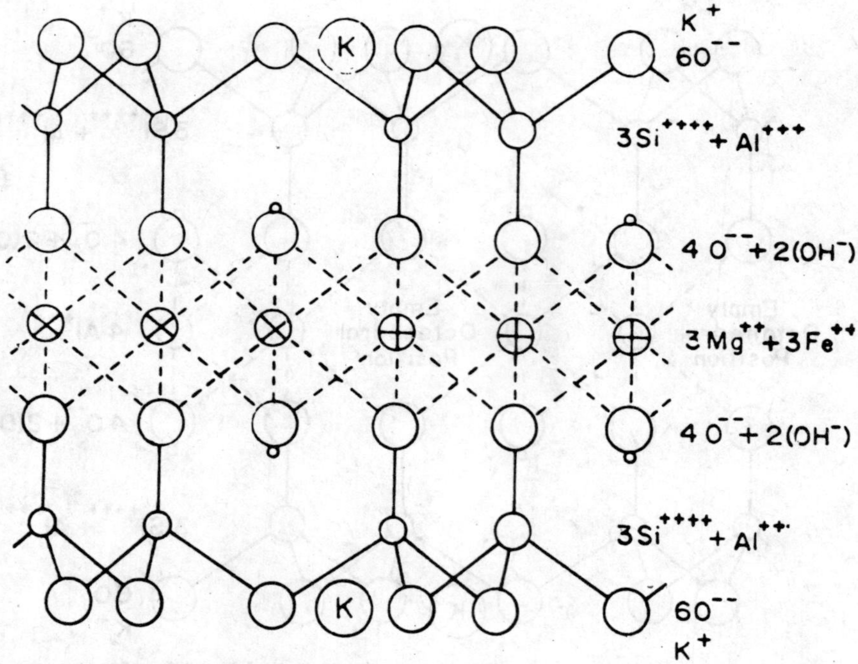

TRIOCTAHEDRAL ILLITE $K_2(Si_6Al_2)(Mg_3Fe_3)O_{20}(OH)_4$

Fig. 7.4. Structure of trioctahedral illite.

ions are isomorphously substituted by aluminium ions and therefore magnesium ions occur between the sheets of vermiculite to balance the structure electrically. Most vermiculite is of the dioctahedral type where the aluminium ions is isomorphously substituted by Mg^{++}, Fe^{++} etc. (Fig. 7.5). Magnesium occupies all the octahedral position in the trioctahedral vermiculite (Fig. 7.6). The basal oxygen surface of the tetrahedral layer has a weak negative charge. A negative charge also develops due to isomorphous ion substitution in the octahedral layer and in the tetrahedral layer. Consequently the negatively charged vermiculite units actually repel each other. Hence the space between vermiculite units increases. But the magnesium ions which occur between units of vermiculite holds them together, so that the expansion of vermiculite is somewhat limited. Hence it expands less than the smectite group of clay minerals. Each magnesium ions is surrounded by six outer molecules which are called bound water. The rest of the water molecules occurring between units of vermiculite are called unbound water, as shown in Figs. 7.5 and 7.6. The Cation Exchange Capacity of Vermiculite is 100 to 150 milliequivalent per 100 gm. The total external and internal surface areas are 780 to 900, 80 to 100 and 700 to 800 square metres gm. respectively.

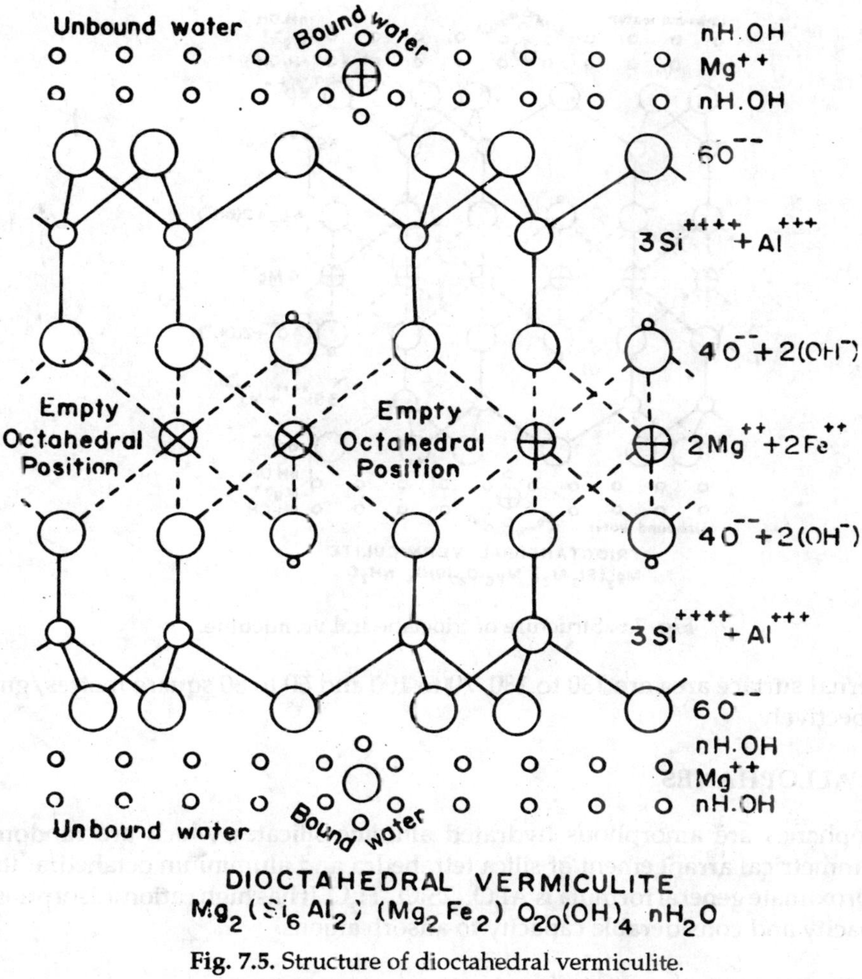

Fig. 7.5. Structure of dioctahedral vermiculite.

7.2.3 TYPES OF CLAY MINERALS

Chlorite: The most common member is chlorite. Chlorite consists of one unit of a vermiculite layer and one unit of a magnesium octahedral layer called Brucite as shown in Fig. 7.7. In the vermiculite layer, one central magnesium and ferrous dominated octahedral layer is attached with two silica tetrahedral layers as shown in Fig. 7.7. The top oxygen of the silica tetrahedral layer occupies the position of the hydroxyl of the Aluminium octahedral layer and is common to both the tetrahedral and octahedral layers. Some silicon ions are isomorphously substituted by aluminium ions. All the octahedral positions are occupied by magnesium, and ferrous ions as shown in Fig. 7.7. All the octahedral positions are occupied by magnesium ions in the brucite layer as shown in Fig. 7.7. Chlorite is also called a 2 : 2 type of clay mineral. The Cation Exchange capacity is 15 to 40 milliequivalents per 100 gms; Total, external and

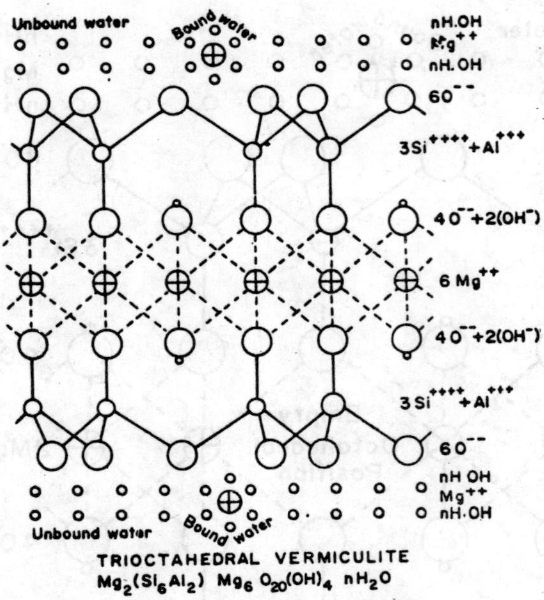

Fig. 7.6. Structure of trioctahedral vermiculite.

internal surface area are 130 to 180, 70 to 100 and 60 to 80 square metres/gm respectively.

7.3. ALLOPHANES

Allophanes are amorphous hydrated alumino-silicates which are random asymmetrical arrangement of silica tetrahedra and aluminium octahedra. Its approximate general formula is $Al_2O_3 . 2SiO_2 . H_2O$. It has high cation adsorption capacity and considerable capacity to adsorb anions.

7.4. HYDROUS OXIDES OF IRON AND ALUMINIUM

The most common examples are gibbsite ($Al_2O_3.3H_2O$) and goethite ($Fe_2O_3.H_2O$). Some of them have crystalline structures. They carry a small negative charge at a high pH value, which attracts cations. Their capacity to absorb cations is even less than that of kaolinite. Some acid soils, have positively charged particles which tend to reduce the electronegativity of the silicate clays. Most of them are not as sticky, plastic or cohesive as are the silicate clays. Soils dominated by kaolinite and hydrous oxides of iron and aluminium have much better physical structure than soils dominated by the smectite group of clay minerals.

7.5. COLLOIDAL STATE

When substances like sodium chloride, ammonium chloride, urea, glucose etc. are added to water and are thoroughly stirred, they break down to their molecular state i.e. their particle diameters are in the order of 10^{-8} to 10^{-7} cm. On the other hand when finely powdered sand, glass, charcoal etc. are put to water and are thoroughly stirred, they do not dissolve in water but settle down to the

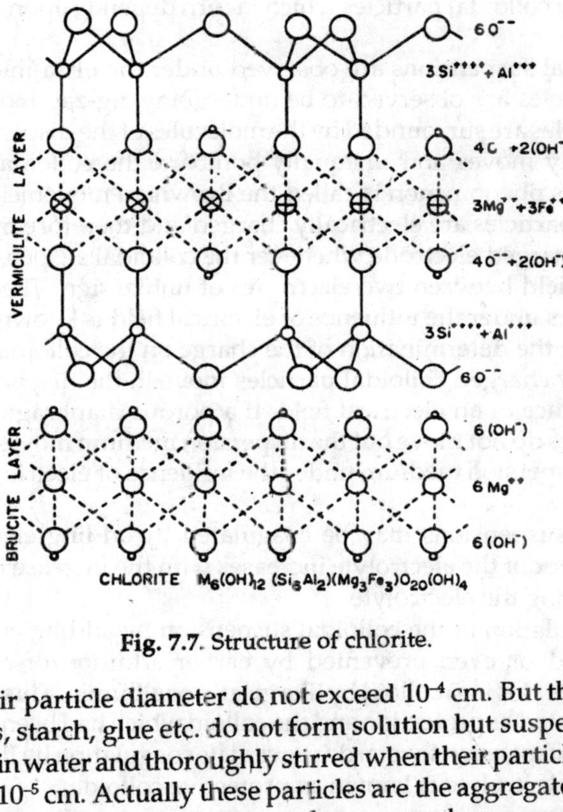

Fig. 7.7. Structure of chlorite.

bottom. Their particle diameter do not exceed 10^{-4} cm. But the substances like egg albumes, starch, glue etc. do not form solution but suspensions whenever they are put in water and thoroughly stirred when their particle diameter varies from 10^{-7} to 10^{-5} cm. Actually these particles are the aggregate (union) of 100 to 10000 molecules. These suspensions often appear homogeneous to the naked eye. But they are not true solutions. When a substance can be reduced to a state of subdivision line enough to be dispersed in another medium, it is considered to exist in the colloidal state when its particle size varies from 100 to 10000 Angstrom (Å) (1Å = 10^{-8} cm). So a colloidal system is a two-phase heterogeneous system in which one phase has been dispersed in a fine state of subdivision (1 micron to 50 millimicrons) in another medium called the dispersion medium. A few important properties of colloids are being mentioned below.

1. The colloidal suspension is constituted of extremely minute particles (dispersed phase that have been dispred in a liquid (dispersion medium).

2. As colloidal particles are actually aggregates (union) of large number of molecules, so the colloidal suspension possess very low osmotic pressure.

3. Colloidal particles easily passes through the ordinary filter paper but not through filters made up of unglazed porcelain.

4. Whenever a strong beam of light falls on a colloidal suspension, the colloidal particles scatter the light enough to illuminate the path ot the beam of light by a bluish light that can be observed from sides. This was discovered by Faraday and later studied by Tyndall. So this is known as Faraday and Tyndall effect.

5. The colour of a colloidal suspension depends upon the wavelength of the light scattered by colloidal particles which in turn depends upon the size of the particles.

6. If a colloidal suspensions are observed under the ultra-microscope, the suspended particles are observed to be undergoing zig-zag motion continuously. The particles are surrounded by the molecules of the dispersion medium which irregularly moves and unequally bombard the colloidal particles to knock them. This phenomenon is called the Brownian movement.

7. Colloidal particles are electrically charged and therefore move towards the oppositely charged electrode whenever the colloidal suspension is placed in an electrical field between two electrodes of unlike sign. The migration of colloidal particles under the influence of electrical field is known as cataphoresis. It facilitates the determination of the charge on the colloidal particles.

8. Electrically charged colloidal particles move in the dispersion medium under the influence of an electrical field. If a porous diaphragm is fixed, the colloidal particles do not move but the dispersion medium moves. This kind of movement of dispersion medium under the influence of electric field is called electro osmosis.

9. Colloidal suspensions may be coagulated by adding electrolyte. This coagulating power of the electrolyte increases with the increase in the valency of ions constituting the electrolyte.

10. The coagulation of the colloidal suspension by adding electrolyte to it may be retarded or even prevented by earlier addition of a more stable hydrophilic (water loving) colloid by like gelatin or albumin. This phenomenon has been designated as protection and the colloid which had been earlier added to the colloidal suspension to retard / prevent its coagulation by the subsequent addition of the electrolyte is known as protective colloid.

7.6. SOIL COLLOIDS

The seat of chemical activity in soils resides in the soil colloidal particles. All soils excluding pure sands contain particles of colloidal size. Soil colloids possess properties very much like those of typical colloidal substance like gelatin, starch etc. Soil colloids are of two kinds (i) Inorganic and (ii) Organic. The former is represented by the clay of size smaller than 0.001 mm or 1 micron (μ) while the latter is represented by the humus.

Clays are complex alumino silicates which are constituted of plate like units (sec 7.2) and therefore possess considerable amount of surface area, both external and internal. Considerable amount of internal surface occurs between the plate like units especially of smectites and vermiculites. The minute (size 1 micron or less) are known as micelle (microcell). So each clay micelle is surrounded by cations. The concentration of these cations is greatest near the negatively charged surface of the clay micelle and gradually decreases with the increase in the distance in the distance between clay micelle surface and the cation as illustrated in Fig. 7.8. Some species of cation especially Na^{++}, H^{+} and Mg^{++} are hydrated (water molecules have not been shown in Fig. 7.8.

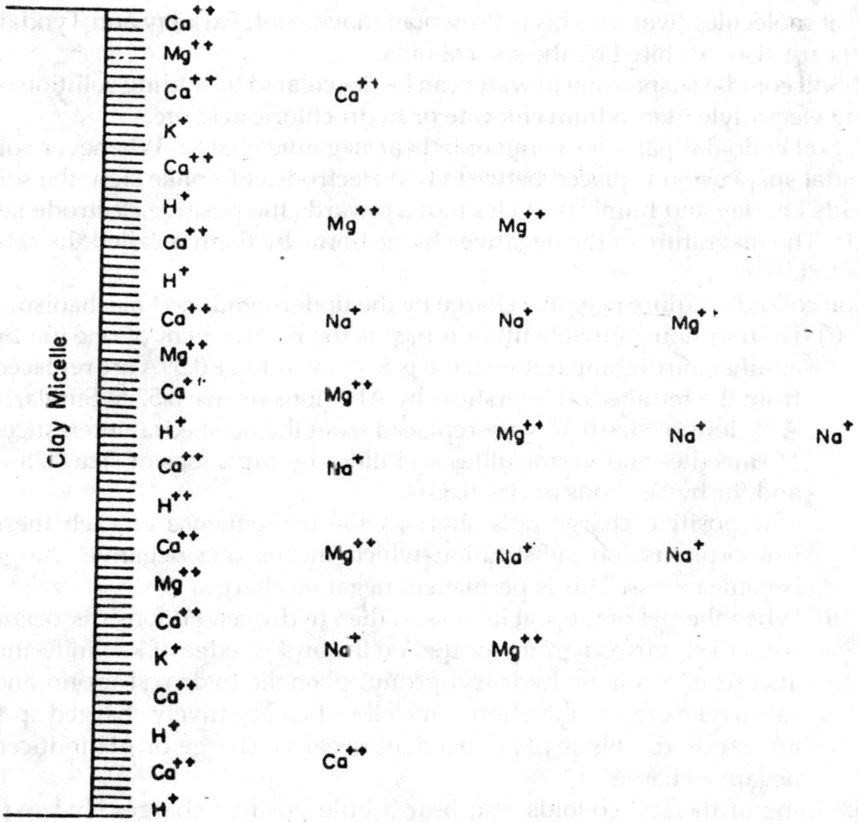

Fig. 7.8. Diagramatic representation of the diffuse double layer around the clay micelle (only one side of it has been shown) water molecules associated with the cations have not been shown.

The organic colloid is represented by the humus. The humic micelle [M] is constituted mainly of carbon and oxygen and subordinately of hydrogen and nitrogen and possesses reactive (functional) groups which are protruded from the main mass of it as mentioned below

(*i*) Phenolic Hydroxyl group [M] * OH or [M] * O^-
(*ii*) Carboxylic group [M]—COOH or [M]—COO^-
(*iii*) Alcoholic Hydroxyl group [M]—OH or [M]-O^-
(*iv*) Keto or Ester Group [M] = O or [M]—O^-
(*v*) Amide group [M]—NH_2 or [M]—NH^- or [M]—NH_2^+

*[M] refers to the main mass (micelle) of the humus.

7.7. PROPERTIES OF SOIL COLLOID

All the properties of colloids, as mentioned earlier are also shown by soil colloids whose important properties are given below.

1. Soil colloidal particles appear to oscillate in their suspension in water when the soil colloid-water suspension is observed under the microscope. The soil colloidal particles appear to colloide with one another and also with the solvent molecules (water). This is Brownian movement. Faraday and Tyndall effects are also exhibited by the soil colloids.

2. Soil collidal suspension in water can be flocculated by adding solution of strong electrolyte like sodium chloride or hydrochloric acid etc.

3. Soil colloidal particles commonly bear negative charge. Whenever soil colloidal suspension is placed between two electrodes of unlike sign, the soil colloids i.e. clay and humic particles move towards the positive clectrode i.e. anode. The magniture of the negative charge borne by them is called the zeta potential.

Soil colloids acquire negative charge by the undermentioned mechanisms.

(*i*) Isomorphous Ion Substitution means the replacement of one ion by another ion of comparable size. e.g. S_2^{+++} ions of size 0.39Å are replaced from the tetrahedral interstices by Al^{+++} ions of size 0.57Å. Similarly Al^{+++} ions of size 0.57Å are replaced from the octahedral interestices of smectites and vermiculites and illites by Mg^{++} ions of size 0.78 Å and/or by Fe^{++} ions of size 0.83Å.

One positive charge falls short as the consequence of each these isomorphous ion substitution which means one negative charge become excess. This is permanent negative charge.

(*ii*) When the pH of the soil increases, then hydrogen H^+ ions dissociate from the hydroxyl group located on the broken edge of kaolinite and also from alcoholic hydroxyl group, phenolic hydroxyl group and carboxylic group of the humic micelle when negatively charged spot are exposed. This is pH dependent negative charge or pH induced negative charge.

5. Some of the soil colloids also bear a little positive charge. Hydroxyl groups occurring at the broken edge of kaolinite and those of hydrous oxides of iron and aluminium as well as the amine groups of humus get protonated. The positive charge arises in this way, on them.

6. Soil colloidal particles selectively adsorb ions. They adsorb some ions more readily than others e.g. NH_4^+, K^+, Mg^{++} and $H_2PO_4^-$ ions edsorbed more readily while Ca^{++}, Na^+ and SO_4^{--} ions are adsorbed less readily and NO_3^-, Cl^- and HCO_3^- ions are not adsorbed or very little if at all. This adsorption of ions mostly takes place by exchange of ions.

The adsorption of ions by colloids depend both on the type of colloid and type of ions as well as on the ionic concentration. The strength of adsorption of cations by the soil colloid increases with the increases in the valency of the cations. So aluminium ions are held most tightly. Calcium and magnesium ions are adsorbed more tightly than sodium and potassium ions. But hydrogen ions, even though they are monovalent are held more tightly than the divalent cations Ca^{++}, Mg^{++} etc. If the valency of the cations is same, then their strength of adsorption by soil collidal particles increases with the increase in the atomic weight of the concerned element. The adsorption of cations also increases with the increase in their concentration in the soil solution. The adsorption of cations also depend on the nature of the soil colloids e.g. smectites adsorb divalent cations more tightly than kaolinites.

The adsorption of anions by the soil colloids depends on the type of soil colloids, the type of anions and the soil pH. Anions are adsorbed more readily by hydrous oxides of iron and aluminium than by kaolinite. They possess considerable capacity to adrosb anions. Phosphates are adsorbed by the maximum possible extent by them Sulphates are also adsorbed to a great extent. Anion adsorption increases with the decrease in soil pH.

The property of adsorption helps in the maintenance of soil fertility. Soil colloids adsorb nutrient ions and water molecules which remains available to the plant roots. Hence the soil colloids serve as a reservoir of plant nutrients. Soluble nutrients which are applied to the soil in the form of fertilisers are retained in it on account of this property of soil colloid. Plant roots assimilate these nutrients.

7. If the zeta potential of the soil colloidal particles e.g. clay particles are very high, they continue to repel one another in their suspension enough to stabilize their suspension, when they remain deflocculated or dispersed. When their zeta potential (magnitude of negative charge) goes down, they unite together to form units called floccules. This phenomenon is known as flocculation. Acids e.g. hydrochloric acid or neutral salts e.g. calcium chloride or sodium chloride whenever added to a deflocculated clay suspension flocculates it. The acid is better in this repsect than its salt. Flocculation is the prerequisite to the aggregation of colloidal particles for the development of granular and crumb structure for maintaining proper tilth for crop growth.

8. A colloid, whenever it is brought in contact with water absorb water and swells up. This phenomenon has been designated as imbibition. It actually causes the disappearance of certain amount of water. Similarly when soil colloidal particles come in contact with water, they imbibe water to and swells up. Smectitic clay imbibes water and swells up more than kaolinitic clay.

9. When water is absorbed by soil colloids, they liberate certain quantity of heat which is known as heat of wetting. Smectitic clay evolves more heat than kaolinitic clay.

10. Whenever the soil moisture content decreases, clay particles adhere (stick) together and also with sand particles. These phenomena are called cohesion in the former case and adhesion in the latter case. Cohesion is primarily due to the attraction of the clay particles for the remaining water molecules. Hydrogen bonding develop between the clay particles and the remaining water molecules. Hydrogen bonding is responsible for cohesion. Smectitic clay exhibit greater cohesion and adhesion than kaolinitic clay. These phenomena ultimately results in the development of stable soil structure.

11. Plasticity is the ability of the soil to be moulded to any desired shape and

to retain that shape when the external force is no longer being applied to the soil. This property is due to plate like nature of the clay particles. Sodium saturated clay exhibit plasticity more than the calcium saturated clay.

7.8. CATION EXCHANGE OR BASE EXCHANGE IN SOILS

Cations surrounding the soil colloidal complex i.e. clay and humic micelle are called exchangeable cations because they can be reversibly replaced equivalently by other cations. For example when ammonium chloride is applied to the soil, then ammonium ions gradually and reversibly replace other cations in equivalent amount as illustrated in the undermentioned equation:

$$\begin{matrix} & Ca^{++}Mg^{++}Ca^{++}K^{+}Ca^{++} & \\ \begin{matrix} H^{+} \\ Na^{+} \end{matrix} & \boxed{\text{MICELLE}} & Ca^{++} \\ & Ca^{++}Mg^{++}H^{+}Ca^{++} & \end{matrix} + (NH_4)_2SO_4 \rightleftharpoons \begin{matrix} & Ca^{++}Mg^{++}Ca^{++}K^{+}Ca^{++} & \\ \begin{matrix} H^{+} \\ Na^{+} \end{matrix} & \boxed{\text{MICELLE}} & \begin{matrix} NH_4^{+} \\ NH_4^{+} \end{matrix} \\ & Ca^{++}Mg^{++}H^{+}Ca^{++} & \end{matrix} + CaSO_4$$

This phenomenon is known as cation exchange which is a reversible reaction because if after some time, sufficient quantities of calcium salt e.g. calcium chloride is added to the soil, then calcium ions will replace other cations from the clay and humic micelles. So the reaction will proceed in the backward direction.

The electrokinetic theory of ion exchange facilitates the explanation of cation exchange in soils. According to this theory, cations which have formed the diffuse layer of cations around the clay/humic micelle, are in a state of oscillation when some of these cations move away from the surface of the clay/humic micelle and some of the cations of the applied salt e.g. ammonium chloride, NH_4Cl slip in between the surface of the clay/humic micelle and the oscillating cation occurring at the outermost part of the diffuse layer of cations. Then the cation i.e. NH_4^+ ion of the applied salt gets adsorbed by the clay/humic micelle and the cation which was earlier held by the clay/humic micelle, now remains in the solution as the exchanged cation.

Monovalent cation are more easily replaced from the clay/humic micelle surface than the divalent cations which in turn are replaced more easily than the trivalent cations. More highly hydrated cation e.g. Na^+ is more easily replaced than less hydrated one e.g. K^+ ion because when the cation is hydrated, its effective size is increased and so it is less tightly held on the surface of the clay/humic micelle.

7.9. CATION EXCHANGE CAPACITY (C.E.C.) OF SOILS

The total number of cation adsorption sites per unit weight of soils is called the cation exchange capacity of soils. The cation exchange capacity of the soil may also be defined as the total number of negative charges per unit weight of the soil. It is expressed in Milliequivalent* per 100 gms of soil.

$$\text{*Milliequivalent} = \frac{\text{Milligram}}{\text{Equivalent weight}}$$

$$\text{or Milligram} = \text{Milliequivalent} \times \text{Equivalent weight.}$$

7.10. FACTORS AFFECTING CATION EXCHANGE CAPACITY OF SOILS

(*i*) *Soil texture:* Since the negatively charged clay micelle attracts positively charged cations, and holds them, the cation exchange capacity of soils increases when their percentage of clay increases i.e. their texture becomes finer.

(*ii*) *Soil humus content:* Since a negatively charged humic micelle attracts cations and holds them, the cation exchange capacity of soils increases when their percentage of humus increases. The cation exchange capacity of humus varies from 200 to 400 milliequivalent per 100 gms. So C.E.C. of soils also depends on the nature of humus they contain.

(*iii*) *Nature of clay:* The cation exchange capacity of soil also depends on the nature of clay minerals which constitute them. The cation exchange capacities of different clay minerals are as follows:

Clay Mineral	*Cation Exchange Capacity*
Montmorillonite	80 to 150 meg/100 gms
Vermiculite	100 to 150 meg/100 gms
Illite	15 to 40 meg/100 gms
Kaolinite	3 to 15 meg/100 gms

Hence the cation exchange capacity of a soil dominated by montmorillonite or vermiculite is much higher than the cation exchange capacity of another soil dominated by Illite or Kaolinite.

(*iv*) *Soil reaction:* When the pH (See Sec. 7.14) of soils increases, more hydrogen ions dissociate from the hydroxyl group located on the broken edge of clay minerals especially kaolinite exposing more hydroxyl groups. Hence the cation exchange capacity of soils dominated especially by Kaolinite increased when the pH of the soil increased.

7.11. PERCENTAGE BASE SATURATION (P.B.S.)

The percentage base saturation of a soil is the proportion of its total negative charge or cation exchange capacity which is balanced by basic cations:

$$\text{Percentage base saturation} = \frac{\text{Milli equi. of basic cations}}{\text{Cation exch. capacity in mill. equiv.}} \times 100$$

For example if C.E.C. of the soil is 10 meg/100 gms and it has absorbed 8 milliequivalents of basic cations then P.B.S. = $\frac{8}{10} \times 100 = 80\%$

Basic cations are replaced from the clay and the humic micelle by hydrogen ions. Then the basic cations are washed down by high rainfall in humic regions. Hence the percentage base saturation of the clay and humic micelle decreases and their percentage hydrogen saturation increases. Hence the soil pH decrease or soil acidity increases.

7.12. CATION EXCHANGE AND AVAILABILITY OF NUTRIENTS TO CROPS

Soil water usually contains soluble salts and nutrients ions. The soil water containing soluble salts and nutrient ions is known as the soil solution. Hydrogen ions are generated by the roots hairs which are given up to the soil solution from which nutrient cations are taken up by root hairs. If the root hairs are in close contract with the clay and the humic micelle under natural condition, then the nutrient cations pass directly from the clay and humic micelle to the root hairs and hydrogen ions without either of the ions coming in contact with the soil soultion. This phenomenon is called contact exchange. The availability of nutrient cations to the root hairs depends upon the following factors:

(*i*) Percentage Saturation of the Clay and Humic Micelle by the Cations

When the percentage saturation of the clay and humic micelle by cations increases, the nutrient cation become more easily available to the root hairs.

(*ii*) Complementary Cations held by the Micelle

Since Hydrogen ion are held most strongly by the clay and humic micelle, the calcium ions, in turn are held more strongly than the sodium ions. Therefore, calcium is more easily available to root hairs when the clay and humic micelles have adsorbed calcium and hydrogen ions than when they have absorbed calcium and sodium ions, because calcium ions are held less strongly by them than hydrogen ions is the first case and calcium ions are held more strongly than sodium ions in the second case.

(*iii*) Nature of the Clay

Divalent cations are held more strongly by the high cation exchange capacity clay minerals like montmorillonite, and monovalent cations are held more strongly by the low cation exchange capacity clay minerals like kaolinite. Hence divalent cations like calcium will be more easily available to root hairs growing in kaolinitic soil and monovalent cations like potassium will be more easily available to root hairs growing in montmorillonitic soil.

"If the cation exchange capacity is high, then relatively larger doses of nutrients can be applied to the soil at one time than if the cation exchange capacity is low because the applied nutrients will not easily be leached down in the former case in contrast with the latter case when the applied nutrients will easily be leached down.

7.13. ANION EXCHANGE

Soil Colloids i.e. clay and humux and hydrous oxides of iron and aluminium are capable of taking up anions from the acidic soil solution. The hydroxyl groups which have been attached to aluminium at the broken edge of 1 : 1 clay i.e. kaolinite or at amorphous clay i.e. allophane and the amine group of humus accept protons i.e. hydrogen ions from the acidic soil solution to become positively charged as illustrated in the undermentioned equation:

$$\underset{\text{Neutral Kaolinite surface}}{(Si, Al) - OH} + \underset{\text{Proton}}{H^{-}} \xrightarrow{\text{Protonated}} \underset{\text{Positively charged kaolinite surface}}{(Si, Al) - OH_2^{+}}$$

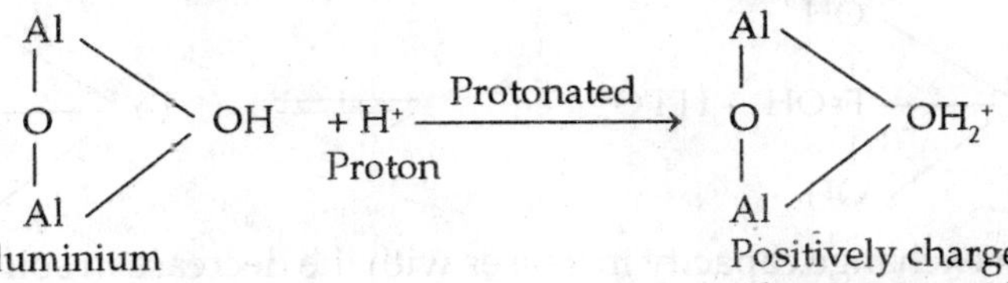

Neutral Aluminium Oxide Surface

Positively charged Aluminium oxide surface.

$$\boxed{\text{Humus}} - NH_2 + H^{+} \xrightarrow{\text{Protonated}} \boxed{\text{Humus}} - NH_3^{+}$$

Neutral Humus surface — Proton — Positively charged Humus surface

These positively charged sites attract and hold anions from acidic soil solution as illustrated in the undermentioned equations.

$$(Si, Al) - OH_2^{+} + NO_3^{-} \rightleftharpoons (Si, Al) - OH_2^{+} NO_3^{-}$$

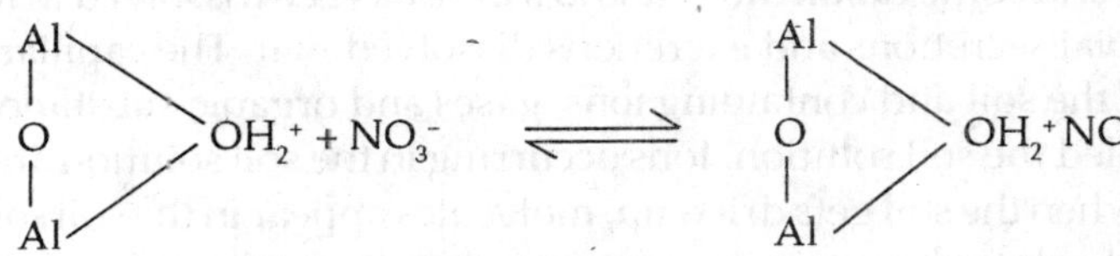

$$\boxed{\text{Humus}} - NH_3^{+} + NO_3^{-} \rightleftharpoons \boxed{\text{Humus}} - NH_3^{+} NO_3^{-}$$

The anion thus held, may subsequently be replaced by another anion as illustrated in the under anion as illustrated in the undermentioned equation.

$$(Si, Al) - OH^{-} + NO_3^{-} + Cl^{-} \rightleftharpoons (Si, Al) - OH^{+} Cl^{-} + NO_3^{-}$$

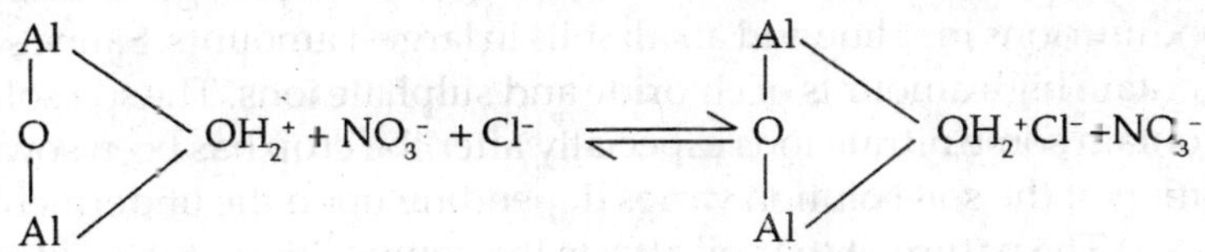

$$\boxed{\text{Humus}} - NH_3^{+} + NO_3^{-} + Cl^{-} \rightleftharpoons \boxed{\text{Humus}} - NH_3^{+}Cl^{-} - NO_3^{-}$$

Some other anions e.g. phosphate $H_2PO_4^{-}$ and sulphate SO_4^{--} ions are held and exchanged in a somewhat complex manner, because specific reactions takes place between them and the soil constituents when they may react with the protonated hydroxyl ion as illustrated in the undermentioned equation.

$$\underset{\text{Soil Solid}}{> Al - OH_2^{+}} + \underset{\text{Soil solution}}{H_2PO_4^{-}} \longrightarrow \underset{\text{Soil Solid}}{> Al - H_2PO_4} + \underset{\text{Soil solution}}{H_2O}$$

Thus the positive charge on the soil colloid is reduced on account of this reaction taking place.

Anions also replace hydroxyl ions of hydrous oxides of iron and aluminium, kaolinite and amorphous clays as illustrated in the undermentioned equation:

$$-(\text{Si, Al})-\text{OH} + \text{H}_2\text{PO}_4^- \rightleftharpoons -(\text{Si, Al})-\text{H}_2\text{PO}_4 + \text{OH}^-$$

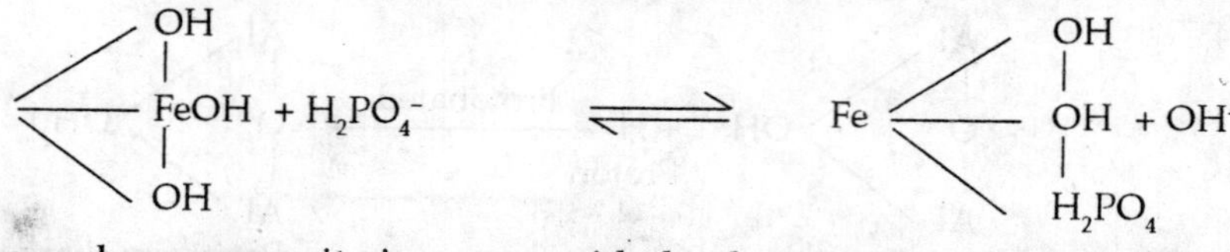

Anion exchange capacity increases with the decrease in soil pH. Soils with net positively charged colloids adsorb anions like phosphate, sulphate etc. and repel cations like calcium, magnesium etc. which are then lost by leaching.

Anion exchange capacity of the soil is usually very low i.e., a few tenth of a milliequivalent per 100 gms of a soil.

7.14. THE SOIL SOLUTION

Plant roots commonly absorb the capillary water which usually contain the soluble products of weathering of minerals and decomposition of organic matter and some cabon dioxide and a little oxygen dissolved in it and also some microbial secretions and excretions dissolved in it. The capillary water occurring in the soil and containing ions, gases and organic substances dissolved in it is called the soil solution. Ions occurring in the soil solution are called soluble ions. When the soil gets dried up, molecules appear in the soil solution. Usually the soil solution contains much more ions than molecules if the soil is moist enough for crop growth.

Usually the soil solution contain calcium, hydrogen, magnesium, sodium and potassium as important cations and bicarbonate, sulphate and chloride as the important anions. Average fertile soils contain calcium as the predominant ion in their solutions. The soil solution contain hydrogen ions in acid soils and sodium ions in saline and alkali soils in largest amounts. Saline soils commonly contain high amounts of chloride and sulphate ions. The soil solution may also contain some nitrate ions especially after the crop has been sown. The composition of the soil solution varies depending upon the undermentioned factors.

(*i*) The nature of the soil affects the composition of the soil solution and its reaction. The soil solution contains calcium as its predominant cation in case of smectic soil (e.g. black cotton soils of India) and hydrogen as the predominant cation in the case of kaolinitic soils (e.g. red soils of India). So the soil solution is alkaline in reaction in the former case and acidic in reaction in the latter case. The soil solution of saline soils possess sodium, magnesium and calcium as the predominant cation and chloride and sulphate as the predominant anions in their soil solution which is strongly alkaline in reaction.

(*ii*) The soil moisture contents affects the composition of the soil solution. The soil solution gets more and more diluted with the increase in soil moisture content as observed during the rainy season and more and more concentrated with the decrease in the soil moisture content as observed during the summer.

The dilution of the soil moisture decreases its sodium, chloride and nitrate ion contents and *vice versa*.

(*iii*) Plant roots of different species absorb ions from the soil solution in a selective manner. For example, nitrates are almost completely absorbed while other anions are less absorbed. Plant roots respire to liberate carbon dioxide which reacts to produce carbonic acid that gradually dissolve insoluble primary minerals to increase the concentration of the soil solution. Roots absorb nutrient ions from the soil solution which is therefore less concentrated in cropped soil than in uncropped soil.

(*iv*) Soil micro-organisms are partially responsible for increasing the concentration of certain ions especially anions in the soil solution. These anions are usually nitrate, sulphate and bicarbonate in the soil solution. Micro-organisms assimilate some of the salts from the soil solution which are released again in the soil solution after their death and decomposition of their dead bodies.

(*v*) Inorganic and organic colloid also adsorb nutrient ions from the soil solution. They do not retain chlorides. They also do not retain sulphates and nitrates for long periods of time. Exchangeable ions held on the surface of the soil colloidal complex, remain in equilibirum with the respective ions in the soil solution. When the ions are absorbed from the soil solution, the equilibrium is disturbed. So the respective ions are transferred from the surface of the soil colloidal complex to the soil solution in order to restore the equilibrium between the exchangeable cations and soluble cations.

(*vi*) When the soil solution contains ions far in excess of the saturation capacity of the soil colloidal complex, then the excess amount is leached down in high rainfall area if no crop is growing.

(*vii*) Application of manures and fertilisers to the soil increases the concentration of nutrient ions in the soil solution.

(*viii*) The productive soil possess more concentrated soil solution than the unproductive soil does. In the former case, the concentration of soil solution varies from 0.05 to 0.2 per cent, its osmotic pressure varies from 0.2 to 2.0 atmosphere and its Ca : Na and Ca : Ma cations exceeds one.

7.15. PHOSPHATE RETENTION AND FIXATION

Kaolinite has a strong affinity for phosphate anions. Phosphate anions can be attracted to soil constituent with the force strong enough to make them insoluble and difficulty available to plant roots. Tisdale and Nelson (1975) have differentiated phosphate retention from phosphate fixation by the fact in the first case, the phosphate anion held by the soil clay can be removed by treatment with the soil clay with dilute acids e.g. 0.005 N sulphuric acid in contrast with the latter case when the phosphate ion held by the soil clay cannot be removed by the treatment of the soil clay with the same acid.

Phosphorus retention may be caused by anion exchange mechanism when the phosphate anion replaces the hydroxyl anion held by the aluminium ion of the clay mineral especially of 1 : 1 type as illustrated in the undermentioned equation:

$$-(S > Al) - OH + H_2PO_4^- \rightleftharpoons -(S >, Al) - H_2PO_4 + OH^-$$

Phosphate fixation is caused by the reaction of the orthophosphate $H_2PO_4^-$ anions with (*i*) soluble aluminium and ferric ions (*ii*) exchangeable aluminium ion (*iii*) protonated aluminium occurring at the broken edge of kaolinite (*iv*) hydrous oxides of iron and aluminium (*v*) exchangeable and soluble calcium ions and (*vi*) calcium carbonate.

The 1 : 1 clay i.e. kaolinite decomposes with the decrease in pH in acid soils to liberate aluminium ions in the soil solution. The aluminium ions thus liberated in the soil solution react with water.

$$Al^{+++} + 3\,H\,O.H \longrightarrow Al\,(OH)_3 + 3\,H^+$$

The increase in soil acidity caused by the above-mentioned liberation of hydrogen ions in the soil solution leads to further decomposition of clay that further liberate more aluminium ions in the soil solution.

The aluminium hydroxide thus formed, reacts with the orthophosphate anions to form insoluble aluminium hydroxy phosphate

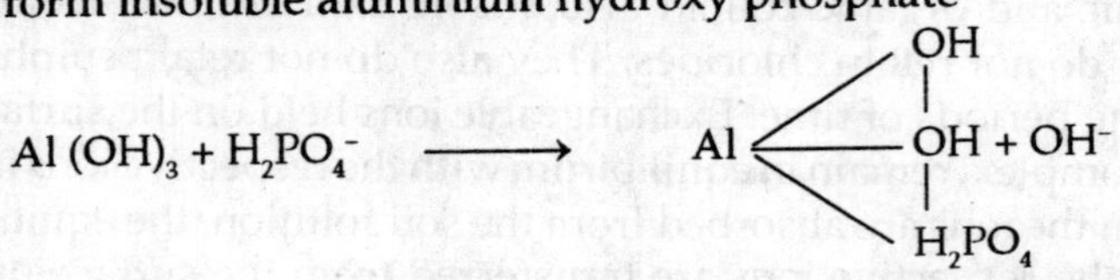

Aluminium Hydroxy Phosphate

Acid soils also contain high amounts of hydrous oxides of iron and aluminium which react with the soluble orthophosphate anions and precipitate them as insoluble iron and aluminium hydroxy phosphates.

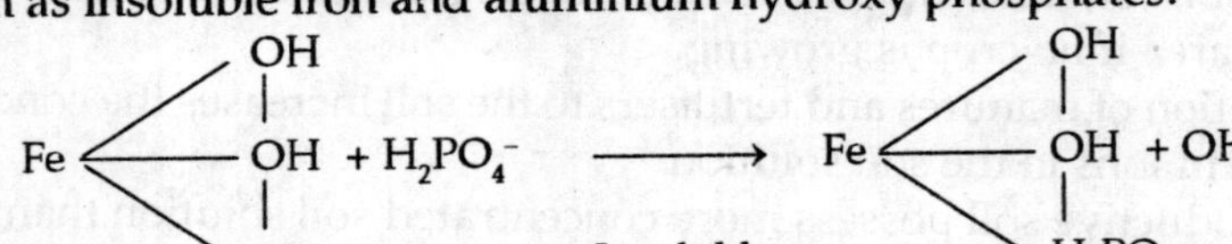

Insoluble Ferric Hydroxy Phosphate

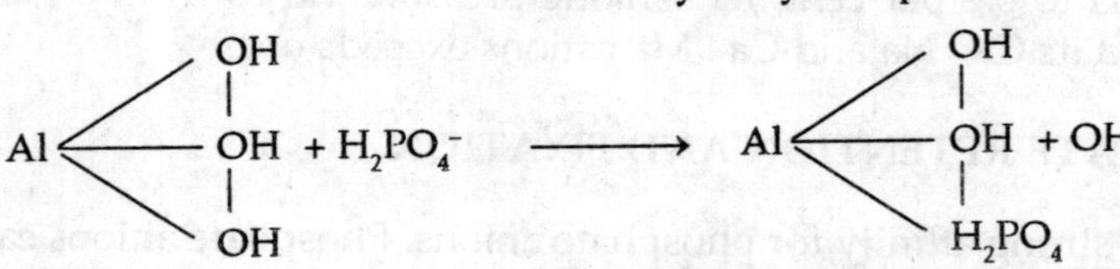

Insoluble Aluminium Hydroxy Phosphate

The phosphate anion also replaces hydroxyl ion held by the aluminium ion of kaolinite

$$> Al - OH + H_2PO_4^- \longrightarrow > Al - H_2PO_4 + OH^-$$

Hydroxyl ions attached with aluminium ions located at the broken edge of kaolinite gets protonated.

$$\underset{\text{Neutral Kaolinite}}{> Al - OH} + H^+ \xrightarrow[\text{Proton}]{\text{Protonated}} \underset{\text{Positively charged Kaolinite surface.}}{> Al - OH_2^+}$$

The orthophosphate anions react with these protonated hydroxyl anion :

$$> Al - OH_2^+ + H_2PO_4^- \longrightarrow > Al - H_2PO_4 + H_2O$$

The exchangeable calcium ions occurring in soils of alkaline reaction acts as bridge to link the orthophosphate anion with the clay as shown below.

$$\boxed{\text{Clay}} - Ca - H_2PO_4$$

The soluble calcium ion precipitate the orthophosphate anion as insoluble tricalcium phosphate.

$$3Ca^{++} + 2H_2PO_4^- + 4\,(OH)^- \longrightarrow Ca_3\,(PO_4)_2 + 4\,H_2O$$

The undermentioned reaction takes place in presence of calcium carbonate.

$$CaCO_3 + 2\,Ca^{++} + 2H_2PO_4 + 2(OH)^- \longrightarrow Ca_3(PO_4)_2 + CO_2 + 4\,H_2O$$

7.16. FIXATION OF POTASSIUM IONS

Hexagonal cavities occur between the two silica tetrahedral sheets of illite, vermiculite, etc. that lie one above the other under the natural condition. The size of the potassium ion is approximately equal to these hexagonal cavity formed between 12 oxygen ions of two silica tetrahedral sheets of illite, vermiculite that lie one above the other under natural condition. When these clay minerals are moist i.e. the soil is moist, they expand. Potassium ions enter in these cavity. When the soil dries up and the clay minerals contract, the potassium ion is entrapped within the cavity. The bonding energy between the potassium ion and the entrapping silica sheets is great enough to prevent the immediate re-expansion of all the clay mineral layers.

7.17. FIXATION OF AMMONIUM IONS

Ammonium ions can replace cations like calcium, magnesium, sodium etc. from the interlayers of vermiculite and smectite when they are in the expanded state. The ammonium ions get entrapped in the interlayer of vermiculite, smectite etc. when they contract e.g. when the soil dries up. The total quantity of ammonium ions thus fixed depend on the quantity of expanding type of clay occurring in the soil. The fixed ammonium can be slowly released by calcium, magnesium and sodium ions when these ions expand the lattice. Potassium ions contract the lattice and therefore cannot replace the fixed ammonium ions.

7.18. SOIL REACTION—pH

When soils contain more hydrogen ions than hydroxyl ions, they are acidic in reaction. If they contain more hydroxyl ions than hydrogen ions, they are alkaline in reaction. If they contain equal number of hydrogen and hydroxyl ions, they are neutral in reaction.

The word "pH" has originated from the French word "pouvoir hydrogen" which means powers of hydrogen. pH is the logarithm* to the base ten of the reciprocal of hydrogen ion concentration.

$$\text{Mathematically pH} = \log_{10} \frac{1}{[H^+]} = \log_{10} [H^+]^{-1} - \log_{10} [H^+]$$

Hence pH of a soil is also defined as the negative logarithm to the base ten of its hydrogen ion cencentration.

The product of hydrogen ion concentration and hydroxyl ion concentration of a solution is called the ionic product of water which has a constant value of 10^{-14} gms per litre for all aqueous solution. Therefore when the hydrogen ion concentration of an aqueous solution increases, its hydroxyl ion concentration decreases in order to maintain the value of the ionic product of water constant i.e. 10^{-14} gms per litre e.g. if hydrogen ion concentration is 10^{-6} gms per litre, then hydroxyl ion constration will be 10^{-8} gms of water when the ionic product of water will be $10^{-6} \times 10^{-8} = 10^{-14}$ gms/litre.

7.19. FACTORS AFFECTING SOIL REACTION, pH

Amount of hydrogen and aluminium ions present in the soil.

When the soil contains more hydrogen ions than hydroxyl ions then it becomes acidic in reaction. When aluminium ions are present in the soil, they react with water to liberate hydrogen ions, which increase the soil acidity

$$Al^{+++} + 3H.OH \longrightarrow Al(OH)_3 + 3H^+$$

Percentage Base Saturation

A low percentage base saturation of soils means soil acidity. Basic elements have been leached down from the soils of a humid region, the percentage base saturation decreases much below 80 and they become acidic in reaction. If it is above 80 then it is neutral in reaction. If it is 90, then it is alkaline reaction.

Nature of the Micelle

More hydrogen ions are absorbed and released by the high cation exchange capacity micelle than by the low cation exchange capacity micelle. For example, if the cation exchange capacities of montmorillonite and kaolinite are 80 meq/100 gms and 3 meq/100 gms respectively, then at 50 per cent base saturation montmorillonite and kaolinite holds 40 meq/100 and 1.5 meq/100 gm. of hydrogen respectively. The montmorillonitic soil solution will contain more hydrogen ions than the kaolinitic soil solution at the same percentage base saturation, since hydrogen ions absorbed by the micelle are in equilibrium with the hydrogen ions in the soil solution. Hence the pH of the montmorillonitic soil will be less than that of the kaolinitic soil.

*Logarithm of a number with respect to any base other than one is the index to which the base must be raised in order to get the concerned number. For example we know that $2^3 = 8$. Here 8 is the concerned number, 2 is called the base and 3 is called the index. Hence logarithm of 8 with respect to the base 2 is 3. This we mathematically write as $\log_2 8 = 3$.
Similarly $10^2 = 100$. So $\log_{10}100 = 2$.
and $10^3 = 1000$. So $\log_{10}100 = 33$.
and so on.

Kind of Exchangeable Ions

Sodium ions are held less strongly by the micelle calcium ions and are therefore more easily hydrolysed than calcium ions as shown below:

$$Na^+ \underset{Na^+\,Na^+\,Na^+\,Na^+\,Na^+}{\overset{Na^+\,Na^+\,Na^+\,Na^+\,Na^+}{\boxed{\text{MICELLE}}}} Na^+ + 6H.OH \rightleftharpoons H^+ \underset{H^+Na^+Na^+Na^+H^+}{\overset{H^+Na^+Na^+Na^-H^+}{\boxed{\text{MICELLE}}}} H^+ + 6Na^+ + 6(OH)^- \quad \ldots (1)$$

$$Ca^+ \underset{Ca^{++}\,Ca^{++}}{\overset{Ca^{++}\,Ca^{++}}{\boxed{\text{MICELLE}}}} Ca^{++} + 2H.OH \rightleftharpoons Ca^{++} \underset{Ca^{++}\,Ca^{+}}{\overset{Ca^{++}\,Ca^{++}}{\boxed{\text{MICELLE}}}} H^+ + Ca^{++} + 2(OH)^- \quad \ldots (2)$$

Hence more hydroxyl ions are released from the hydrolysis of sodium icns than from the hydrolysis of calcium ions as shown in the Equations (1) and (2) respectively. Hence the pH of sodium saturated soils is higher than that of calcium saturated soils.

Formation of Carbon dioxide under Natural Condition

Soil micro organisms and plant roots respire to produce carbon dioxide which is also liberated from the decomposition of organic matter in the soil. Carbon dioxide reacts with water to form carbonic acid, which, in turn react with the clay and humic micelle (suppose they contain Ca^{++}, Mg^{++}, K^+, Na^+ and H^+ ions in the ratios of (54 : 8 : 3 : 5 : 30) and insoluble primary minerals present in the soil as shown in the following equations:

$$CO_2 + H_2O \longrightarrow H_2CO_3$$

$$K_3^+ \underset{Na^+_5\,H^+_{30}}{\overset{Ca_{54}\,Mg_8}{\boxed{\text{MICELLE}}}} + 4H_2CO_3 \rightleftharpoons K_3^+ \underset{Na^+_5\,H^+_{34}}{\overset{Ca_{53}\,Mg_7}{\boxed{\text{MICELLE}}}} + \underset{\text{soluble}}{Ca(HCO_3)_2} + \underset{\text{soluble}}{Mg(HCO_3)_2}$$

$$\underset{\text{Insoluble}}{CaCO_3} + H_2CO_3 \longrightarrow \underset{\text{Soluble}}{Ca(HCO_3)_2}$$

The soluble bicarbonates thus formed are washed down by rain in the high rainfall areas. So the above reversible reactions proceeds in the forward direction. Hence more basic elements are gradually replaced from the clay and humic micelle to form soluble bicarbonates. Insoluble primary minerals are gradually converted to soluble bicarbonates. These soluble bicarbonates are washed down by high rainfall and the pH of soils gradually decreases. Ultimately soils become acidic in reaction in humid region.

Drying of soil slightly increases soil acidity.

On accout of the decrease in the thickness of the diffuse layer of cations around the clay micelle caused by drying as explained below.

The clay micelle possess a diffuse layer of cation around its negatively charged surface. The concentration of cations including hydrogen ions is greatest just near its negatively charged surface and gradually decreases towards the outer side of the diffuse cationic layer. So more hydrogen ion occur in the inner portion of the diffuse layer of cation than in its outer portion. The soil pH which is commonly measured, indicates the quantity of hydrogen ions occurring at the outer portion of the diffuse layer of cation (including hydrogen ions). The increase in the soil moisture content increases the thickness of the diffuse layer when less hydrogen ion would be present in the outer portion of the diffuse layer than in its inner portion. So little higher pH would be indicated by the pH measuring device. When the soil dries up, the thickness of the diffuse layer decreases. The concentration of cations including hydrogen would increase at its outer portion. So the pH measuring device would indicate a little less pH than when the soil was moist.

Ammonium ions are oxidized to form nitric acid in the soil. Nitric acid reacts with calcium carbonate and clay and humic micelle as shown below:

$$\underset{\text{Insoluble}}{CaCO_3 + 2HNO_3} \rightleftharpoons \underset{\text{Soluble}}{Ca(NO_3)_2} + H_2CO_3$$

$$\underset{\text{Insoluble}}{CaCO_3 + H_2CO_3} \rightleftharpoons \underset{\text{Soluble}}{Ca(HCO_3)_2}$$

$$\text{Clay Micelle}\left]\begin{array}{l}Ca^{++}\\Ca^{++}\\H^{+}\\Ca^{++}\\Mg^{++}\\H^{+}\\Ca^{++}\\Mg^{++}\end{array}\right. + 2HNO_3 \rightleftharpoons \text{Clay Micelle}\left]\begin{array}{l}H^{+}\\H^{+}\\Ca^{++}\\H^{+}\\Ca^{++}\\Mg^{++}\\H^{+}\\Ca^{++}\\Mg^{++}\end{array}\right. + \underset{\text{Soluble}}{Ca(NO_3)_2}$$

The soluble calcium nitrates and calcium bicarbonates thus formed, are leached down by high rainfall resulting in the increase in soil acidity. Hence continuous application of ammoniacal fertiliser increases soil acidity. Continuous application of urea also increases soil acidity because urea undergoes transformation in the soil to form nitric acid in the soil. As shown below:

$$\begin{array}{l}NH_2\\ \;|\\ CO + 3O_2 \longrightarrow 2\,HNO_3 + H_2O\\ \;|\\ NH_2\end{array}$$

Soil aeration status also affect the soil reaction. In poorly aerated soil, electrons are available from the microbial oxidation of organic compounds. These electrons are utilized in the process of reduction of ferric iron to ferrous iron. When H^+ ions are consumed as shown below:

$$Fe^{+++} + 2HOH \rightleftharpoons Fe(OH)_2^{+} + 2H^{+}$$

$$Fe(OH)_2^{+} + 2H^{+} + e^{-} \rightleftharpoons Fe^{++} + 2H_2O$$

Summing up $\quad Fe^{+++} + e^{-} \rightleftharpoons Fe^{++}$

Hydrogen H^+ ions are also consumed when sulphates is reduced to sulphide in poorly aerated soil and also when carbon dioxide is reduced to glucose in poorly aerated soil as shown below:

$$SO_4^{--} + 10H^+ + 8e^- \rightleftharpoons H_2S + 4H_2O$$

$$CO_2 + 4H^+ + 4e^- \rightleftharpoons \underset{\text{glucose}}{CH_2O} + H_2O$$

So poorly aerated soils are of lower pH than well aerated soil:

7.20. ACTIVE AND RESERVE ACIDITY OF SOILS

The active acidity of soils is the amount of hydrogen and aluminium ions present in the soil solution. Reserve acidity is the amount of hydrogen and aluminium ions absorbed by the clay and the humic micelle. Both are in equilibrium with each other as shown below:

H^+ and Al^{+++} adsorbed on the micelle $\rightleftharpoons$ H^+ and Al^{+++} Present in the soil solution.

Reserve acidity of soils — Active acidity of soil

When H^+ and Al^{+++} ions present in the soil solution are neutralised, H^+ and Al^{+++} ions move into the soil solution. Therefore the H^+ and Al^{+++} ions adsorbed on the micelle is also called the exchange acidity of the soil. So reserve or exchange acidity of soils must be neutralised before their pH can be increased. It is a thousand or million times greater than the active acidity.

7.21. BUFFERING OF SOILS

Buffering of soils is their resistance against any change in the concentration of hydrogen or other cation in the soil solution. For example, the clay micelle is almost completely saturated with hydrogen ions in strongly acid soils which also contain some in the soil solution. The hydrogen ions absorbed by the micelle (Reserve of Exchange acidity) are in equilibrium with the hydrogen ions present in the soil solution as shown below:

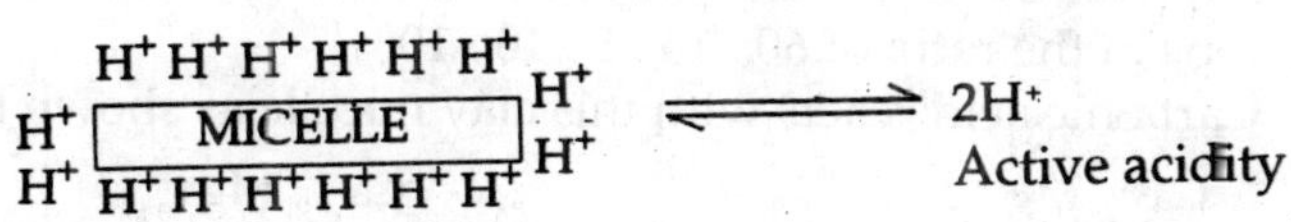

Reserve or Exchange acidity

When limestone is added to the soil it first neutralises the active acidity as shown below:

$$CaCO_3 \rightleftharpoons Ca^{++} + CO_3^{--}$$

$$2H^+ + CO_3 \rightleftharpoons H_2CO_3 \longrightarrow H_2O + CO_2\uparrow$$

The Ca^{++} ion then replace the H^+ ion from the clay micelle to restore the original level of active acidity as shown below:

$$\begin{array}{c} H^+H^+H^+H^+H^+H^+ \\ H^+ \boxed{\text{MICELLE}} H^+ \\ H^+ \qquad\qquad\quad H^+ \\ H^+H^+H^+H^+H^+H^+ \end{array} + Ca^{++} \rightleftharpoons \begin{array}{c} H^+H^+H^+H^+H^+Ca^{++} \\ H^+ \boxed{\text{MICELLE}} H^+ \\ H^+ \qquad\qquad\quad H^+ \\ H^+H^+H^+H^+H^+ \end{array} + \underset{\text{Active acidity}}{H^+}$$

Exchange or Reserve acidity — Exchange or Reserve acidity

The buffering capacity or inorganic or mineral soils depends mainly on their texture. Since clay can adsorb ions and not sands. The buffering capacity of soils increases when their percentage of clay increases *i.e.* their texture becomes finer. The buffering capacity of soils also depends upon the nature of clay and humus of soils. The buffering capacity of black cotton soils, dominated by smectite group of clay minerals, is much higher than those of red soils which are dominated by the kaolinitic group of clay minerals and hydrous oxides of iron and aluminium.

When the buffering capacity of the soil is very high, larger quantities of nutrients in the form of fertilizers can be added to soil at longer intervals because they will be retained by the clay micelle and not washed down. Similarly more lime is to be added to clayey soils of a high buffering capacity, in order to control the reserve of exchange acidity of soils.

Soil pH is very stable in the soils of high buffering capacity.

7.22. FORMATION OF ACID SOILS

Soil microorganisms and plant roots respire to form carbon dioxide, which is also produced when organic matter decomposes. This carbon dioxide reacts with water to form carbonic acid that reacts with the insoluble primary minerals present in the soil as well as the clay and humic micelle.

$$CO_2 + H_2O \rightleftharpoons H_2CO_3$$

$$\underset{\text{Insoluble}}{CaCO_3} + H_2CO_3 \rightleftharpoons \underset{\text{Soluble}}{Ca(HCO_3)_2}$$

Suppose the clay and humic micelle have absorbed Ca^{++}, Mg^{++}, K^+, Na^+ and H^+ ions in the ratio of 60 : 15 : 5 : 10 : 10.

Carbonic acid reacts with this clay micelle as shown below:

$$\begin{array}{c} Ca^{++}_{60}\,Mg^{++}_{15} \\ K^+_5 \boxed{\text{MICLLE}} \\ Na^+_{10}H^+_{10} \end{array} + 6H_2CO_3 \rightleftharpoons \begin{array}{c} Ca^{++}_{58}\,Mg^{++}_{14} \\ K_5 \boxed{\text{MICELLE}} \\ Na^+_{10}H^+_{16} \end{array} + \underset{\text{Soluble}}{2Ca(HCO_3)_2} + \underset{\text{Soluble}}{Mg(HCO_3)_2}$$

These soluble bicarbonates of calcium, magnesium, potassium and sodium are washed down by high rainfall in humid regions. So the above reversible reactions proceeds in the forward direction. This means that basic cations are gradually released from the primary minerals and clay and humic micelle in the form of soluble bicarbonates, which are gradually washed down by high rainfall in the humid regions where the soils ultimately become strongly acidic in reaction.

7.23. CHARACTERISTICS OF ACID SOILS HARMFUL FOR PLANT GROWTH

Acids soils are formed when the basic elements like calcium, magnesium, potassium etc. have been wahsed down. So, strongly acidic soils are always deficient soils are in calcium, magnesium and potassium. Highly leached acid soils are also deficient in sulphur and boron.

Strongly acidic soils contain excessive amounts of aluminium, iron and manganese in the soluble form, which is very harmful for crop growth. Acid soils may also contain toxic amounts of copper and zinc. Phosphate, if present as inorganic orthophosphate ions, H_2PO_4, precipitated as insoluble hydroxy phosphates as shown below:

$$Al^{+++} + H_2PO_4^- + 2H.OH \rightleftharpoons Al\begin{cases}OH\\OH\\H_2PO_4\end{cases} + 2H^+$$

Insoluble Aluminium Hydroxy Phospnate

$$Fe^{++} + H_2PO_4^- + 2H.OH \rightleftharpoons Fe\begin{cases}OH\\OH\\H_2PO_4\end{cases} + 2H^+$$

Insoluble Ferric Hydroxy Phosphate

Strongly acidic soils are therefore deficient in available phosphorus and also available molybdenum because it reacts with the silicates and iron and aluminium compounds. Bacteria and actinomycetes cannot function properly in strongly acidic soils, where fungi dominates the soil micro organisms.

TABLE 7.1. Optimum pH requirement of some important plants

Cereals		*Vegetables*		*Fruits*	
Paddy	5.0-6.5	Potato	5.0-5.5	Watermelon	5.0-5.0
Sorghum	5.5-7.0	Sweet Patato	5.0-6.0	Pineapple	5.5-6.5
Maize	5.5-7.0	Carrot	5.5-6.5	Mango	5.5-6.5
Wheat	5.5-7.5	Bitter Gourd	5.5-6.5	Apple	5.5-6.5
Oats	6.0-7.0	Brinjal	5.5-6.5	Pummelo	5.5-6.5
Barley	6.5-8.0	Pea	5.5-6.5	Guava	5.5-7.0
		Raddish	5.5-6.5		
		Chillies	5.5-6.5		
		Turnip	5.5-6.5		
		Tomato	5.5-6.0		
Oil Seeds					
Groundnut	5.0-6.0	Cowpea	5.5-7.0	Lemon	6.0-7.0
Mustard	6.0-7.0	Onion	6.0-7.5	Fig	6.0-7.0
Sugar Crops		Lettuce	6.0-7.0	Banana	6.0-7.0
		Cauliflower	6.0-7.0	Papaya	6.0-7.5
Sugarcane	6.0-7.0	Cabbage	6.0-7.0	Grape fruit	6.0-7.5
Sugarbeet	6.0-7.0	Guard	6.0-7.0	Lime	6.0-7.5
		Beans	6.0-7.5	Pomogranate	6.0-7.5
		Ladies finger	6.0-7.5	Cucumber	6.0-8.0
				Grapes	6.0-8.0

(Contd.)

Cereals		*Vegetables*		*Fruits*	
Fibre Crop					
Cotton	5.0-6.0	Beans	6.0-7.5	Pear	6.0-8.0
		Ginger	6.0-7.5	Walnut	6.0-8.0
		Spinach	6.5-7.0	Date	6.5-8.0
		Mint	6.0-8.0		
Fodder Crops		*Flowers*		*Plantation Crops*	
Lucerne	6.0-7.0	Orchids	4.0-5.0	Tea	5.0-5.5
Clover	6.0-7.0	Crotons	5.5-6.5	Coffee	5.0-6.0
		Bougainvillea	5.5-7.0	Pepper	6.0-6.5
		Rose	6.6-7.0	Coconut	
		Sunflower	6.0-7.0		
		Chysanthemum	6.0-7.5		

7.24. CONTROL OF SOIL ACIDITY

Soil acidity is controlled by adding lime to the soil. The different forms of lime are as follows:

Calcium oxide, CaO, called quick lime or burnt lime because limestone $CaCO_3$ is burnt to produce it.

$$\underset{\text{Calcite}}{CaCO_3} \xrightarrow{\text{heat}} CaO + 2CO_2\uparrow$$

$$\underset{\text{Dolomite}}{CaMg(CO_3)_2} \xrightarrow{\text{heat}} CaO + MgO + 2CO_2\uparrow$$

Calcium hydroxide $Ca(OH)_2$ is called the hydroxide of lime. It is formed when quick lime reacts with water.

$$CaO + H.OH \longrightarrow Ca(OH)_2$$

Calcium carbonate or carbonate of lime occurs under natural conditions. When applied to the acid soil, the soil acidity is neutralised as shown in the following equation.

$$CaCO_3 \rightleftharpoons Ca^{++} + CO_3^{--} \quad \ldots (i)$$

$$2H^+ + CO_3^{--} \longrightarrow H_2CO_3 \longrightarrow H_2O + CO_2\uparrow$$

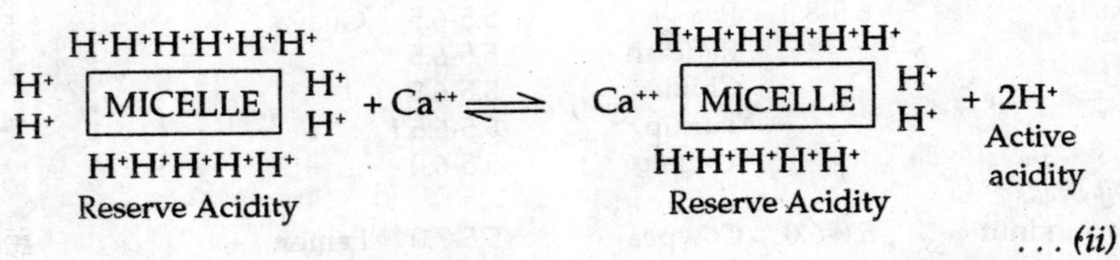

. . . (*ii*)

Lime first neutralizes the acitve acidity as shown in Equation (*i*). The Ca^{++} ion replaces the H^+ ion from the micelle to restore the original level of active acidity as shown in Equation (*ii*). So reversible reactions proceed in the forward direction. In this way soil acidity is ultimately controlled.

7.25. LIME REQUIREMENT OF ACID SOILS

Take 100 gms of finely powered soil in a series of beakers, to which powdered limestone is added at an increasing rate of 440, 880, 1320...kg/ha, which is equivalent to 20, 40, 60, 80...mg/100 gms. The limestone powder is thoroughly mixed with soils after they are, moistened.

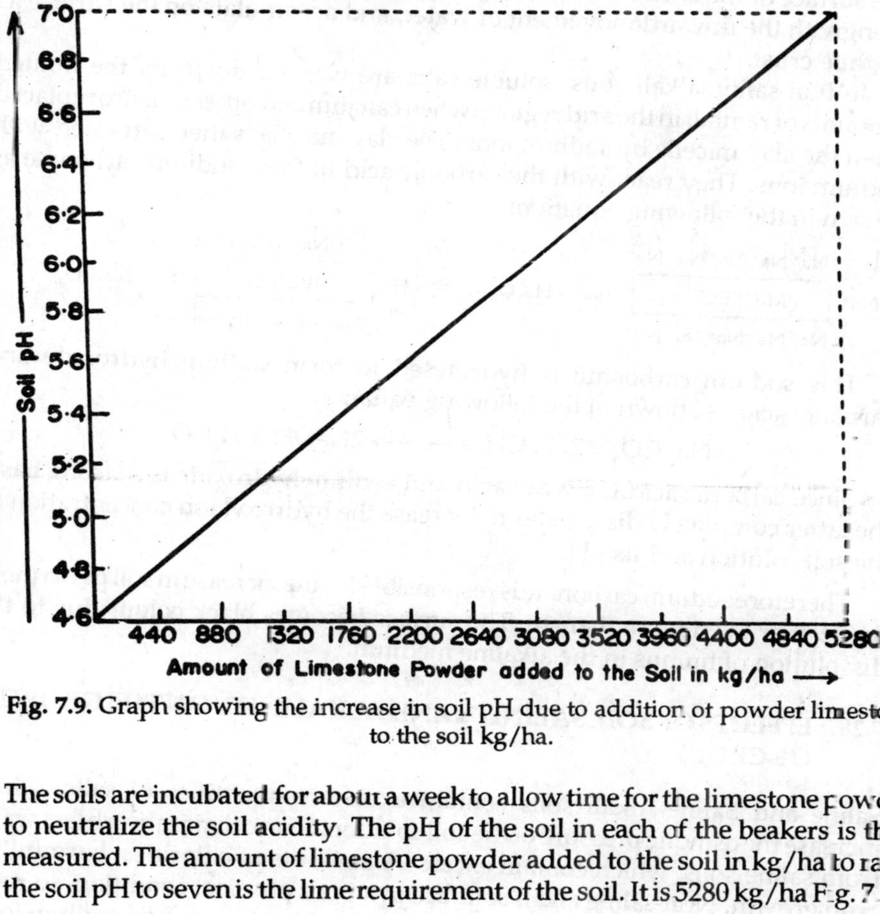

Fig. 7.9. Graph showing the increase in soil pH due to addition of powder limestone to the soil kg/ha.

The soils are incubated for about a week to allow time for the limestone powder to neutralize the soil acidity. The pH of the soil in each of the beakers is then measured. The amount of limestone powder added to the soil in kg/ha to raise the soil pH to seven is the lime requirement of the soil. It is 5280 kg/ha Fig. 7.10.

7.26. CHARACTERISTICS OF SALINE AND ALKALI SOILS

Saline soils contain excessive amounts of neutral salts like chloride and sulphate of sodium, calcium and magnesium, and less than 15 per cent exchangeable sodium. Their pH is less than 8.5, due to the presence of excessive amounts of neutral salts. Saline soils are in excellent physical condition and are permeable to water. Saline-alkali soils also contain enough neutral salts like chloride and sulphate of sodium and calcium and magnesium to maintain their pH value below 8.5, even when they contain more than 15 per cent exchangeable sodium.

Non-saline alkali soils contain more than 15 per cent exchangeable sodium. Their pH is more than 8.5 because they do not contain any neutral soluble salts.

They are sticky and plastic when wet and very hard when dry. Water does not percolate through them.

7.27. FORMATION OF SALINE AND ALKALI SOILS

Soluble salts originate in the soil from the decomposition of primary minerals. Whenever drainage is restricted and excess amounts of water evaporate from the surface of the land in the arid regions, neutral soluble salts move upward, alongwith the upward movement of water, and accumulate on the surface as a white crust.

In non-saline alkali soils, soluble salts are washed down by the limited amounts of rainfall in the arid regions when calcium and other ions are replaced from the clay micelle by sodium ions. The clay micelle is then saturated with sodium ions. They react with the carbonic acid to form sodium carbonate as shown in the following equation:

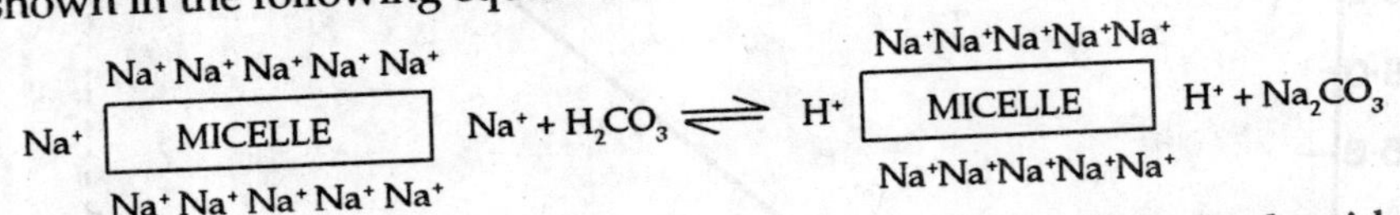

This sodium carbonate is hydrolysed to form sodium hydroxide and carbonic acid as shown in the following equation

$$Na_2\,CO_3 + 2H\,.\,CH \longrightarrow 2NaOH + H_2CO_3$$

Since carbonic acid is a weak acid and sodium hydroxide is a strong base, the latter completely dissociates to increase the hydroxyl ion concentration of the soil solution and its pH.

Therefore sodium carbonate is responsible for the increase in soil pH to more than 9.0 or even more than 10. The surface becomes black colour due to the dissolution of humus in the alkaline medium.

7.28. EFFECTS OF SOIL SALINITY AND ALKALINITY ON THE GROWTH OF CROPS

Saline and Saline-Alkali soils contain excessive quantities of salts, which increase the osmotic pressure of the soil solution. So root hairs absorb less water from saline soils, which contain excessive amounts of chlorides, harmful for crop growth. Non-saline alkali soils are in a poor physical condition, so roots suffer from poor aeration. They contain borate, bicarbonate and sodium ions Even small amounts of borates are toxic to crops. Excessive amounts of bicarbonate and sodium ions are also harmful for crops; they make phosphorus and all micronutrients except molybdenum unavailable to crops.

7.29. CONTROL OF SOIL SALINITY AND ALKALINITY

Underground drainage system should be established. As saline soils are permeable to water heavy irrigation can be used to wash down the soluble salts

By adding gypsum to alkali soils, the sodium ions from the clay micelles are gradually replaced by calcium ions, as shown in the following equation:

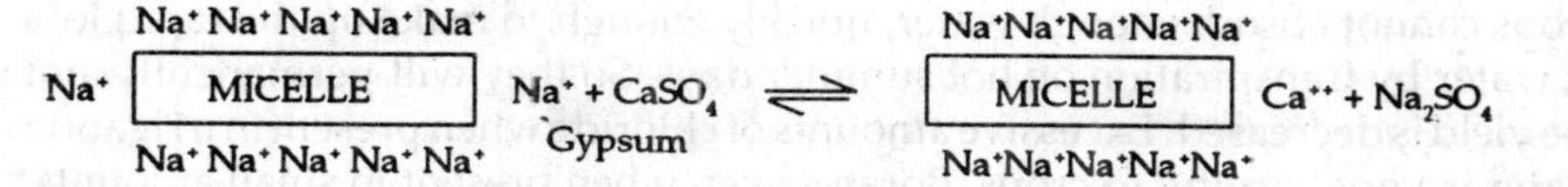

Soluble sodium sulphate is washed down by heavily irrigating the alkali soils, when the above reversible equation proceeds in the forward direction, which means that the sodium ions are gradually replaced from the clay micelle by the calcium ions, which ultimately saturate the clay micelle when soil alkalinity is controlled.

If the alkali soil contains free calcium carbonate, then powdered sulphur of ferrous sulphate may be added to the alkali soil, first producing sulphuric acid and then, calcium sulphate, as shown in the following equations:

$$2S + 3O_2 \longrightarrow 2SO_3$$
$$2SO_3 + 2H_2O \longrightarrow 2H_2SO_4$$

Summing up: $2S + 3O_2 + 2H_2O \longrightarrow 2H_2SO_4$

$$FeSO_4 + H_2O \longrightarrow H_2SO_4 + FeO$$
$$H_2SO_4 + CaCO_3 \longrightarrow CaSO_4 + H_2O + CO_2\uparrow$$

The calcium ions from this calcium sulphate gradually replace sodium ions from the clay micelle as shown in the following equation:

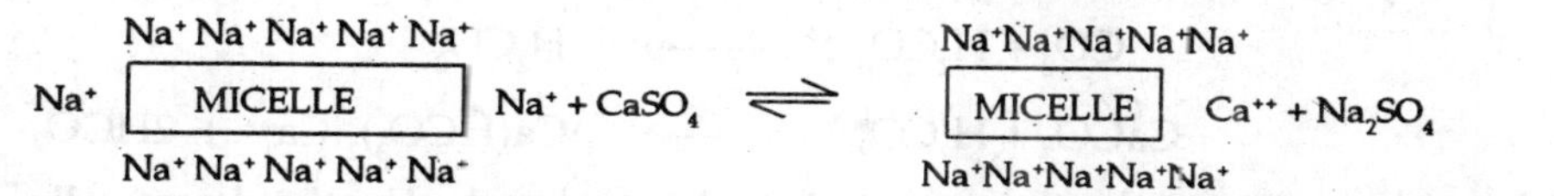

Sodium sulphate is washed down by heavily irrigating the field. Ultimately soil alkalinity is controlled.

7.30. GYPSUM REQUIREMENT OF ALKALI SOILS

To determine the gypsum requirement of alkali soils, the same process used to determine the lime requirement of acid soil using gypsum powder is used. The amount of gypsum required to reduce soil pH to seven is defined as its gypsum requirement.

7.31. QUALITY OF IRRIGATION WATER

Nature of irrigation Water

In arid regions of India, *i.e.* Uttar Pradesh, Haryana, Punjab, Rajasthan, Gujarat etc. water available for irrigation usually contains soluble salts which include ions of calcium, magnesium, sodium, chloride, sulphate, bicarbonate and sometimes borate and nitrate.

Their relative amounts depend on the source of water. Well water from drier areas of Uttar Pradesh, Haryana, Gujarat and Rajasthan are very saline.

Harmful Effect of Poor Quality of Irrigation Water

If water which contains excessive amounts of soluble salts is applied to the field, these salts increase the osmotic pressure of the soil solution. Consequently, crops cannot absorb enough water, quickly enough to make up the rapid loss of water by transpiration on hot summer days. So they wilt permanently and the yield is decreased. Excessive amounts of chloride when present in irrigation water is very harmful to crops. Borates even when present in small amounts, are also very harmful to most crops. If excessive amount of sodium ions are present in irrigation water, these will be adsorbed by the clay micelle; the clay will deflocculate, consequently destroying the soil structure and water and air will not able to circulate properly in the soil.

Principles on which The Criteria for Suitability of Irrigation Water is Based

Irrigation water may contain soluble salts which may include carbonates, bicarbonates, chlorides and sulphates of sodium, magnesium and calcium.

When saline irrigation water is applied to the fields, soluble salts, if present in the soil, dissolve to increase the concentration of salts in the soil solution, and sodium ions replace calcium ions from the clay micelle to destroy the soil structure. So irrigation water should contain enough calcium salts when calcium ions instead of sodium ions will be adsorbed by the clay micelle because calcium ions are held more tightly than sodium by the clay micelle. If the soil contains free calcium carbonate, this is dissolved in the irrigation water to release enough calcium ions into the water as shown in the following equation:

$$CO_2 + H_2CO_3 \longrightarrow H_2CO_3$$

$$CaCO_3 + H_2CO_3 \longrightarrow Ca(HCO_3)_2\ Ca^{++} + 2HCO_3^-$$

Calcium ions prevent the sodium ions from being absorbed by the clay micelle. So irrigation water containing higher amounts of soluble salts may be used if the soil contains enough lime. If not, only low salinity water should be used. Irrigation water also contains suspended materials.

Criteria of Suitability of the Water for Irrigation

On the basis of the above ideas, the suitability of any water for irrigation depends on (*i*) total soluble salts in the water as determined by its electrical conductivity (*ii*) total solids in the water (*iii*) the ratio of sodium to calcium in the water as determined by its (*a*) sodium absorption ratio (*b*) soluble sodium percentage and (*c*) residual sodium carbonate value (*iv*) the quantity of sodium salts present in the soil as determined by its exchangeable sodium percentage, and (*v*) presence of borates in the water

$$\text{Sodium Absorption Ratio (S.A.R)} = \frac{Na^+}{\sqrt{Ca^{++} + Mg^{++}}}$$

$$\text{Soluble Sodium Percentage (S.S.P)} = \frac{Na^+ \times 100}{Ca^{++} + Mg^{++} + Na^+}$$

Na^+, Ca^{++} and Mg^{++} are the concentration of the sodium, calcium and magnesium in the water in milliequivalents per litre. All carbonates (CO_3^{--}) and (HCO_3^-) in the water in milliequivalents per litre in excess of calcium (Ca^{++}) and magnesium

(Mg^{++}) in milliequivalent per litre in water, occur in the form of sodium carbonate and bicarbonate and therefore increase the pH of the water.

Residual Sodium Carbonate (R.S.C.) = (CO_3^{--} + HCO_3^-) - (Ca^{++} + Mg^{++})

Irrigation waters have been grouped into classes of low, medium, high and very high, depending upon their suitability for irrigations as shown below:

Water Class	Electrical conductivity in micro per cm. at 25°C	S.A.R.	P.P.M. of Boron in Water for			R.S.C.
			Sensitive crops e.g. cabbage	Semi tolerant crops e.g. (Rice)	Tolerant crops, e.g. sugarbeet	
Low	< 250	< 10	0.33	< 0.67	< 1.0	< 1.25
Medium	250 - 750	10 - 18	0.33 - 0.67	0.67-1.0	1.0 - 2.0	1.25 – 2.50
High	750 - 2250	18 -26	0.67 - 1.0	1.0 - 2.0	2.0–3.0	> 2.50
Very High	> 2250	> 26	> 1.0	2.0	> 3.0	

Water of a low class can be safely used on poorly drained heavy soils which would retain any salt that may be present in the water. Medium class should be used on permeable light soils which contain enough calcium carbonate and have low water table. Water of high class should not be used.

The pH of most of Indian rivers varies from 7.8 to 8.4. On the basis of the above standards, waters of rivers Kosi, Brahmaputra and Gandak are excellent for irrigation, due to their low salt content (Electrical Conductivity varies from 142 to 253 micro mhos/cm at 25°C) and low sodium content (Sodium Absorption Ratio varies from 0.17 to 0.26). Waters of a few other important rivers like Beas, Kaveri, Ganga, Sutlej and Narmada are also fairly good for irrigation due to their medium content of soluble salt (Electrical conductivity varies from 269 to 339 micro mhos per am at 25°C) and low sodium content (S.A.R.) varies from 0.46 to 0.99). Three south Indian rivers Krishna, Hagari and Tungabhadra contain very saline water (Electrical conductivity varies from 1392 to 1730 micro mhos/am at 25°C) which should not be used for irrigation, even when the S.A.R.'s of the first two rivers are below ten.

Improvement in the Quality of Irrigation Water

A little saline irrigation water can be improved by adding gypsum or farmyard manure to water.

A saturated solution of gypsum should be mechanically added at a constant rate to flowing water when sodium bicarbonate is converted to calcium carbonate.

$$2NaHCO_3 + CaSO_4 \rightleftharpoons Ca(HCO_3)_2 + Na_2SO_4$$

$$Ca(HCO_3)_2 \rightleftharpoons CaCO_3 + CO_2\uparrow + H_2O$$

Sodium ions may also be removed from irrigation waters by passing them through high cation exchange capacity calcium and hydrogen charcoal, which absorb sodium ions from them as indicated in the following equations:

$$\text{Ca Charcoal} + Na_2CO_3 \rightleftharpoons {}^{Na}_{Na}\!\!>\text{Charcoal} + CaCO_3$$

$$\text{H Charcoal} + Na_2HCO_3 \longrightarrow \text{Na Charcoal} + CO_2 + H_2O$$

The used charcoal may be converted again to hydrogen charcoal by treating it with dilute acids.

SELECTED READINGS

1. Agarwal, R.R. and Gupta, N.N. (1968), *Saline and Alkali Soils in India.* I.C.A.R. New Delhi.
2. Biswas, T.D. and Mukherjee, S.K. (1987): Textbook of Soil Science. Tata McGraw Hill Publishing Co., Ltd. Asaf Ali Road, New Delhi
3. Brady, N.C. (1984) *Nature and Properties of Soils*, 9th Ed. Macmillian Publishing Co., New York.
4. Bohn, H., McNeal, B. and O' Connor, G. (1985), *Soil Chemistry* John Wiley & Sons. 2nd Ed. p. 283.
5. Breeman, N van and Brinkman, R. (1978), *Chemical Equilibrium and Soil Formation.* In Soil Chemistry. A Basic Elements. Editors, G.H. Bolt and M.G.M. Bruggenwerts. Elsevier Scientific Publishing Co. 160.
6. Foth, H.D. and Turk L.M. (1973), *Fundamentals of Soil Science*, Wiley Eastern Private Ltd., New Delhi-110049.
7. Kanwar, J.S. and Grewal, J.S. (1971), A review of Phosphorus Fixation in Soils of India. I.C.A.R. *Technical Bulletin* (*Agril*) No. 29.
8. Mandal, S.C. (1971), *Acidic Soils of Indian and their Management.* Presidential Address to the section of Agricultural Science, 58th Session of Indian Science Congress, Bangalore.
9. Ray Choudhary, S.P. (1958), Properties of Colloidal Soil Constituents. I.C.A.R. *Research Series* No. 17.
10. Russel E.W. (1973), *Soil Conditions and Plant Growth*, 10th Ed. Eng. Book Society & Longman.

8
BIOLOGICAL PROPERTIES OF SOILS

Ecology is the science that deals with the study of the interrelationships between organisms and their environment. The ecosphere is the total number of living organisms in the global environment. It has been divided into ecosystems which are numerous, self-sustaining communities of organisms and their inorganic environments and resources. Each ecosystem comprises of a unique combination of living organisms and biotic resources that interact to maintain a continuous supply of nutrients and energy to the concerned organisms, both fauna (animals) and for (plants). Biological properties of soils comprise of those transformation in soils caused by the activities of the living organisms that live in them.

8.1. SOIL AS A LIVING SYSTEM

The soil is inhabitated by diverse groups of living organisms which include both macro and micro fauna and flora. The total number of living organisms in soil may be as much as billions per gm of soil and their live weight may go up to five tons per acre furrow slice of soil. The soil ecosystems include the macro and the micro organisms that live in the soil as well as the inorganic and the organic constituents of soils. All ecosystems comprise of two kinds of living organisms *i.e.* Autotrophs and Heterotrophs. The former use inorganic carbon, principally from carbon dioxide and are the producers. But the latter use organic carbon dioxide and are the producers. But the latter use organic carbon and are the consumer and the decomposers. Heterotrophs and Autotrophs have been sub-divided into two sub-groups photo heterotrophs, chemoheterotrophs, photoautotrophs and chemoautotrophs. The phototypes get their energy from sunlight and the chemo types get their energy from oxidation of inorganic compounds. Photoautotrohs, chemoautotrophs and chemoheterotrophs are most important in soils. Higher plants and alga are the photoautotrophs; chemoautotrophs include sulphur oxidizing and nitrifying bacteria. Animals, protozoa, fungi and most bacteria are chemoheterotrophs.

The major primary producers are higher plants and algae that photosynthesise carbohydrates form carbondioxide and water. Animals eat the top of the plants and excrete dung. The underground parts are consumed and decomposed by soil organisms. Unconsumed plant materials and dead bodies of soil organisms are decomposed by micro-organisms to release the nutrients for use in another cycle of growth and decay. The major function of soil organisms in the ecosystem centres around the flow or cycling of energy and nutrients.

Soil organisms depend mainly upon the climate and the resultant vegetation. Soil flora and fauna also depend upon soil factors like temperature, moisture-air relationships, soil reaction and humus and nutrient content of soils. Forests usually support more diverse fauna than grasslands do. But the fauna of grasslands are metabolically more active and their live weight per hectare is greater than forest fauna. Usually soils organisms are fewer in number and lower in weight in cultivated land than in uncultivated virgin land. Their population decreases as soil acidity is increased.

Microflora is responsible for about 80 per cent of the total soil metabolism. It decomposes organic residues to form the humus, from which nutrients can be assimilated by crops.

8.2 CLASSIFICATION OF SOIL MICROBES

The *A* horizon contains more organic debris and is more favourable for the growth of micro-organisms than the *B* horizon. Although other factors influence the activity and population of microbes besides nutrients and energy supplies, the greatest number of microbes offers in the *A* horizon or surface soil. Soil microfauna (animal) comprises of the following:

1. Nematodes: They are also called threadworms or eelworms. They are mainly microscopic in size and are the most abundant animals (maximum number is 50 per gm). Nematodes live mainly in the water film that surrounds soil particles on or in plant roots. They form cysts when conditions become unfavourable and are active when conditions become favourable. These organisms are round and spindle shaped. Their caudal end is pointed. They have been divided into three groups on the basis of their food requirements.

(*i*) Nematodes that live on decaying organic matter.
(*ii*) Predatory Nematodes feed on other nematodes, plant parasites protozoa, bacteria and the like.
(*iii*) Parasitic Nematodes that attack that roots of crops. These have a needle-like anterior end (stylet) with which they pierce the roots of the host, which develops galls or deformed roots.

2. Protozoa: They are much lager than bacteria although both are unicellular soil micro-organisms. Phylum protozoa have been divided into five classes on the basis of their means of locomotion as shown below:

(*a*) *Mastigophera*: Which moves by means of long whip-like appendages called flagella, e.g. Cerconodo in Fig. 8.1A.
(*b*) *Sarcodina*: They are also called rhizopods. Actually they consist of a naked mass of protoplasm which moves by pseudopodia e.g. Euglypus in Fig. 8.1B.
(*c*) *Ciliate*: They bear hairs on their body which help them to move e.g. Colpoda in Fig. 8.1C.

Protozoa encyst when the environment is unfavourable. The majority of them depend upon performed organic matter either as sporozoic feeders, obtaining their nutrition from soluble organic and inorganic substances, or by a holozoic nutrition characterized by direct feeding upon microbial cell Although they are the most varied and numerous because more than 250 species have been identified, their function in soil is still not fully known except that they serve to regulate the bacterial population.

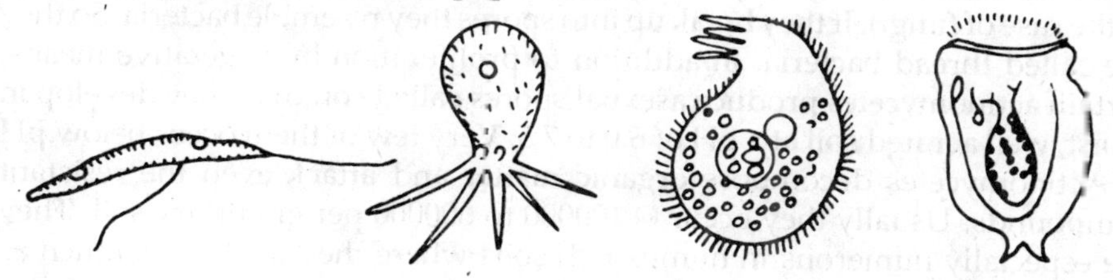

Fig. 8.1. Soil Protozoa *A*, *B* and *C*, *A* common Rotifera *D*.

3. Rotifera: They occur in moist soils. Their anterior side is modified into a retractile disc that bears circles of hairs called cilia, which, while in motion, give the appearance of a moving wheel. These hairs sweep floating food materials into the animal. The posterior end of a rotifer can attach itself to a convenient object (Fig.8.1D). Microflora include algae, fungi, actinomycetes and bacteria. Physiologically they may be heterotrophic or aurtotrophic in nutrition.

1. Algae: Morphologically, algae may be unicellular or may occur in short filaments. They may be divided into the following four classes:

(*i*) *Chlorophyceae*: They contain chlorophyll in chromatophores. So they are called green algae. *e.g.* chlorella.

(*ii*) *Cyanophyceae or myxophyceae*: Chlorophyll is uniformly distributed in the cytoplasm of this class which also contains a blue pigment called phycocyanin in addition to chlorophyll. So they are called blue-green algae, *e.g.* Nostoc.

(*iii*) *Xanthophyceae*: They contain a yellow-green pigment called xanthophyl. So they are called yellow-green algae, *e.g.* Bortrydium.

(*iv*) *Bacillariaceae or Diatomaceae*: They are unicellular or collonial algae which are surrounded by a highly silicified layer. They are golden brown in colour. So they are called golden brown algae or diatoms *e.g.* Nevicula.

2. Fungi: The vegetative body of a fungi consists of elongate thread-like structures which are individually called hyphae. The entire vegetative body or groups of hyphae are called mycelium. The hyphae of fungi are much broader than those of actionmyceters (about 0.5 to 1.2 micron in diameter). Soil fungi belongs to either of the three following classes:

(*a*) *Class fungi imperfecti*: They possess non-septate mycelium which bear sexual spores called conidia, on specialised structures called conidiophores.

(*b*) *Class phycomycetes:* They possess non-septate and filamentous hyphae and bear asexual spores called sporangiospores in Sparangia.

(*c*) *Class ascomycetes:* They posses hyphae and form sexual spares called as cospores in a sac-like structure called an ascus. Fungi belonging to the classes Phycomycetes and Ascomycetes produce both asexual and sexual spores.

Fungi usually occur in the soil @ 8000 to 100000 per gm. of dry soil. Fungi can decompose most complex organic molecules like starch, pectin, cellulose, fast and even lignins which cannot attacked by bacteria.

3. Actinomycetes are intermediate between fungi and bacteria. They are unicellular micro-organisms that consist of slender branched mycelia which are narrower than the fungal mycelium (0.5 to 1.2 microns in contrast to 5 microns in the case of fungi). If they break up into spores they resemble bacteria. So they are called thread bacteria. In addition to proliferation by vegetative means, certain actinomycetes produce asexual spores called conidia. They develop in moist, well aerated soil at a pH of 6.0 to 7.5. Very few of them occur below pH 5. Actinomycetes decompose organic matter and attack even the resistant compounds. Usually they occur @ 100000 to 600000 per gm of dry soil. They are especially numerous in humus rich soils where they may be as much as 0.25 tonnes per hectare.

4. Bacteria: They are unicellular micro-organisms. Larger ones seldom exceed 4 to 5 microns, and smaller bacteria are the size of clay particles *i.e.* less than 2 microns. A rod shaped bacterium is called Bacillus (Plural Bacilli) a spherical shaped bacterium is called a coccus (Plural, Cocci) and a spiral shaped bacterium is called spirillum (Plural Spirilla). Bacteria occur in soil @ 0.1 to 20 millions per gm of soil. Some produce spores which allow them to survive during unfavourable conditions. They occur in the greatest number near the surface of the land where favourable conditions occur most of the time. Most of the soil bacteria are heterotrophic *i.e.* they obtain their carbon and energy from organic matter. A few of them may get their energy from sunlight when they are called photoheterotrophic bacteria. Some bacteria are autotrophic, and obtain their carbon from carbon dioxide and energy sunlight when they are called photoautotrophic bacteria or from oxidation of inorganic compounds when they are called chemoautotrophic bacteria.

Bacteria are the most important soil micro-organisms. They are responsible for basic enzymatic transformation like nitrification, nitrogen fixation and sulphur oxidation. Aerobic bacteria can use only oxygen gas. Anaerobic bacteria can only use oxygen in the combined form . They grow best at 50 to 75 per cent of the maximum water holding capacity of soils, between 20 to 40°C and at a pH of 6.0 to 8.0. Most of them require abundant calcium.

8.3. MYCORRHIZAE

The symbiotic association between the roots of some plants and some fungi is called Mycorrhiza. The fungus is highly habitat limited and is usually found in the immediate vicinity of or within the roots. The fungal hyphae on the exterior of roots usually serve as an extension of roots and store large amounts of carbohydrates. Roots supply carbohydrates to the fungi which absorb nutrients from the soil and supply them to the crop. The mycorrhizae are of two kinds (1) Ectomycorrhizae and (2) Endomycorrhizae. In the former case, the fungus forms a mantle around the exterior of the roots a network composed of a mass of hyphae entering the intercellular spaces while in the latter case it penetrates the root cells. Endotrophic mycorrhizae are quite common among Ericaceae and Orchidaceae as well as in fruit trees especially citrus, coffee, rubber and various legumes. Coils of hyphae or branched hyphal structures formed within the root cells are called arbuscles. The fungus benefits the host in the following way:

1. The effective root surface is increased, so absorption of nutrients, especially phosphorus, by the roots, increases.
2. The rootlets function for a longer time.
3. The capacity of the host plant to tolerate heat and drought increases.
4. Soil nutrients become more easily available.
5. The host plant becomes more disease resistant. This last benefit occurs only in the case of ectomycorrhizae.

Mycorrhizae develop well in soils low in phosphorus and nitrogen. They are most vigorous on roots which have a large reserve of available carbohydrates.

8.4. LICHENS

Symbiotic association between some algae and fungi are called Lichens. The algae benefits in part because the fungal hyphae envelop it and protect it from environmental stresses, while the fungus gains by absorption of carbohydrates that have been photosynthesised by the algae. If a blue green is part of the lichen, the fungus additionally gains nitrogen because the blue green algae fixes atmospheric nitrogen. Lichens release compounds that cause weathering of rocks.

8.5. BENEFICIAL EFFECTS OF SOIL MICRO-ORGANISMS

Plant and animal residues are broken down by the soil fauna and flora which prevent unwanted accumulation of fresh organic material. Secondary complex organic forms of nutrients are converted to the simple inorganic forms. The stability of soil aggregates are increased both by the slimy intermediate products of decay and by the humus. When sulphur is added to alkaline soils, it gets oxidized by the activity of sulphur oxidising bacteria to sulphuric acid that neutralizes the alkalinity. Soil alkalinity is quickly neutralized if sulphur is added along with bulky organic manures.

Proteins are decomposed to ammonium compounds by the soil micro-organisms, which further oxidise the ammonium compounds to nitrites and nitrates. In well drained soils, divalent ferrous ions are oxidized to trivalent ferric ions and divalent manganous ion to trivalent and tetravalent manganic ions. These higher valent forms are insoluble. Hence the greater part of iron and manganese are kept in the insoluble non-toxic form by the soil micro-organisms. Molecular nitrogen is assimilated by the symbiotic and non-symbiotic bacteria and also by certain Blue Green Algae. Insoluble phosphorus compounds are dissolved in soil by organic acids produced by certain genera of soil micro-organisms to release the phosphorus contained in them in the soluble form for crop growth. Hence seeds of crops are inoculated with cultures or preparations of specific micro-organisms to increase the amount of water soluble nitrogen and phosphorus in soil for crop growth.

8.6. HARMFUL EFFECTS OF SOIL ORGANISMS TO HIGHER PLANTS

Some soil fauna such as snails and slugs damage and destroy crops. Ants transfer aphids to certain crops. Some crop roots are infested with nematodes sometimes seriously enough to affect the growth of these crops adversely. Bacteria, fungi and actinomycetes cause plant diseases. However, fungi are responsible for most of the common plant diseases. For example, species of Pythium, Fusarium and Rhizoctonia are responsible for the "damping off" of seedlings.

Injurious organisms live in the soil for variable periods, soil. Some disappear within a few years if their host plants are not grown, but some can survive on other hosts. Once a soil is infested, it is likely to remain so for a long time. Organisms may be carried from infested fields to other fields on implements, plants or debris of any kind.

Soil micro-organisms may compete with crops for available nutrients, especially when their supply is limited. Micro-organisms assimilate appreciable amounts of nitrogen, phosphorus, potassium and calcium and growth of crops would suffer due to their non-availability. Competition for trace elements may be even more serious. Soil micro-organisms take their quota of soil nutrients first and crops survive on the remaining soluble nutrients.

If the drainage of the soil is somewhat restricted, the aerobic micro-organisms may deplete the limited oxygen supply of the soil. This may adversely affect growth in two ways: The minimum amounts of oxygen necessary for the respiration of the root would not be available. Crops need respiratory energy for absorbing nutrients from the soil. Nitrates and sulphates would be reduced to nitrous and nitric oxides and nitrogen gas, and hydrogen sulphide respectively and lost to the atmosphere. The reduced forms of iron and manganese are excessively soluble and become toxic for crop growth. Thus micro-organisms can induce the deficiencies of some nutrients as well as toxicities of some others.

8.7. PARTIAL STERILISATION OF SOILS

Harmful soil organisms can be eliminated form the soil by partial, sterilisation, which can be achieved by either heating the soil by passing steam through it, or by treating it with sterilising solutions like carbon disulphide or formaline. Steam is passed through the soil for sometime. Thereafter, the field is flooded and left undisturbed for about a month. The field is then ploughed and harrowed and crops are sown or transplanted. The crop usually grows better after this partial sterilisation of the soil, though the exact mechanism is unknown. Probably organic nitrogenous compounds are rapidly decomposed to ammonium compounds. Soluble humic compounds are also increased.

Initially the population of both bacteria and protozoa decreases, but soon reverts to the original number. The fungal population is decreased. The effect of partial sterilisation with chemicals is similar to that with steam, except that most of the fungi are killed by formaline and carbon disulphide.

Indian farmers have known from time immemorial the heating the soil by burning stubbles on it increases the crop yield. When the soils are heated and dried and again rewetted, organic nitrogenous compounds decompose in 10 days to ammonium compounds which are oxidised to nitrate if some fresh soil has been added to the heated and dried soil before it is rewet. More ammonium compounds and nitrates are formed if soils rich in organic matter are heated to higher temperatures and kept dry for longer periods. This happens in hilly areas where tribal people burn the forest when the soil is partially sterilised and plant nutrients are added to the soil in the form of plant ash. Thereafter the field is ploughed and the crops sown. The tribal people get a good yield in the first year immediately after burning the forest. In northeastern India, farmers clear the forest by burning it, grow crops on the land for 4 to 4 years, and then abandon the land and move to another area. This practice is known as shifting cultivation or Jhum cultivation.

Farmers in some other parts of India, especially in the heavy rainfall areas along the west coast, spread a layer of cowdung on the field which is then covered with a layer of waste organic material. They then cover the waste organic material with a layer of dry soil and set the organic material on fire. After burning it, they plough the land and grow crops on it; the ploughed land produces a higher crop yield than the untreated land. This higher yield of the crop has been attributed to the increase in the availability of nutrients especially potash and phosphoric acid. This practice is known as "rab" to the farmers of the west coast of India.

8.8. SOURCE OF ORGANIC MATTER

The organic matter which is decomposed in the soil by micro-organisms comes form several sources. Forest plants add organic matter to the soils mainly in the form of fallen leaves. Dead plant, roots and soil organisms like insects, millipedes, nematodes etc. also contribute some organic matter to the soil. The mixture of leaves, roots and bodies of soil organisms is decomposed by micro-organisms to form humus. If this humus lies just above the mineral horizon, it is called mor. If the organic matter has been intimately mixed with mineral matter and then decomposed, this humus incorporated mineral layer is called mull.

8.9. ORGANIC MATTER VERSUS HUMUS

Soil organic matter may be defined to be any substance of organic origin. This means that soil organic matter comprises of living or dead plant and animal tissues. Humus is the substance which was formed from microbial decomposition of plant and animal tissue. The organic form of plant and animal material can be identified in the soil organic matter while the structure of the original plant and animal material cannot be traced in the humus. Actually the humus is more complex than the original plant and animal materials. Both the decomposition of the original organic compounds and the synthesis of new organic compounds are involved in the formation of humus.

8.10. THE CHEMICAL NATURE OF THE ORGANIC MATTER

The plant material which goes into the soil mainly consists of cellulose (15 to 16%), hemicellulose (10 to 30%) and lignin (5 to 30%). It also contains fats, waxes and resins, sugars, starches, proteins, amino acids and minerals. When plants age, water soluble constituents like sugars, starches and proteins decrease whereas cellulose hemicellulose, and lignins increase.

8.11. DECOMPOSITION OF ORGANIC MATTER TO FORM HUMUS

Whenever organic matter decomposes, it is acted upon by varieties of micro-organisms. Fungi dominates over others in the initial stages while bacteria are the important agents of decomposition during the later stages. At first, the easily decomposable substances like sugars, starches and water soluble proteins are acted upon by the micro-organisms and decomposition and digestion is rapid. Crude protein is next in order, followed by hemicellulose. Cellulose, which is more resistant to microbial attack than hemicellulose, decomposes much more rapidly than oils, fats, waxes, resins etc. Lignin decomposes very slowly and continues to dominate soil organic matter when the decay precess slows down.

The sugars, starches, hemicelluloses and celluloses are ultimately decomposed to carbon dioxide and water, and energy is liberated which is utilized by micro-organisms. Some oils, fats, waxes and resins are also slowly decomposed to carbon dioxide and water some energy liberated for use by micro-organisms. A small portion of lignin may be slowly decomposed to form aromatic compounds. Other portions may be chemically altered. Some other portion may chemically unite with protein to form part of the soil humus.

Proteins are gradually decomposed to amino acids and amides which are further decomposed to ammonium compounds by micro-organisms. Ammonium compounds are oxidised to nitrites by Nitrosomonas bacteria. Nitrites are further oxidized to Nitrates by Nitrobacter bacteria. Phosphorus is present in the organic matter as phytin, nucleic acid and phospholipids, which are decomposed to liberate the phosphorus present in them as orthophoshate ions, $H_2PO_4^-$. Similarly, sulphur containing amino acids like methionine, cysteine etc. are decomposed by micro-organisms to liberate the sulphur contained in them as sulphate. When organic matter decomposes, other complex organic forms of nutrients are converted to simple ionic forms like K^+, Ca^{++}, Mg^{++}, etc. This process of conversion of complex organic forms of nutrients to simple inorganic forms by micro-organisms is called the mineralisation of nutrients. A portion of the nutrients thus mineralized is assimilated by the micro-organisms themselves for synthesis of their cell protoplasm. Thus the simple inorganic form of nutrient is recovered to the complex organic form of nutrients. This process of conversion of the simple inorganic form of nutrients to the complex organic form of nutrients is called immobilization of nutrients. During the earlier stages of decomposition of organic matter all the simple inorganic forms of nutrients are assimilated by micro-organisms which multiply rapidly and continue to decompose the organic matter. When almost all the carbon compounds have been decomposed the micro-organisms die due to the lack of sufficient amounts of energy giving carbon compounds. Their bodies decay when the complex organic forms of the nutrients are reconverted to simple inorganic forms. Some proteins combine with organic compounds like lignins, tannis, humic acids etc. Some proteins are adsorbed by the clay, especially the expanding ones. All these reactions protect the proteins from microbial decomposition. At this stage, almost all the original organic material has been converted to a dark heterogeneous mass called humus.

Humus is a resistant complex mixture of dark brown to black coloured colloidal and amorphous substances, synthesized or modified from the original organic materials by various micro-organism.

8.12. CHARACTERISTICS AND PROPERTIES OF HUMUS

Humic materials include the following three chemical groupings:

(*i*) Fulvic acids, of low molecular weight light coloured and soluble in both acid and alkali. Fulvic acids are the least resistant to decomposition by micro-organisms.

(*ii*) Humic acids are of medium molecular weight and colour, are soluble in alkali but insoluble in acids. They have a medium resistance to decomposition by micro-organisms.

(*iii*) Humins have the highest molecular weight are the darkest in colour and insoluble in both acid and alkali.

All these three groups of materials, in spite of the difference in their chemical and physical makeups have some similarities and can be termed humus.

The nature of the humus is exceeding complex. The soil humus comprises additionally of nitrogenous compounds, of humic acids, phospholipids, vitamins, sulpholipids and polysaccharide, hormones etc. These additional substances may be residue from plant tissue, synthesized by micro-organisms or residues of microbial degradation.

The nitrogen content of humus varies widely. Mostly, humus contains 3-6 per cent nitrogen and about 58 per cent carbon. On the basis of the 58 per cent of organic carbon contained in humus, the organic matter be determined by multiplying the percentage of organic matter by 1.724. The Carbon: Nitrogen ratio of humus is from 10 : 1 to 12 : 1. This C/N ratio varies with the nature of the humus, the stage of its decomposition and the soil and climatic conditions under which it has been formed. Humus also contains appreciable amounts of phosphorus and sulphur. The Carbon : Nitrogen : Phosphorus : Sulphur ratio is 120 or 100 : 10 : 1 : 1. The cation exchange capacity of humus varies form 200 to 400 milliequivalents per 100 gms. This cation exchange is due to several chemically active groups. During the decomposition of organic matter, lignin is altered in such a way that both the carboxylic group (—COOH) and the phenolic hydroxyl group($\Rightarrow$ OH) increase. The hydrogen of these groups can be replaced by cations. The humus actually behaves as a huge negative anion which is made up of carbon, oxygen and hydrogen. These carboxylic and phenolic hydroxylic groups are attached with this central unit. The metallic cation like Ca^{++}, Mg^{++}, K^{+} etc. are absorbed on the negative sites as a result of which these ions are not leached down from the soil. They can be taken up roots by cation exchange. Humus absorbs large amounts of water and swells up.

8.13. CARBON: NITROGEN RATIO

A close relationship exists between the organic matter and nitrogen content of soils. The ratio of organic carbon to total nitrogen in soil is called the Carbon: Nitrogen ratio or C/N ratio of soils C/N of surface soils varies form 8 : 1 to 15 : 1.

Importance of C/N Ratio

The C/N ratio helps to conserve both the nitrogen and organic matter content of soils. When the C/N ratio is great, the carbon compounds decompose to release energy which is utilized by the heterotrophic micro-organisms. They multiply rapidly and assimilate all the ammoniacal and nitrate nitrogen to synthesize their cell protoplasm. This nitrogen is thus immobilised and conserved, it is not lost in drainage. When the organic matter is fairly decomposed, the C/N ratio becomes lower; the micro-organisms die and decay when the complex organic forms of nitrogen are again reconverted to simple inorganic forms. Since the C/N ratios of soils are fairly constant *i.e.* about 10 : 1 to 12 : 1, whenever organic matter that contains a high amount of nitrogen is added to the soil, less of the originally added organic matter is decomposed to attain the fairly constant ratio of 10 : 1 to 12 : 1. Hence organic matter is conserved.

Stabilisation of C/N Ratio

The C/N ratio of organic matter usually ranges from 40 : 1 to 60 : 1. Micro-organism decomposes this organic matter to convert carbon compounds to carbon dioxide and water, and nitrogenous compounds to ammonium compounds which are further oxidized to nitrates. In the initial stages of decomposition of organic matter, carbon dioxides is lost and ammonium compounds and nitrates are assimilated by the micro-organisms to synthesize their cell protoplasm. Hence the C/N ratio is narrowed down. The micro-organisms live only for short time, and when they die their bodies decay; complex organic forms of nitrogen are converted to simple inorganic forms, on which another crop of micro-organisms grow, and continue to decompose the organic matter when more carbon is lost as carbon dioxide. Ammonium and nitrate however, are assimilated to synthesise their cell protoplasm, and the Carbon : Nitrogen ratio is further narrowed down. This above-mentioned process is repeated till the constant C/N ratio of 10 : 1 to 12 : 1 is attained.

8.14. NITROGEN CYCLE

Nitrogen is added to the soil and lost from it when it undergoes a number of transformations, almost simultaneously as shown in Fig. 8.1. A small portion of nitrogen gas is directly assimilated by symbiotic and non- symbiotic nitrogen fixing bacteria and also by blue-green algae, to synthesize their cell protoplasm. Proteins present in organic matter as well as in the bodies of soil micro-organisms are decomposed to amino acids which are further oxidized to nitrates. Considerable amounts of nitrate may be leached down during the rainy season. A portion of nitrates may be further reduced to nitrous oxide and nitrogen gas by the denitrifying bacteria and lost to the atmosphere (Fig. 8.2). Plant assimilate nitrates, grow and are eaten up by animals. Plant material and animal dung is decomposed to form the humus. The above sequence of biochemical reactions, by which forms of nitrogen change under natural conditions, is called the Nitrogen Cycle as shown in Fig. 8.2.

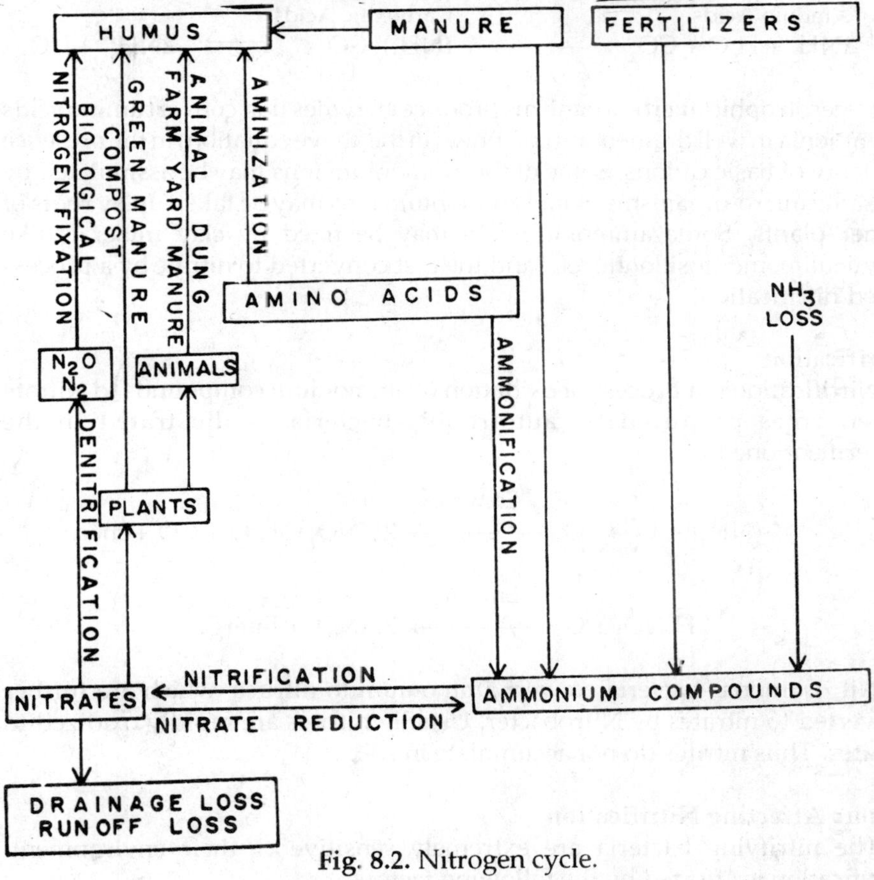

Fig. 8.2. Nitrogen cycle.

8.15. TRANSFORMATION OF NITROGEN IN SOILS

A large number of heterotrophic micro-organisms bring about complex transformations of nitrogen as shown below:

Proteins → Amino Acids → Ammonium Compounds → Nitrites → Nitrates

Aminization

Proteins are decomposed by the proteinase enzyme to amino acids

$$\text{Proteins} \xrightarrow[\text{enzymes}]{\text{Proteinase}} \text{Amino Acids} + CO_2 + \text{Energy}$$

This process is called aminization.

Ammonification

Amino acids are converted to ammonium compounds by a process called ammonification

$$\underset{\text{Amonio Acids}}{R.C.H_2\,\overset{\overset{NH_2}{|}}{C}H.COOH} \xrightarrow{\text{Enzyme}} \underset{\text{Carboxilic Acids}}{R.C.H. = CH.\,COOH+NH_3.}$$

$$2NH_3 + H_2O + CO_2 \longrightarrow (NH_4)_2\,CO_3 \rightleftharpoons 2NH_4^+ + CO_3^{--}$$

Heterotrophic micro-organisms produce enzymes that convert amino acids to ammonia in well drained soil (as shown in the above equation) in the presence of plenty of basic cations. Some of the ammonium ions may be assimilated by the same micro-organisms. Some ammonium ions may be taken up by roots of higher plants. Some ammonium ion may be fixed by clay minerals like vermiculite, montosillonite, etc, and the rest converted to nitrate by a process called nitrification.

Nitrification

Nitrification is a process of oxidation of ammonium compounds to Nitrate by enzymes produced by autotrophic bacteria as illustrated in the undermentioned

$$(NH_4)_2\,CO_3 + 3O_2 \xrightarrow{\text{Enzyme}} 2HNO_2 + 3H_2O + CO_2 + \text{Energy}$$

$$2H.NO_2 + O_2 \xrightarrow{\text{Enzyme}} 2HNO_3 + \text{Energy}$$

Nitrosomonas bacteria convert ammonium to nitrites, which are further converted to nitrates by Nitrobacter. Usually nitrites are rapidly oxidized to nitrates. Thus nitrites do not accumulate in soil.

Factors Affecting Nitrification

The nitrifying bacteria are extremely sensitive to their environment. Nitrification is affected by the following factors:

(*i*) *Aeration*: As nitrification is an oxidative process it is increased by tillage operations like ploughing and cultivations, which increase the supply of oxygen of soils, provided other conditions, are favourable.

(*ii*) *Temperature*: 30 to 32°C is most favourable for nitrification. It takes place very slowly at 5°C or less and at 40°C or more.

(*iii*) *Moisture*: Nitrates usually appear most readily at one half to two-thirds of the maximum available water holding Capacity of Soils.

(*iv*) *Soil reaction*: The optimum pH for nitrification appears to be between 6.5

to 7.8. Bacterial strains which occur in acidic soil continue this process best at a pH of 6.5 while those that occur in alkaline soils continue this process best at pH 7.8.

(*v*) *Season*: The interaction of moisture, temperature, cropping and nutrients is the seasonal effect.

8.16. SYMBIOTIC NITROGEN FIXATION

Symbiosis means the physiological partnership between two living organisms. Bacteria belonging to the genus Rhizobium live in the nodules that occur on the roots of leguminous plants (Legumes) in most cases and assimilate nitrogen gas from the atmosphere to synthesise amino acids which bacteria supply to the host plant, which supplies carbohydrates to the Rhizobium bacteria. The carbohydrates are oxidized to carbon dioxides and water and energy in liberated which is utilized by the Rhizobium bacteria. Bacteria which belong to the genus Rhizobium have been classified into several cross-inoculation groups on the basis of their capacity to infect the roots of a restricted number of leguminous plants and form nodules in addition to the roots of the legume from which the bacterial strain were first obtained. Seven cross-inoculation groups recognised so far are as follows:

Cross-inoculation	*Rhizobium species*	*Host legumes*
1. Alfalfa Group	Rhizobium meliloti	Alfalfa, Sweet clover
2. Clover Group	R. trifolii	Clovers (Barseem)
3. Pea Group	R. leguminosarum	Pea, Lentil, Sweet pea
4. Bean Group	R. phaseoli	Beans
5. Lupin Group	R. lupini	Lupines
6. Soybean Group	R. japonicum	Soybean
7. Cowpea Group	R. species not identified so far	Cowpea, Groundnut, Pidgeon pea, Sunhemp.

Rhizobium Bacteria which infect roots of cowpea and sunhemp and form nodules on them have been observed at I.A.R.I., New Delhi to fix about 90 kg of nitrogen per hectare per year. It has also been observed at I.A.R.I, New Delhi, that Rhizobium trifolii, that forms nodules on the roots of barseem (Trifolium alexandrium), fixed about 120 kgs of nitrogen per hectare per year.

The weight and numbers of nodules decrease when excessive amounts of ammonium compounds or nitrates accumulate in the soil.

A high Carbohydrate: Nitrogen ratio within the host plant increases the amount of nitrogen fixed by the bacteria, whereas an even higher Carbohydrate: Nitrogen ratio resulting from very high photosynthesis decreases the amount of nitrogen fixed.

Rhizobium bacteria do not infect the roots of leguminous plants below. pH 5.0 Molybdenum and calcium are essential for nitrogen fixation.

8.17. NON-SYMBIOTIC NITROGEN FIXATION

Certain free living organisms like Azotobacter, Beijerinekia, Clostridium, blue-green algae etc. can directly assimilate nitrogen gas from the atmosphere.to synthesize their cell protoplasm. When these organisms die, their bodies decompose and nitrogen becomes available to higher plants. The addition of readily decomposable organic matter increases nitrogen fixation for which an adequate supply of molybdenum and calcium is also essential. Non-symbiotic nitrogen fixation by Azotobactor stops below pH 5.0.

8.18. MECHANISM OF SYMBIOTIC AND NON-SYMBIOTIC NITROGEN FIXATION

The exact mechanism of symbiotic and non-symbiotic nitrogen fixation is not clear. The enzyme nitrogenase activates the nitrogen gas, which combines with the hydrogen to form ammonia, that reacts with the carboxylic acids to form into amino acids, which combine with each other to form proteins as shown in the following equation:

$$N_2 + 3H_2 \xrightarrow{\text{Nitrogenase}} 2NH_3$$

$$NH_3 + \underset{\substack{|\\(CH_2)_2\\|\\COOH}}{\overset{\substack{COOH\\|}}{C}} = O \longrightarrow \underset{\substack{|\\(CH_2)_2\\|\\COOH}}{\overset{\substack{COOH\\|}}{C}} = NH + H_2O$$

Imino Glutaric Acid (Carboxylic Acid) → Glutamic Acid (Amino Acid)

Then amino acids combine together to form proteins.

8.19. DENITRIFICATION

The sequence of step that result in the gaseous loss of nitrogen is called denitrification. Certain micro-organisms like Pseudomonas, Clastridium etc. use the combined oxygen in nitrates under anaerobic conditions to oxidise organic compounds when nitrates are reduced to nitrates, oxides of nitrogen and nitrogen gas as shown in the following equations:

$$2HNO_3 \xrightarrow{-2[O]} 2HNO_2 \xrightarrow{-2[O]} HON = NOH \xrightarrow{+2[H]} N_2 + 2H_2O$$

$$2HNO_2 \xrightarrow{-[o]} 2NO + H_2O \longrightarrow N_2O + H_2O$$

$$HON = NOH \xrightarrow{-H.OH} N_2O + H_2O$$

8.20. MICROBIAL TRANSFORMATION OF PHOSPHORUS AND SULPHUR

Bacteria, for example species Pseudomonas and Bacillus megaterium var.

Phosphaticum produce organic acids that dissolve insoluble tricalcium phosphate, making the phosphorus contained in it available for plant growth. When organic matter decomposes, the phosphatase enzyme breaks up the

$$\text{Humus—O—}\overset{\overset{\displaystyle \text{O}}{\|}}{\text{P}}\text{—(OH)}_2 \text{ bond to Form Humus—OH} + H_3PO_4$$

When organic matter is decomposed, sulphur containing proteins are decomposed to sulphur containing amino acids, which are further decomposed to form sulphate. In well aerated soils, sulphur is oxidised to sulphate by thiobacillus thiooxidans bacteria, respectively as represented in the following equations:

$$2S + 3O_2 + 2H_2O \quad \rightarrow \quad 2H_2SO_4$$

In poorly drained soils desulphovibro and desulpho maculum bacteria use combined oxygen in sulphate to oxidize organic compounds when sulphates are reduced to sulphides as has been illustrated in the undermentioned equation:

$$2R - CH_2OH + H_2SO_4 \quad \rightarrow \quad 2R - COOH + 2H_2O + H_2S.$$

The soil pH falls.

8.21. SOIL AS A STORAGE PLACE FOR NUTRIENTS

The nutrient storage capacities of soils depends upon the amount and nature of clay and humus present in them.

Humus possess a very high cation exchange capacity. Montmorillonite and vermiculite also possess a high cation exchange capacity, whereas the cation exchange capacity of Kaolinite is very low. Organic soils can store high amounts of nutrients delivered to the soil in the form of fallen leaves and dead roots containing nutrients in the complex organic form. Soil micro-organisms decompose them to release the nutrients contained in them in the soluble form, which are loosely held by the humus. Micro-organisms also assimilate some of these nutrients to synthesise their cell protoplasm.

Hence some of the nutrients remain stored in their cell protoplasm in the complex organic form. As soon as the energy supplying organic materials have been exhausted, the micro-organisms die and their bodies are decomposed to release the nutrients in the simple inorganic water-soluble form which can be assimilated by roots.

The nutrient storage capacity of the black cotton soil in India, in which montmorillonite predominates is very high, due to the high cation exchange capacity of montmorillonite. Black cotton soils are considered fertile where varieties of crops like cotton, sorghum, wheat etc. are grown. The red soil of India, which is dominated by kaolinite, possesses a poor storage capacity due to the low cation exchange capacity of kaolinite.

SELECTED READINGS

1. Alexander, M (1961), *Introduction to soil microbiology*. John Wiley & Sons, New York.
2. Biswas, T.D. and Mukherjee, S.K. (1987): *Textbook of Soil Science*, Tata McGraw Hill Publishing Co. Ltd, Asaf Ali Road, New Delhi.

3. Brady, N.C. (1984), *Nature and properties of soil*. Macmillan Publishing Co., New York.
4. Foth, H.D. (1984), *Fundamentals of Soil Science*. 7th Ed. John Wiley & Sons, New York.
5. Khambata, S. R. and Bhat, J.V. (1957), Final Report on the Survey of the Microflora of Soils in relation to their Structure and, the role of Earth worms in the alteration of the Microflora of Soils.
6. Russel, E.W. (1973), *Soil Conditions and Plant Growth*, 10th Ed., Eng. Lang. Book Society & Longman.
7. Stefferud, A. Editor (1957), *Soils: Year Book of Agriculture* 1957, U.S.D.A.
8. Waksman, S.A. (1957), *Soil Microbiology*, John Wiley & Sons, New York.

9
SOIL CLASSIFICATION AND SURVEY

Soil taxonomy means the orderly arrangement or classification of soils in order to study their properties, to increase their productivity. Soils are classified for the following purposes:

1. To enable us systematic study of soils.
2. To determine the relationship between the different kinds of soils.
3. To know how they can be best used.
4. To estimate the productivity of soils under different management practices.

There are various systems of soil classification. Only the Genetic system, which was in use up to 1949 and the Modern system of soil classification will be dealt with here.

9.1. GENETIC SYSTEM OF SOIL CLASSIFICATION

Russian scientist V.V. Docuchaev first suggested the genetic system of soil classification, which was further developed by the American Soil Scientist C.F. Marbut. According to this system, each soil possesses certain properties which develop as a consequence of the effects of a particular combination of factors affecting soil formation. The soils inherit their properties from the parent material in the initial stages of soil formation. For example, soil developed from basic parent material are rich in basic elements and plant nutrients and are alkaline in reaction in the initials stages of development. But as soils mature, the inherited character gradually disappear and the soils are dominated by the character acquired from the climate and vegetation. For example, as time passes, basic elements are gradually washed down and ultimately soils which had developed from basic parent materials become poor in basic elements, plant nutrients and acidic in reaction.

According to the genetic system, soils of the world have been classified into three orders : Zonal, Intrazonal and Azonal.

(1) **Zonal soils**: The characteristics of soils belonging this order are mainly determined by the climate and vegetation. This means that soils that have developed from different parent material under the same climatic and vegetational conditions, possess same general properties.

(2) **Intrazonal soils**: General characteristics of soils belonging to this order are mainly determined by certain local conditions like poor drainage, soluble salts etc. The soil characteristics due to these conditions, dominate over the characteristics due to climate and vegetation.

(3) **Azonal soils**: Soils belonging to this order to do not possess well developed profiles because they have not been subjected to the influence of the climate and the resultant vegetation long enough for the soil profile to develop.

Zonals soils have been divided into six suborders on the basic of wider ranges of variation in climate and the resultant vegetation. Each of the sub-orders have been subdivided into great groups on the basis of narrower ranges of variation of climate and the resultant vegetation.

Intrazonal soils have been divided into three sub-orders:

(*i*) Hydromorphic soils that contain excessive amounts of water.

(*ii*) Halomorphic soils that contain excessive amounts of soluble salts.

(*iii*) Calciomorphic soils that contain excessive amounts of lime or gypsum.

Hydromorphic soils have been subdivided into great groups on the basis of increasing degrees of accumulation of water in soils.

Halomorphic soils have been subdivided into the two great groups, Saline and Alkali soils.

Saline soils contain excessive amounts of neutral salts *i.e.* sulphates and chlorides of sodium, calcium and magnesium.

Alkali soils contain excessive amounts of sodium saturated clays and sodium carbonate.

Calciomorphic soils are divided into two great groups;

(*i*) Brown forest soils that are dark brown at the surface and gradually grade into gray calcareous parent material.

(*ii*) Black earth or Grumosols that are dark brown grassland soils developed from limestone.

The order Azonal doesn't have any suborders, but the three following great groups have been recognised under it:

(*i*) *Lithosols*: They are commonly found on steep slopes and consist of very shallow stony soils over bed rock.

(*ii*) *Regosols*: They have developed from deep soft mineral deposits. They are also very young soils like lithosols and differ from Lithosols in that they are not stony.

(*iii*) *Alluvial soils*: The main characteristics of alluvial soils have been derived by the deposition of products of weathering of rocks by the great river systems like the Ganges, the Brahmaputra and the Indus etc.

Great groups have been divided into families, each of which includes a certain number of identical soil series possessing certain common properties. Each family has been divided into a number of series. Soils belonging to a series have originated from the same parent material, under same climate and resultant vegetation and therefore possess the same profile characteristics. Each series has been divided into a number of soil types on the basis of the texture of the surface soil. Soil types have been divided into phases on the basis of some important deviations like slope, erosion (removal of top soil by running water or blowing wind is called soil erosion), wetness, stoniness etc.

9.2. MODERN SYSTEM OF SOIL CLASSIFICATION

The Genetic system of soil classification is based entirely on soil genesis. Some soils which have been cultivated for centuries have greatly changed. Yet they are still classified on the basis of the properties presumed to be possessed by them when they were not cultivated but under natural vegetation. For these reasons, the soil survey staff of the United States, Department of Agriculture have proposed the new system of soil classification called the seventh Approximation because it has been modified or approximated seven times. This system of soil classification is based on the present properties of soils. However, these properties are also the consequence of soil genesis. According to the modern system of soil classification, soils of the world have been divided into ten orders based on the variation in the major soil forming processes as indicated by the presence or absence of major diagnostic horizons. These ten orders are as follows:

Orders	*Main Characteristics*
1. *Entisols*:	Very young soils that do not possess a well developed soil profile.
2. *Inceptisols*:	Soil in which the horizons have just started to develop. They are relatively low in organic matter or base saturation or both and contain some weatherable minerals.
3. *Aridisols*:	Soils of dry regions e.g. Deserts. They are relatively low in organic matter and possess a few pedogenic horizons.
4. *Vertisols*:	Soils that are very rich in the expanding type of clay-minerals and therefore have swollen and shrunk a number of times. As a consequence, they possess deep wide cracks at some stage in most of the years.
5. *Mollisols*:	Grassland soils that possess a thick soft dark coloured surface horizon, relatively high in organic matter and high base saturation in all horizons.
6. *Spodosols*:	Soils that possess ash grey horizons near the surface due to eluviation of clay and sequioxide and the subsurface illuvial horizon called the spodic horizon, rich in sesquioxide and humus.
7. *Alfisols*:	Cool humid region soils containing appreciable amounts of primary minerals from which basic elements are being slowly leached. They have gray to brown surface horizon and are medium to high in base saturation and possess an illuvial horizon of silicate clays that is more 35 per cent base saturated.
8. *Ultisols*:	Warm humid region, highly leached acidic soils containing low amounts of basic elements and plant nutrients. So the base saturation is low. They possess an illuvial horizon of

silicate clays with base saturation lower than 35 per cent.

9. *Oxisols*: Hot humid climate soils that possess an oxic horizon (this horizon contains high amounts of low cation exchange capacity clays, which are highly resistant to dispersion in water). These clays are very inactive and do not possess an illuvial horizon of silicate clays.

10. *Histosols*: Organic soils that consist of plant materials.

Orders (1 to 9) have been divided into sub-orders on the basis of properties associated with wetness or dryness, major parent material and vegetation, degree of decomposition of organic matter is the basis of classification of the order Histosols.

The sub orders have been divided into great groups on the basis of differentiating soil horizons and soil features. Differentiating soil horizons include those that have accumulated clay, sesquioxide and/or humus and those that have pans (hard layers). The differentiating soil features include expansion and contraction properties of clays, soil temperature and the presence of basic elements and soluble salts in the soil.

Great groups have been divided into not more than the following three sub-groups:

(*i*) The first sub-group possess only the properties of the concerned great group.

(*ii*) The second sub-group possesses some of the properties of other sub-groups, great groups, sub-orders and orders in addition to the properties of the concerned great groups.

(*iii*) The third sub-group possess some properties that have not been found in any other great groups, sub-orders and orders in addition to the properties of the concerned great group.

Sub-groups have been divided into families on the basis of these properties important to plant growth, including soil texture, clay minerals, soil reaction, mean annual soil temperature and mean rainfall. Families have been divided into series on the basis of profile characterstics which means that soils belonging to a series have developed from the same parent material under the same climate and resultant vegetation and therefore possess identical profile characteristics. Soil series have been divided into phases which describes soil texture, percentage slope, stoniness, wetness, erosion, soluble salts etc.

9.3. SOIL SURVEY

Definition

A soil survey is the systematic description of soils in the field, their grouping into well defined mapping units like soil series, phases etc. to identify their best use and show their location on the map.

Purpose

1. To classify soils into well defined mapping units *i.e.* soil series, phases etc.
2. To shows their distribution in the field on the map.

3. To find out their best use.
4. To predict their performance under different management practices *i.e.* yields of crops under different management practices.

Soil surveys are of three types:

(*i*) *Detail soil survey*: Soils are studies in detail. Soil boundaries are demarcated on a large scale (1 : 10,000 to 1 : 5,000) base map called cadastral map. Mapping units is soil series.

(*ii*) *Reconnaissance soil survey*: Soils are not studied in detail *i.e.* observations are taken at a longer interval and noted down on the Survey of India topo sheets of scale 1 : 2,50,000 to 1 : 50,000. The mapping unit is soil association.

(*iii*) *Detailed Reconnaissance*: First reconnaissance soil survey is carried out and subsequently about 15 per cent of the area where research projects are to be established, are surveyed in detail.

Method of Detail Soil Survey

Soils are grouped into units possessing certain common physical properties and also morphological properties, which can be readily recognized in the field. Physical and morphological properties are texture, structure, colour, pH, carbonate, natural vegetation, slope, erosion, depth of soil, natural vegetation etc. The soil units are usually defined on the basis of the above-mentioned characteristics of the surface soil, soil depth, slope and erosion and the soil profile. The soil units are the soil series. Cadastral (village) maps are the base maps (Scale 1 : 10,000 to 1 : 5,000).

Select a convenient starting point like a bench mark, or building or pond, or anything in the field and identify the same on the cadastral map (1 : 10,000 to 1 : 5000). The Soil Surveyor moves up or down the slope because usually the soil properties change in that direction. Start walking down the slope from the starting point *i.e.* the bench mark in the Fig. 9.1 and continue observing the natural vegetation, slope of the land, soil erosion, soil depth, soil colour, texture, by feeling the Soil; pH with the help of universal indicator; and carbonate with the help of dilute acid at an interval of about 100 to 200 metres. As you are walking down the slope and studying these properties, you suddenly find that a number of soil characteristics change; For example, slope erosion, soil depth, colour, texture etc. Then, follow the line of change in the soil properties as shown in Fig. 9.1 and demarcate it. This is the boundary between the soil units *A* and *B*. Resume walking down (Traversing) the slope till you again find that number of soil characteristics change. Follow the line of change of soil properties between the soil units *B* and *C* and demarcate it on the field. In a similar way, find out and demarcate the boundaries between soil units *C* and *D*, *D*, and *E* and so on as shown in Fig. 7.1. Then dig profiles in each of the soil units *A*, *B*, *C*, *D* and *E* the number of profiles depend on the relative area of the soil units. Study profiles, the characteristics of which become the basis of identifying soil series. If the characteristics of the profile in soil units *A*, *B*, *C*, *D* and *E* are different then the soul units *A*, *B*, *C*, *D* and *E* become the soil series *A*, *B*, *C*, *D* and *E*. Then a soil survey report is written, describing the soil series and providing other useful information about the area.

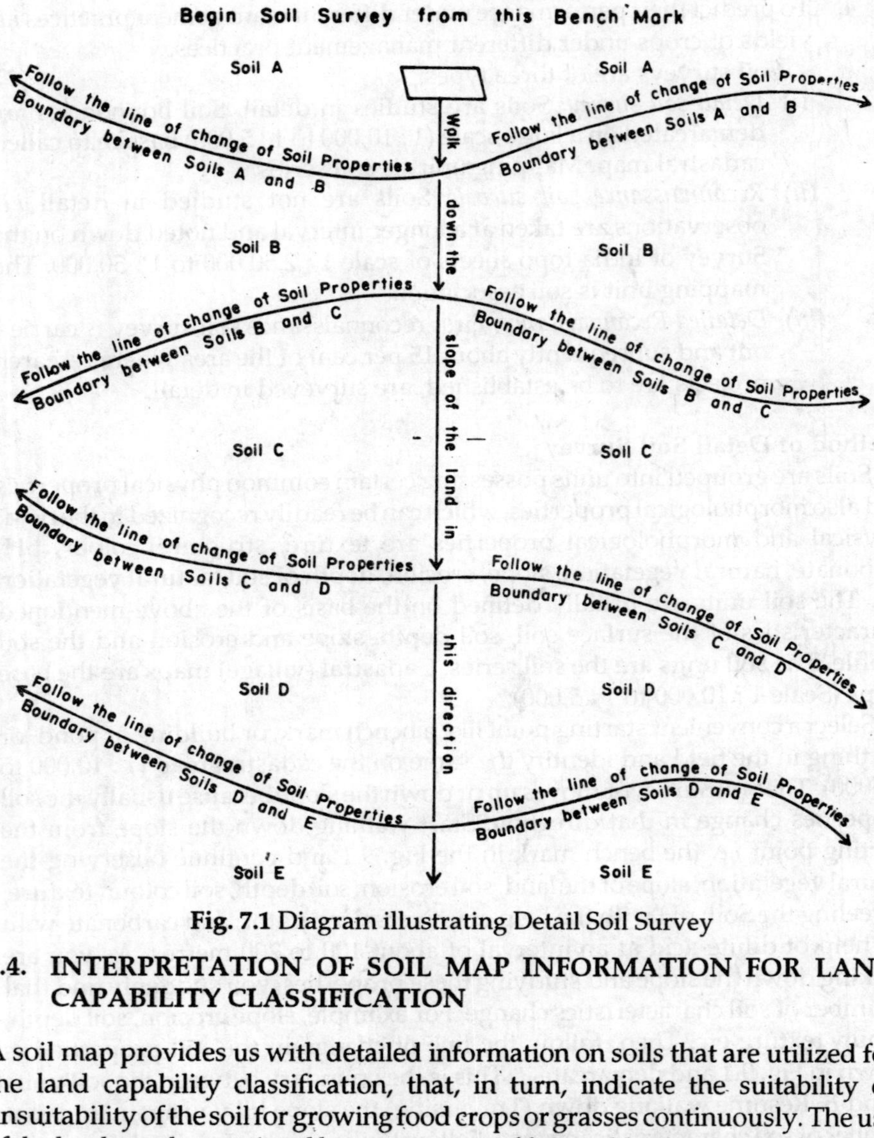

Fig. 7.1 Diagram illustrating Detail Soil Survey

9.4. INTERPRETATION OF SOIL MAP INFORMATION FOR LAND CAPABILITY CLASSIFICATION

A soil map provides us with detailed information on soils that are utilized for the land capability classification, that, in turn, indicate the suitability or unsuitability of the soil for growing food crops or grasses continuously. The use of the land may be restricted by one or more of factors like the nature of the soil, the land form or the climate, if these are unfavourable for growing crops. Land capability classification emphasises that to get the maximum possible return from the land without any loss in its productivity, it must be used only for what it is best suited, and adequate precautions must be simultaneously taken to prevent any damage of the land or decrease in its productivity.

Land capability classification is an interpretive grouping of soils based on inherent soil characteristics, external land features and environmental factors that may restrict the use of the land for growing varieties of crops. The soil survey of an area is undertaken to establish the soil series of that area which has a definite range of morphological, physical, chemical and mineralogical properties, each of which influences land productivity. Morphological properties include the depth of the soil, and the presence or absence of impermeable layers like clay pan, fragi pan. The soil must be deep enough to encourage the development of a deeper root system. Impermeable layers like clay pan, and fragi pan prevent the root system from going deeper into the soil. Important physical properties include soil texture, structure, permeability, and colour. The nutrient and moisture retention capacity of the soil depends mainly on the soil texture. The soils structure and texture effects the circulation of air and water through it. Enough rain water must percolate into the soil to reduce the volume of water running down the slope of the land and causing soil erosion. The chemical properties of soil include soil reaction, cation exchange capacity and total and available nutrient content. Soil reaction influences the availability of nutrients. Soils of a high cation exchange capacity can store more nutrients than the soils of a low cation exchange capacity. Fertility of soils is their capacity to supply nutrients in the soluble and available form to crops. The mineralogical properties of soil include the nature of the parent material and primary and secondary minerals present. If the parent material is rich in basic elements and plant nutrients, the resultant soil is expected to be rich in plant nutrients and alkaline in reaction. Fertile soils with a high moisture retension capacity usually contain easily decomposable ferromagnesian minerals and the smectite group of clay minerals. Unfavourable landscape and climatic features may severely restrict the use of land. If the land slope increases, the soil erosion also increases, which, in turn, decreases the productivity of the land. Many varieties of crops cannot be grown on the land throughout the year in the arid region. Other features like soil salinity and alkalinity, wetness, stoniness and rockiness may also limit the use of the land.

We need the following two kinds of information for land capability classification (*i*) the susceptibility of the soil to various factors that cause soil damage and decrease in its productivity and (*ii*) its potentiality for crop production. We get the first group of information from the soil map but it is not so easy to get the second group, which can only be obtained by growing crops on soils. A soil with capacity to grow varieties of crops for giving high yields, shall qualify to be placed in a better class than another soil which is a little limited in this respect.

The most important characteristic of class I land is that it must be suitable for growing varieties of crops using ordinary good farming practices like use of good manures and fertilizers, crop rotation etc. Paddy is the most important crop in India. It is perferably grown on lowland clayey soils. Low earthen embankments called bunds are to be erected around the filed which must be puddled to destroy the soil structure in order to retain water within the field for paddy cultivation. So once paddy is cultivated in the field, the field cannot be easily prepared for growing other arable crops like wheat.

Hence a paddy field which gives very high yield of paddy is not placed in class I land, because special farming practices are required for paddy cultivation, and this land not suitable for growing varieties of crops. Thus paddy lands are usually placed in class III land or at the most, in class II land but never in class I land.

Lands are first tentatively placed in different land capability groups on the basis of slope of the land, erosion and depth of soil which are considered individually as mentioned below:

Slope class	Slope range%	Description	Land Capability Classification		
			Red soil moderate depth	Red soil deep	Black soil
A	0 - 01	Nearly level	I	I	II
B	1 - 03	Very gently sloping	II	II	II & II
C	3 - 05	Gently sloping II	II	II	IV
D	5 - 10	Moderately sloping	III	II	VI & VII
E	10 - 15	Strongly sloping	IV	III	
F	15 - 25	Moderately steep	VI	IV	
G	25 - 33	Steep	VII	VI	
H	33 - 50	Very steep	VIII	VII	
I	>50			VIII	

Lands belonging to slope class *A* is put in class I lands in black, red and allevial soil, provided other Factors are favourable. As the black soils are extremely susceptible to erosion, so lands belonging to slope *B* and *C* are put in II, III and IV respectively.

Cultivation of lands of more than 5-10 per cent slope is discouraged, and due to their increasing degree of susceptibility to erosion therefore put in classes VI and VII. Black soils of more than 10 per cent slope are usually not found. As red soils are usually resistant to erosion, the moderately deep and deep red soils can be regularly cultivated for raising grain crops up to *D* (5 to 10 per cent slope) and *E* (10-15 *E* slope) slope class respectively. Lands belonging to *D* and *E* slope classes belong to class III lands in moderately deep and deep red soils respectively. Lands belonging to E & F slope classes are put in class IV in red soil of moderate and good depths respectively as they can be occasionally cultivated for growing grain crop. Lands belonging to slope classes *F* and *G* and immoderately deep red soil and lands belonging to *G* and *H* slope classes in deep red soil are put in class IV and VII respectively.

Classification of effect of erosion on land capability.

Erosion symbol	Erosion class	Description	Capability class
e_1	Slight erosion	—Mild sheet erosion	I, II
e_2	Moderate erosion	—Sheet & rill erosion	III
e_3	Severe erosion	—Excessive sheet & rill erosion exposing sub-soil and small gullies are formed	IV
e_4	Very severe erosion	—Extensive gully formation	VI, VII

Classification on the basis of effective soil depth.

Depth range cm	*Depth range inches*	*Depth class symbol*	*Land capability classification*
> 90 cm	>36 inches	d_5	I
45 - 90 cm	18 - 36 inches	d_4	II
22.5 - 45 cm	9 - 18 inches	d_3	III
7.5 - 22.5 cm	3 - 9 inches	d_2	IV
< 7.5 cm	< 3 inches	d_1	VI, VII

The effect of climate on land capability classification may be considered only when all other factors are favourable as shown in Table 9.1.

TABLE 9.1

Nature of climate	*Land capability classification*
1. Humid climate. Rainfall is evenly distributed	can start from class I
2. Humid climate. Dry spells occasionally occur	can start from class II
3. Crop yields are frequently reduced by drought in sub-humid climate	can start from class II to III
4. Semi arid Climate	can start from class III to IV
5. Arid climate	can start from class IV

Other factors which are to be considered for land capability classification include the nutrient and water retention capacity of soil, its salinity and alkalinity, the available nutrient status of the soil, wetness, stoniness, rockiness, toxicity of certain nutrients.

Each of the above factors considered separately for the tentative placement of lands in capability classes which are then superimposed on each another to get the final land capability classification.

9.5. CHARACTERISTICS OF LANDS BELONGING TO DIFFERENT CAPABILITY CLASSES

Class I: *Lands* are nearly level lands with deep, well drained soils of a satisfactory texture and structure. They are free from the harmful effects of salinity and alkalinity. They are subjected to not more than slight water or wind erosion. They are suitable for growing varieties of crops by adopting ordinary good farming practices. They respond to manuring and possess a high moisture and nutrient retention capacity. They are very fertile lands.

Class II: *Lands* can also regularly cultivated for growing grain crops although some easily applied special conservation practices like contour farming are to be followed for maintaining the productivity of lands. They are gently sloping lands with deep soil a satisfactory texture and structure. They are slightly affected by a salinity, alkalinity, wetness erosion etc. which can be easily corrected. They have poor quantity of available nutrients.

Class III: *Lands* can also be regular cultivated for raising grain crops by following some intensive soil conservation practices like terracing. They are moderately sloping lands with a moderately deep soil. They are subjected to frequent wetness. They are more severely affected by salinity and alkalinity and soil erosion than class II lands. They have a poor nutrient content and water retention capacity.

Class IV: *Lands* are used to cultivate grain crops once in four years. These lands should be left under grasses during rest of the time. They are strongly sloping lands with shallow soils.
They are severely affected by soil salinity and alkalinity, soil erosion and wetness. Class IV lands usually possess sandy or concretionery soil of a poor moisture retention capacity.

Class V: *Lands* are nearly level lands which are not subjected to more than a slight erosion. Cultivation of grain crops on this land is not possible due to some permanent obstruction like rockiness, stoniness. These lands are best suited to growing grasses and forest trees.

Class VI: *Lands* are even limited for growing grasses or forest trees. The limitations of class VI land include steep land of shallow soils which are highly susceptible to erosion they are usually dry gravelly/ sandy/loamy soils.

Class VII: *Lands* are severely limited even for growing grasses and forest trees. They are steep rough stony lands of extremely shallow soil due to past erosion. Class VII lands might have been extremely damaged by salinity and alkalinity.

Class VIII: *Lands* are not suitable even for growing grasses and forest trees because all the soil has been eroded away. So they are steep rough stony mountains.

9.6. INTERPRETATION OF SOIL MAP INFORMATIONS FOR LAND IRRIGABILITY CLASSIFICATION

Land irrigability maps indicate the suitability of lands belonging to different irrigability classes for irrigation farming, the precautions that have to be taken for growing different crops on lands and the necessity of applying soil amendments like gypsum or lime to irrigated lands. The classification of land according to its irrigability is based on inherent soil characteristics like soil depth, presence of impermocable layer (clay pan, fragi pan) below the surface of the land, soil texture and structure, infiltration permeability of soils, available moisture characteristics, the cation exchange capacity of soil, the soluble salt content and pH of soils and external land features like slope, erosion stoniness, rockiness, presence of saline and alkaline patches of lands.

Each of these factors indicate the suitability of different lands for irrigation farming. Deep root systems develop in deep soils of a favourable texture and structure. Crops develop a shallow root system if any impermeable layers like clay pan, fragi pan etc. are present near its surface. Sufficient water will not infiltrate in the soil if the texture and structure of the surface soil is impermeable. Similarly, drainage of the land depends on the permeability of the sub soil which, in turn, depends on the texture and structure of land. Soils should be of a medium texture, that allows water to percolate down the soil at a reasonable depth, yet stores enough water for crop growth. The soil texture mainly affects the water holding capacity of the soil, which is also modified by soil structure, and humus content and the nature of gravels and stones present in the soil. The available moisture retention capacity of medium textured soils like silty loam and very fine sandy loams is higher than that of the finer textured soils like clay and clay loam. A high cation exchange capacity of a soil is responsible for its high nutrient—storage and buffering capacity. Such soils are placed in a better land irrigibiality class than the soils of a low nutrient storage and buffering capacity, because they are highly productive and will not be damaged even if moderate saline water is applied to them. Saline lands qualify to be placed in the inferior land irrigability classes due to the higher cost of reclaiming these land than the non saline lands. Similarly, strongly saline and alkaline soils of very high pH and strongly acidic soils of very low pH are placed in land irrigability classes inferior to the neutral soils, due to the high cost of reclamation of saline alkali and acidic soils.

It is very convenient to evenly distribute irrigation water to nearly level land where slope percentage of the land does not exceed 0.2 per cent. The suitability of different lands for land irrigability classification should be determined by the following criteria.

Criteria for land irrigability classification

Land slope per cent	*Effective Soil depth c.m.*	*Soil texture*	*Permeability*	*Available W.H.C. %*	*ECe* mm hos/cm at 25°C*	*Exchangeable*	
						Sodium per cent	*Land irrigability class*
<0.2	100	V. Fine sandy Loam or silty loam or Silty clay loam	Medium Moderately	20%	4	10	I
0.2-2	75-100	Sandy Loam or Sandy clay or Clay loam	Slow or Moderately rapid	5-20	4-8	10-15	II
2-5	50-75	Loamy sand or clay	Slow or rapid	10-15	8-12	15-30	III
5-10	25-50	Clay or sand	Very slow or very rapid	5-10	12-16	30-50	IV
10	25	Silty clay or very coarse sand	Very slow or very rapid	5	16	50	V

*ECe-Electrical of saturation Extract in milliohms/cm at 25°C

Soil permeability classification.

Permeability class	*Rate in cm/hour*
Very slow	0.13
Slow	0.13 to 0.5
Mod. slow	0.5 to 2.0
Moderate	2 to 5
Mod. rapid	5 to 13
Rapid	13 to 25
Very rapid	25

9.7. CHARACTERISTICS OF LANDS BELONGING TO DIFFERENT IRRIGABILITY CLASSES

Class I *Lands* are excellent lands with no limitations. They are deep level lands where the land slope does not exceed 0.2 per cent. They are moderately permeable to water. The soil texture varies from very fine sandy loam to silty loam or silty clay loams which are not adversely affected by soil salinity and alkalinity or erosion. They possess a high available water holding capacity.

Class II *Lands* are slightly limited for irrigation. They may be very gently sloping land of deep soil which possess fairly high available water holding capacity. The soil texture may be sandy loam, sandy clay loam or clay loam which are slightly affected by soil salinity or alkalinity. Permeability may be moderately slow or moderately rapid.

Class III *Lands* are moderately limited for irrigation. They are gently sloping lands with a fairly deep soil which is moderately affected by soil salinity and alkalinity wetness or droughtiness. They possess a low to moderate available water holding capacity.

Class IV *Lands* are severely limited for irrigation. They are moderately sloping lands with shallow soils with low available water holding capacity. they are severely affected by soil salinity, alkalinity, wetness or droughtiness.

Class V *Lands* are not suitable for irrigation because they are strongly sloping and have very shallow soil with a low available water holding capacity. They are very severely affected by soil salinity and alkalinity.

9.8. PRINCIPLES OF LAND UTILIZATION

Land utilization comprises of surface utilization of all developed, undeveloped or vacant land at a specific point, at a given time and space. The land on the surface of the earth, has some utility. Land utilization at any place is governed by the undermentioned principles:

(*i*) The principle of maximization.
(*ii*) The equimarginal principles.
(*iii*) The principle of diminishing marginal rate of substitution.
(*iv*) The principle of diminishing marginal productivity.
(*v*) The principle of specialization and comparative advantages.
(*vi*) The principle of first choice.

(*vii*) The principle of mapping and search for alternative better utilization of the land.

(*viii*) The optimization of the quality of the land in relation to the environment.

(*ix*) The Principle of increasing the efficiency and capability of land.

(*x*) The fixation of cropping systems around human habitation for least effort and less cost travel.

(*xi*) The conservation of the land for future judicious use.

These principles are being elaborated herewith.

The Principle of Maximization

The principle of maximization points out that the user of the land tend to maximize his value returns because he is motivated by a desire to make the best utilization of as many resources as possible that are lying at his disposal. He always tries to use his land to the maximum possible extent subject to the abundance or scarcity of other factors of production e.g. capital, labour, irrigation and other facilities. For example if a huge amount of capital has accumulated through saving and thrift in a densely populated area where cheap labour and irrigation and other facilities are available but only a small fraction of land is available for agricultural use, then this small area would be intensively cultivated. Consequently the yield per hectare would be very high. On the other hand, much land may be available in a sparsely populated undeveloped area where capital, labour and irrigation are severely restricted, the land would be extensively cultivated. Thus under both the condition, the user of land attempts to maximize his value return.

The Equimarginal Principles

Every piece of land has potentialities for many kinds of use; e.g. the land may be utilized for growing jute, paddy or ladies finger near Calcutta. Inputs are assigned to every kind of use in proportion to return expected from them. As ladies finger fetches good return from Calcutta market, so more input is assigned to it. The transfer of input from one use to another shall equal the loss in return from the rest of uses.

The Principle of Diminishing Marginal Rate of Substitution

The increasing amount of production and consumption, the increment of satisfaction for any type of land use decreases. It is valid when two goods are substituted for each other. The smaller quantities of greater good are necessary to compensate for the substitution of another goods. As for example the use of the land near Calcutta for production of ladies finger is less in terms of production of rice as the farmer substitutes the former for the later because the former is a perishable commodity. Every-day, it is to be harvested and disposed off.

The Principle of Diminishing Marginal Productivity

According to Malthus, the rate of increase in food production lags behind the rate of increase in population. Usually it is not possible to get the same proportion of return consistently with the increasing rate of substitution since the cost of labour and inputs also increases. Rate of substitution since the cost of labour and inputs also increases.

The Principle of Specializaion and Comparative Advantage

No crop can be grown in all the agroclimatic zones of India with the same degree of return. For example cotton is being grown most profitably in black soils under semi arid alimate of Maharashtra, Gujarat, Western parts of Madhya Pradesh and smaller areas in Andhra Pradesh, Tamil Nadu and Karnataka. It cannot be profitably grown in the alluvial or other soils under the humid climate prevailing in the north eastern India and the eastern India, otherwise the cotton crop would suffer from severe attack on insect paste. Measures adopted to control insect pests would increase the cost of production per hectare and reduce the return per hectare. Jute is most profitable grown in the alluvial soils of West Bengal, Assam and Bihar under the humid climate.

The Principle of First Choice

Usually a limited amount of land is available for growing specific crops. For example vegitables are cultivated in well drained soils of medium texture located on a little elevated ground. Paddy is grown on lowland soils of heavier texture. Ladies finger may be grown on lowland soils of heavier texture because it can tolerate waterlogging to some extent as has been observed near Calcutta.

The Principle of Mapping and Search for are Alternative Better Utilization of the Land

Land use maps are produce through field to field survey and analysed. As the land is the base of all types of human decisions, so the search for its alternative utilisation involves:

(*a*) The land is the free gift of the nature to the man for the sustenance of life.

(*b*) The productive capacity of each land must be increased.

(*c*) Each land have fixed boundary according to human needs and physical resistance factor of the nature.

(*d*) Each land has owner under certain customary rights.

(*e*) Each piece of land has some utility e.g. agriculture industry, recreation etc.

(*f*) Each piece of land used by the man has some economic rent.

(*g*) Land use is the product of the demand set by the people of any race, religion and income.

(*h*) Hence with the concept of demand and supply under certain input-output ratio, there is competition between different types of land use. This automatically paves the way for better alternative use of land.

Optimization of the Quality of the Land use in Relation to the Environment

Apple can be grown in the temperate climate of Himachal Pradesh, Kashmir and Arunachal Pradesh. Banana is grown in cool humid climate of eastern, southern and western India. These fruits should be grown under the climatic condition most suitable to them in order to attain their maximum production.

Fixation of Cropping System Around Human Habitation for Least Effort and Less Cost Travel

Land near big cities are utilized for growing vegetable if irrigation water is available because they can be conveniently disposed of in the city market.

Conservation of the Land for the Future Judicious Use

Conservation is the reduction in the process of the rate of loss of land productivity *i.e.* the land must be protected from the hazards of erosion, waterlogging and salinity and alkalinity. Crops must be protected from the attacks of insect pests and diseases. Requisite amounts of manures and fertilizers should be applied to fields in order to build up the soil fertility. Crops should be irrigated and intercultured whenever necessary. Attempts should be made to consolidate the fragmented land holdings to one place if it is possible.

SELECTED READINGS

1. Arakari, H. R. & Donahue, R. L. (1985), *Principles of Soil and Water Conservation and Water Management*, Oxford & IBH Publishing Co., 66 Janpath, New Delhi-1
2. Anonymous (1972), *Soil Survey Mannual*. I.A.R.I., New Delhi.
3. Brady, N. C. (1984), *Nature and Properties of Soil*, Macmillan Publishing Co., New York.
4. Buol, S. W. *et. al.* (1973), *Soil Genesis and Classification*. Oxford & IBH Publishing Co., 66, Janpath, New Delhi.
5. Donahue, R. L. (1964), Soils, *An Introduction to Soils and Plant Growth*. Prentice Hall, Engle wood cliff, New Jersay 076632.
6. Foth, H. D. (1984), *Fundamental of Soil Science*, John Wiley & Sons.
7. Mandal, R. B. (1982), *Land Utilization*: Theory and Practice. Concept Publishing Company, H-13 Bali Nagar, New Delhi-110015.
8. Soil Survey Staff (1966), *Soil Survey Manual*, U.S.D.A. Hand Book No. 18 Oxford & IBH Publishing Co., 66 Janpath, New Delhi.

10
SOILS OF INDIA

The Indian subcontinent extends from 8° to 37° North latitudes and from 69° to 93° East longitudes and covers an area of 32, 74, 90 square kilometers (327.46 million hectares). It has the monsoon type of climate, there being two distinct periods of heavy annual rainfall. The south-west monsoon is active during June to September and the north-east monsoon during the winter. The rainfall, is however, controlled by the relief. The Himalayas and the Western Ghat mountain ranges lying across the path of the south-west monsoon winds are responsible for precipitating their moisture as rain on the wind-ward side. The Himalayas also deflect the south-west monsoon towards upper India. Consequently, the area extending from Assam to Punjab receives goods rainfall. Rajasthan does not experience good rainfall because the south-west monsoon winds pass unobstructed over it as the Aravallis lie along its path. Another dry zone extends from the interior of Tamil Nadu to central Gujarat.

The Indian subcontinent is characterized by a great diversity of parent rocks (Section 2.14 to 2.29) which have weathered under different climate, vegetation and topography to form soils with different characteristics in different regions of the Indian Union.

J.A. Voelekar (1893) and J.W. Leather (1893) recognised four major soil groups; the Indo-Gangetic Alluvium, Black soil, Red soil and Laterite soil. There are four other groups of soils which include forest and Hill soils, the Desert and Semi-Desert soils, the Saline and Alkali soils and Peaty and Marshy soils.

10.1. MAJOR GROUPS OF INDIA

Soils of India have been broadly classified into the major groups: (*i*) Alluvial soils, (*ii*) Black soils, (*iii*) Red soils, (*iv*) Laterite soil, (*v*) Mountain and hill soil, (*vi*) Arid and Desert soil, (*vii*) Saline and Alkali soil, and (*viii*) Peaty and other organic soil.

Alluvial Soils

These soils have been found in Major parts of Punjab, Haryana, U.P., Bengal, Assam, North Bihar and small areas in the other States of the Indian Union. The main features of alluvial soils have been derived as silt deposition laid down by the Indian river systems like the Indus, the Ganges, the Brahmaputra and the rivers like Narmada, Tapti, Mahanadi, Godavari, Krishna and Cauvery. These rivers carry the products of weathering of rocks constituting the mountains and deposit them along their path as they flow down the plain land towards the sea. Most of the alluvial soils do not possess a well developed horizons. Geologically alluvial soils have been divided into two broad groups, namely old and new alluvium, which are also called Bangar and Khadar respectively in North India. Some of the old alluvial soils possess well developed horizons. The surface soil of these old alluvial soil are reddish brown of grey brown sandy loam or loamy in texture, and the sub-soil contain more clay than the surface soil. The texture of the new alluvium is very coarse on the river banks and very fine in the low lying marshy lands. Alluvial soils usually contain lime nodules intermixed with soil at different depths. Frequently drainage is restricted and soluble salts accumulate in the soils of more arid regions *i.e.* Uttar Pradesh, Punjab and Haryana. Alluvial soils are usually deficient in humus and nitrogen, occasionally in phosphorus but not in potash and lime.

Black Soils

Almost the whole of Maharashtra, the western half of Madhya Pradesh, the southern parts of Gujarat and some areas in Andhra Pradesh, Karnataka and Tamil Nadu and Bundi and Tonk districts of Rajasthan and Bundelkhand division of U.P. are covered by black soils which have been divided into three broad groups *i.e.* (*i*) shallow black soil of a depth of 30-50 cms, (*ii*) medium black soils of depth of 50-120 cm, and (*iii*) deep black soil of a depth of more than 120 cms. In many places, black soils often occur in close proximity to red soils. In the vicinity of the hills, the red soils merge interceptibly into the black soil.

Shallow black soils of a texture that varies from silty loam to loam and colour that varies from dark yellowish brown to dark brown, have formed from basalts of the Deccan trap. Parent materials of medium black soils include basaltic trap, Dharwar Schist, basic granite gneisses, hornblende and chlorite schists, whereas deep black soils consist of basaltic trap. The soil texture in both cases varies from silty clay to clay. Both contain 40 to 60 per cent or more clay. Gypsum accumulates in association with lime in some of the soil profiles.

The colour of these soils varies from deep black to light black. The exact cause of the black colour of these soils is unknown. Probably it is due to the presence of titeniferous magnetite, organic compounds of iron and aluminium, humus and colloidal hydrated double iron and aluminium silicate.

Black soils are rich in the smectite group of clay minerals and therefore possess, a high cation exchange capacity (40 to 60 meg/100 gms) and high base status. They are very sticky and plastic when wet and form very wide and deep cracks when dry. Black soils are very rich in iron, lime, magnesia and alumina. They are deficient in organic matter, nitrogen and phosphorus but not in potash. Soluble salts alkaline in reaction accumulate in medium and deep black soils. Black soils, especially deeper ones, can retain a high amount of moisture and are fairly fertile, giving reasonably good yields of crops despite continued cultivation without proper manuring. These soils are well suited for cultivation of cotton, cereals and oil seeds. Deep black soils are often called black cotton soils because cotton is most extensively grown on them.

Red Soils

Red soils cover almost the whole of Tamil Nadu, Karnataka, Andhra Pradesh, Orissa, Nagaland, Manipur, Mizoram and Tripura, eastern and southern parts of Madhya Pradesh, Bundelkhand division and Mirzapur district in Uttar Pradesh, Chhota Nagpur division and Santhal Parganas districts in Bihar, parts of West Bengal, Arunachal Pradesh and Meghalaya, Karbianglong districts of Assam, Thane and Kolaba districts of Maharashtra and also parts of Rajasthan.

Red soils include (*i*) red loam soils, (*ii*) red and yellow soils, and (*iii*) red sandy soils. Red soils have been formed from varieties of rocks which include granite, gneiss, schist, sandstone, hornblende, shales etc. under a forest of broad leaved trees in warm humid climate. As basic elements have been leached from most of these soils, they become acidic in reaction. The soil organic matter has decomposed and they are poor in humus, and nitrogen, phosphorus and lime. Whenever water soluble phosphates are added to these soils, they react with ferric and aluminium ions and their hydrous oxides and are precipitated as insoluble ferric and aluminium hydroxy phosphate. Kaolinite and hydrous oxides of iron and aluminium are the dominant clay minerals in red soils, responsible for the excellent physical condition of red soils. A number of kharif crops are thus grown on these soils, even though they cannot retain enough moisture for their growth. These soils differ greatly in fertility and depth.

Laterite Soils

Laterite soils occur in small patches in almost all the states of north-eastern, eastern, southern and western India. They are formed from several kinds of rocks like basalt under a hot, humid climate when the basic elements and silica have been washed down and iron is precipitated and oxidized. Whenever drainage is restricted, a soft deposit of iron oxide occurs at or near the water table, which hardens when it dries. In some places like Malabar, this is cut in the form of bricks and used as a building material. The word Laterite has originated from the Latin word "Later" which means bricks. This possesses an indurated honey combed mass of iron oxides.

Laterite soils are usually shallow and gravelly at higher lands, but are very deep loam to clay soils in the valleys where good paddy crops are produced on them. Laterite soils are deficient in basic elements like lime, magnesia, potash and nitrogen and available phosphorus. These soils are not sticky and plastic when wet. They are in excellent physical condition due to the presence of hydrous oxides of iron and aluminium and kaolinite as the dominating clay minerals.

Forest and Hill Soils

Forests cover nearly 14 per cent of the total area of this country. They mainly occur in Himachal Pradesh, Uttar Pradesh, West Bengal, Assam, Arunachal

Pradesh, Meghalaya, Mizoram, Nagaland, Tripura, Orissa, Madhya Pradesh, Karnataka, Kerala and Tamil Nadu.

Forest soils are constantly enriched by fallen leaves. Hence they are usually rich in organic matter and total nitrogen. Forest soils developed from acidic parent material, are poor in basic elements and acidic in reaction while some developed from basic parent material, are fairly rich in basic elements, neutral to slightly alkaline in reaction and are fairly fertile.

Desert Soils

Desert soils cover major areas in Rajasthan. They are also found in Haryana, Gujarat and Uttar Pradesh. These soils consist mainly of sands which have been blown mainly from coastal areas. Soil fertility improves in the eastern and north-eastern direction. Some of these soils are saline or alkaline and contain calcium carbonate. The phosphate content of the aeolian sand may compare favourably with some of the alluvial soils. They contain low amounts of organic matter and total nitrogen and are alkaline in reaction. Some of them contain nitrates. These soils may be cultivated if irrigation water is available.

Saline and Alkali Soils

Saline and alkali soils of India occur mainly in the arid and semi arid regions of Uttar Pradesh, Haryana, Punjab, Rajasthan, Gujarat and Maharashtra soluble salts accumulate on the surface, of the saline soils as salt efflorescences which are mainly chlorides and sulphate of sodium, calcium and magnesium. Salts move upwards, along with the capillary water and reach the surface in the case of saline soil in the areas where canal or well irrigation is practised and also near the sea coast. Saline soil or solonchek contain neutral sodium and other salts, whereas alkaline soils contain exchangeable sodium and sodium carbonates, (e.g. in Uttar Pradesh, Punjab, and Haryana) which are impermeable to water in contrast to saline soils, which are permeable to water.

Peaty and Marshy Soils

Large amounts of organic matter have accumulated in low lying areas. This organic matter has decomposed in the presence of excess moisture to form the Peaty and Marshy soils. Marshy soils are blue due to their ferrous iron content. They contain varying amounts of organic matter. Marshy soils of this type are found in the coastal areas of Orissa, West Bengal and Tamil Nadu, the central portions of north Bihar and the Almora district of Uttar Pradesh.

Saline peat soils called the Kari soils in Kerala, which occur in Ambalapuzha, Kunnathanad, Vaikom and Shertallai, generally remain submerged during the rainy season. These black soils are highly acidic in reaction, rich in organic matter (10 to 40 per cent) and sesquioxide and poor in phosphorus. These soils contain high amounts of ferrous and aluminium sulphate which are toxic for plant growth.

10.2. SOILS OF BIHAR

Alluvial soils have covered the area between the Himalayas and the Ganges in

north Bihar and between the Ganges and the Chota Nagpur plateau on the south Bihar. Properties of the alluvial soils of Bihar vary widely because they were formed from different parent materials which were deposited by the tributaries of the Ganges. Alluvial soils of Bihar have been classified into four broad groups on the basis of the differences in the composition of the parent alluvium, climate and vegetation: *(i)* Calcareous Alluvial Soils of North Bihar, *(ii)* Non-Calcareous Alluvial Soils of North Bihar, *(iii)* Heavy Clayey Alluvial Soils of Southern Bihar, and *(iv)* Sone Alluvial Soils of South Bihar.

A belt of calcareous alluvial soils extends from Gorakhpur district in Uttar Pradesh in the west to the Darbhanga district in the east. These soils partially cover the districts of Champaran, Saran, Muzaffarpur and Darbhanga. They are whitish-grey to dark grey sandy loam to silty loam soils, containing high amounts of lime and are strongly alkaline in reaction. They contain low amounts of total nitrogen, available phosphorus and potash. They are saline and saline-alkali soils and are strongly alkaline in reaction in the more arid regions of Saran district.

The non-calcareous soils cover the rest of the area in the north Bihar. They are grey brown to dark brown sandy loam to clay loam soils of good physical condition. They are rich in potash but deficient in phosphorus and are alkaline in reaction except in the eastern parts of Purnea district where they are acidic in reaction (pH 5.0 to 6.0), rich in nitrogen but deficient in phosphorus and potash.

Heavy clayey soils cover part of Patna, Gaya and Sahabad districts in the south Bihar. They are dark grey at the surface and contain high amounts of clay almost uniformly distributed throughout the profile. They are poorly drained and contain a fair amount of total nitrogen and available phosphorus and potash and are usually neutral in reaction.

Some alluvial soils occur in the Valley of the river Sone and other Vindhyan rivers in the Sahabad district. They have been divided into three broad groups *i.e.* sand, sandy clay and clay. They are not fertile and are alkaline in reaction.

Red and yellow soils which cover Chota Nagpur division and parts of Singhbhum and south Sahabad and Mongher districts are usually brownish-yellow to red in colour and lighter in texture, although heavier soils have also been found in some pockets. They are acidic in reaction. Greyish, neutral to slightly alkaline soils occur at the low land. Red and yellow soils contain low amounts of lime and magnesia and moderately amount of potash. They contain low amounts of available phosphorus because water soluble phosphorus have been precipitated as in soluble ferric and aluminium hydroxy phosphate.

Most of Singhbum district is covered by mixed red and black soil which comprises of dark brown loamy soil to black clayey soils and frequently contain lime. Black soils which occurs in pockets of the Singhbhum district are clayey in texture and alkaline in reaction. Laterite soils also occur in small pockets within the red and yellow soil areas, which contain ferruginous concretion which are compacted into blocks of laterite.

10.3. SOILS OF UTTAR PRADESH

Hilly regions of Uttar Pradesh comprising Almora, Tehri Garhwal districts, major area of Dehra Dun district and parts of Garhwal, Uttarkashi and Chamoli districts are covered by brown hill soils which soils which originated from sandstone and shale under forest of Quercus sp and Pinus longifolia and grasses in humid temperate climate. A moderate amount of organic matter is present in the surface soil. B horizon is brown. Sandy soils originated at a slope gradient of about 45° or more while loamy and clayey soils originated in less shaped lands.

Parts of Nainital, Garhwal and Dehra Dun districts are covered by dark grey sandy loam to silty loam Tarai soils which are usually rich in organic matter and acidic in reaction. Some clay has been illuviated in the *B* horizon. Haldwani, where the deep water table occurs, is covered by loamy soils containing gravel. But sandy soils occur at Kichha and Kashipur where the water table occurs at a depth of 4 to 10 feet. Grey loamy soils occur in Dehra Dun while dark grey slightly heavier soils occur at Doiwala.

Alluvial soils of Uttar Pradesh have been grouped as: (*i*) Alluvial soils of western and north-western Uttar Pradesh of a lighter texture (*ii*) Alluvial soils of central Uttar Pradesh of an intermediate texture; and (*iii*) Alluvial soils of eastern Uttar Pradesh of a heavier texture.

Alluvial soils of western and north-western Uttar Pradesh: Saharanpur, Muzaffarnagar and Meerut districts are covered by deep reddish brown to yellowish brown fertile sandy loam soils where the water table occurs at a depth of 8 to 15 feets. Bijnor, Moradabad, Rampur, Bareilly and Pilibhit are covered by light yellow to red coloured heavy loam soils. They are usually slightly alkaline in reaction except at the east of Piliphit where they are slightly acidic in reaction.

Alluvial soils of central Uttar Pradesh: Ukheri and Sitapur districts are covered by sandy loam soils which are slightly acidic in reaction, except in the eastern part of Kheri district and western part of Sitapur district where they are slightly alkaline in reaction. Sandy loam soils of Hardwar district are slightly acidic at the western part and slightly alkaline at the eastern part. Barabanki and Lucknow districts are covered by single grained slightly alkaline sandy loam soils.

Alluvial soils of eastern Uttar Pradesh: There are two district type of soils locally known as Bhat and Bangar. A third type called Dhub occurs near the river banks which are liable to inundation. Single grained sandy loam to clay loam Bhat soils occur in low lying areas where the water table occurs at a depth of 1.5-4.5 metres. These soils of high moisture retention capacity contain a high percentage of lime and are slightly to moderately alkaline in reaction. Bangar soils occur at relatively elevated areas where the water table occurs at a depth of 15-30 feet below the surface. These sandy loam to loam soils of pH 6.0 to 8 contain less lime than the Bhat soil. Sugarcane is grown extensively on these soil under irrigated conditions. Loam soils possessing a high water holding capacity and a zone of accumulation of calcium carbonate are called Dhankar in Varanasi district, which are suitable for paddy cultivation.

Dark clayey soils which resembles the black cotton soils of central India, occur in low lying, basin shaped areas in both old and new alluvium. The rivers which drained the trap rocks of Bundelkhand brought down the products of decomposition of basalt and deposited this basaltic alluvium at these basin shaped areas, where these dark clayey alluvial soils were formed. There is little textural variation in this uniform dark grey to black coloured soil profile. The organic matter content of these soils decreases form 1.14 per cent at the surface to about 0.69 per cent at a depth of 1.3 to 1.8 metres. They are neutral to moderately alkaline in reaction. Calcium forms 75 per cent of the bases, and magnesium forms 20 per cent of the basics. The base exchange capacity of these soils is about 70 milliequivalent per 100 gms.

The Bundelkhand region comprising the districts of Jhansi, Jalaun, Hamirpur and Banda is covered by black and red soils. Black soils are very sticky and plastic when wet but when they dry, wide and deep creacks appear on their surface. Crops can be grown on these soils without applying irrigation water because they can retain a high quantity of moisture. They are locally known as MAR and KABAR. Mar is less clayey than Kabar. Red soils include two types called PARWA AND RAKAR. Parwa is a brownish grey loam to sandy or clay loam, a fairly good soil which responds to manuring and irrigation. Rakar is the eroded soil which is commonly found at higher elevation. Red loam soils have developed on Vindhyan parent materials in Benaras and Mirzapur districts.

About 13 lakh hectares of land are affected by salinity and alkalinity in Uttar Pradesh; over 40,000 hectares of land are affected by salinity and alkalinity in each of the districts of Aligarh, Mainpuri, Etawah, Kanpur, Unnao and Pratapgarh; 20,000 to 40,000 hectares of land are affected by salinity and alkalinity in each of the districts of Bulandshahar, Farrukhabad, Hardoi, Lucknow, Rae Bareili, Fatehpur, Allahabad, Sultanpur, Jaunpur and Azamgarh. The characteristics of Saline, Saline-alkali and alkali soils of Uttar Pradesh are as follows:

Saline soils: The land surface is covered by extensive white or ash coloured salts. These lands occur near canals or rivers where the natural drainage is restricted. These loamy sand to loam soils have an open structure and no hardpan or kankar layer in the sub soil. Soluble salts are mainly chloride and sulphate of sodium and calcium. The soil pH is less than 8.5.

Saline-alkali soils: These lands occur in low lying pockets of the upland and possess a hard indurated clay or kankar pan through which water cannot pass. Sulphates and carbonates of sodium occur in the surface. The international drainage is very restricted if sodium carbonate is the predominant salt when the soils are very compact. But internal drainage can be easily established if sodium sulphate is the important salt.

Alkali soils: Clay to clay loam surface soils occur above the heavier indurated and columnar sub-soil. When wet, soils are very gummy and plastic. Usually a thick and highly cemented bed of bigger sized kankar is to be found in the sub soil which may be cemented like rocks. Water does not pecolate down the profile. The exchangeable sodium percentage of these soils is more than 15 and pH is more than 9.0, often more than 10.

10.4. SOILS OF PUNJAB

Soils of Gurdaspur developed in a hot humid climate. Clayey soils containing kaolinite occur near the Government Agricultural Farm where calcium carbonate is absent. The districts of Amritsar, Jullandhar, Ropar and Patiala and parts of Ludhiana have been covered by new alluvial soils which do not possess a well developed soil profile and are sandy loam to clay loam in texture. Soils of Amritsar, Jullandhar, Patiala and Ludhiana districts were formed in a hot semi arid to semi humid climate. Alluvial soils contain enough phosphorus and potash for crop growth but are deficient in humus and nitrogen. The Ferozepur district is covered by darker, coloured clayey old alluvial soils. Calcareous sierozemic soils occur in Sangrur, Bhatinda and parts of Ludhiana. These soils are also grouped under alluvial soils because of their alluvial region. These are light, coloured clayey soils which are neutral to slightly alkaline in reaction and deficient in phosphorus. 2 : 1 clay minerals dominate the clay mineralogy of these soils.

Almost all the districts of Punjab except perhaps Jullandhar and Ropar are adversely affected by soil salinity and alkalinity. All the salt incrusted lands are called Thur land and the saline efflorescence is called Thur, and consists mainly of sodium sulphate and sodium chloride. These soils are usually lighter in texture and alkaline in reaction, Bari soils are very difficulty to plough, because they are very hard and impermeable. When these soils are dried, a crust rich in clay is formed at the surface. Calcium carbonate occurs at a depth of about four feet. Bari soils are typical saline-alkali soils of Punjab because they contain high amounts of sodium sulphate and sodium chloride. Rakar is the typical non-saline alkali soil which possesses an upper, highly impervious layer and contains calcium carbonate in the sub soil.

10.5. SOILS OF HARYANA

Soils of Haryana were formed in hot humid climates (Ambala) to hot arid climates (Karnal, Hissar, Gurgaon, Rohtak). Ambala, Karnal, parts of Rohtak and Gurgaon districts are covered by new alluvial soil, while calcareous sierozemic soil which is of alluvial origin, occurs in Hissar, Jind and small areas in Rohtak and Gurgaon districts. The majority of alluvial soils are sandy loam or loam in texture, where genetic horizons have not been developed. Soluble salts are usually present in the alluvial soils, the lower horizons also contain lime kankar nodules. They are alkaline in reaction and contain adequate amounts of phosphorus and potash but lack organic matter and nitrogen. Calcareous sierozemic soils have developed from the alluvium brought by the Indus river system. These soils are covered by a mantle of blown sand, which combined with the arid climate, results in poor soil development. They are usually lighter, coloured deeper sandy loam, loam to clay loam soils. Clay and lime have been washed down the profile. A layer of calcium carbonate occurs below 150 cm. They are neutral to alkaline in reaction and are deficient in phosphorus.

Alluvial soils and also the calcareous sierozemic soils are adversely affected by soil salinity and alkalinity. These soils are locally known as Kallar, Reh and Usar. Excessive amounts of sulphate and chloride of calcium, magnesium and sodium and carbonate of sodium and potassium occur in the saline and alkali soils. Soluble salts accumulate in the semi-arid climate with an annual rainfall of less than 500 mm. Application of saline irrigation water on these soils results in the development of soil salinity, as has happened in large areas in Gurgaon, Rohtak and Hissar districts where the water table has come up 150 cms below the surface of lands.

Saline soils contain a high amount of sulphate and chloride of sodium and calcium, and calcium carbonate and are alkaline in reaction. They usually have a lighter texture and do not possess compact layers. Saline-alkali and alkali soils chiefly occur in the Karnal district where a calcareous pan is present that restricts the downward movement of water. Soils of an I.A.R.I. farm at the Karnal district are sandy loam in texture and contain about 1.14 per cent calcium carbonate. Soils of Patti in the Karnal district are clayey in texture and contain about 1.20 per cent calcium carbonate. The clay micelle is saturated by sodium ions (20 to 23 per cent). It is regarded as sodium-magnesium clay because magnesium ions increase to about 28 per cent of the exchangeable cations in the sub soil. Alkali soils of the Karnal district also contain toxic amounts of water soluble boron. Alkali soils are deficient in humus and nitrogen.

10.6. SOILS OF HIMACHAL PRADESH

Himachal Pradesh is covered by a belt of Brown Hill Soil from the Kangra to the Sirmur district. These soils have developed from varieties of parent material like sandstone, grey micaceous sandstone and shale. The surface soils are dark brown loam to silty clays, which are moderately rich in organic matter. These soils do not contain lime and are therefore slightly acidic to neutral in reaction. *B* horizon is fairly thick (50 to 80 cm) and of silty loam to clay loam in texture.

Chamba, Lahul-spiti, Kulu and Kinnaur districts are mostly covered by sub-mountain soils (Podzolic) which have developed from sandstone and shales under conditions of high rainfall which support coniferous vegetation. Dark brown sandy loam soils contain loose undercomposed organic matter up to a depth of 15 cms from the surface. Up to another 15 cm thickness the sub soil is lighter coloured, followed by layer below this brown to reddish brown compact sandy clay up to a depth of another 100 cms. These soils do not contain lime and are therefore acidic in reaction.

In general, soils of Himachal Pradesh contain good amounts of plant nutrients like nitrogen, phosphorus and potash. Soils become heavier in texture and their pH decreases with the increase in the altitude.

10.7. SOILS OF RAJASTHAN

Churu, Bikaner, Jodhpur, Barmer about three-fourths of Jaisalmer, the western half of Nagaur, the eastern half of Jalore and small areas in Jhunjhunu and Ganganagar districts are covered by deep, yellowish brown to pale brown coloured regosolic soils which are low in humus but contains a fair amount of weatherable minerals. Some soils are highly saline in the lower horizon. An appreciable amounts of gypsum is present in the lower horizons of some regosolic soils of Bikaner. The regosolic soils contain free calcium carbonate, soluble salts and are alkaline in reaction. Soils develop salinity, and alkalinity in some of the courses of ancient rivers. Alkali soils are in a poor physical condition.

One fourth of the land area of eastern Jaisalmer district is covered by lithosolic soil, which consists of a thin layer of gravel or sand over bed rocks. Certain xerophytic plants grow wherever there are sandy soils of a depth of about 10 to 15 cm, because these soils retain enough moisture for the growth of xerophytic plants. No soil profile develops, because thin sandy layer is continuously shifted from place to place.

The southern part of Jhunjhunu, Sikar and the eastern half of the Nagaur districts, Pali, Sirohi and one-third of the western part of the Udaipur districts, are covered by grey brown soils of alluvial origin. They consist of lighter coloured sandy loam surface soil and darker sub soil which is of a heavier texture than the surface soil. Calcium carbonate is uniformly distributed in the surface, soil calcium carbonate is also uniformly distributed throughout the profile. Therefore the calcium ion is the major cation absorbed by the clay micelle. Hence these soils are neutral to alkaline in reaction and contain low amounts of nitrogen and phosphorus though enough potash.

A belt of red and yellow soils cover parts of Ajmer, Bhilwara, Udaipur and Dungarpur districts. These soils as studied at Pisanganj in the Ajmer district, are sandy loam in texture, contain calcium carbonate and are alkaline in reaction.

About two-thirds of the Kotah district are covered by medium black soil. The rest of the Kotah district, Bundi, Chitorgarh and the Bhilwara districts are covered by mixed red and black soil. The black soils are fairly deep where some clay has been illuviated in the *B* horizon. These soils crack during the summer and are fairly well drained. The *B* horizon contains calcium carbonate. Grey brown to dark grey brown clay with an alkaline reaction occur at the surface. In a predominant red soil area, black soils occur in patches in the low lying area. Alternately in the predominantly black soil area, red soils have originated from granite, granitoid gneiss, while black soils have developed from basalt.

10.8. SOILS OF GUJARAT

Kheda, Ahmedabad, major areas in Surendranagar and a small area in Banaskantha and Vadodara districts are covered by grey-brown soils which were developed from alluvial parent material in the semi-arid climate of Gujarat.

The *B* horizon soil is heavier in texture than *A* horizon soil. Calcium carbonate is uniformly distributed in the soil profile. So calcium is the most important cation absorbed by the clay micelle. Hence the soils are neutral to alkaline in reaction. These soils, known as "GORADU" in Ahmedabad, Mehsana and Kaira districts are usually fertile and respond to manuring and irrigation. In the Vadodara district, soils developed from older deposits of alluvium are called Bhata. Both Gorat and Bhata soils are deep, poor in organic matter and nitrogen but fairly rich in phosphate and potash. Deep black soils (120 cms deep) occur in Surat, Bharauch, Vadodara and Bhavanagar districts. Medium black soils occur in Rajkot, Jamnagar and Bhavanagar districts. Black soils are dark coloured clayey soils. They possess a high moisture retension capacity. They contain abundant iron, fairly high quantities of lime, magnesia and alumina. They are poor in humus, nitrogen and phosphorus but not in potash. The smectite group of clay minerals dominate that these soils.

Saline soils that usually contain sodium chloride as the important salt, occur in the Rann of Kutch. They are black clayey to silty clayey soils. The sub soil is of a lighter texture than the surface soil. Gypsum is present in the soil profile. These soils are alkaline in reaction (pH 7.8 to 8.4).

10.9. SOILS OF MAHARASHTRA

Maharashtra is mainly covered by black soils which are of three kinds *i.e.* (*i*) shallow, (*ii*) medium; and (*iii*) deep black soils. Shallow black soils have developed from the basalt of the Deccan trap. They are dark yellowish brown silty loam to dark brown clays of a granular or weakly blocky structure and contain lime grains or nodules. Cracks appear on the surface during the summer. They are well drained and moderately alkaline in reaction. Shallow black soils occur in the districts of Nimar, Wardha and west of Nagpur.

Medium black soils have developed from a variety of rocks which include basaltic trap, basic granite, gneiss and hornblende schist which have weathered to form darker silty clay to clay soils, rich in basic elements, and organic matter and alkaline in reaction. They are fairly well drained and contain lime nodules and low amount of phosphorus. Medium black soils occur at Nagpur and Akola.

Deep (120 cms) black soils occur in the Vidarbha, Jalgaon, Dhule, Parbani and Ahmedanagar districts. These are the result of weathering of Basalt. They are silty clay to clay in texture, contain lime nodules and are moderately alkaline in reaction.

Laterite soils occur in Ratnagiri. They are rich in coarser material and plant nutrients except lime.

Darker, deep clayey coastal alluvium occurs along the west coast, which is frequently inundated with the saline sea water.

10.10. SOILS OF MDAHYA PRADESH

Shallow black oils have developed from the weathering of the Deccan trap in parts of Betul, Chindawara, Balaghat, Seoni and Mandla districts. They are dark yellowish brown silty loam to dark brown clay in texture and are alkaline in reaction. They are poor in humus, nitrogen and phosphorus but not in potash and lime.

Vast areas of northern, western and central Madhya Pradesh are covered by medium black soil (50 to 120 cms). These silty clay to clay soils are moderately rich in humus and lime.

Deep black soils (120 cms) occur in central Madhya Pradesh. They have developed from Basalt. They are rich in lime and alkaline in reaction. They are rich in 2 : 1 type of clay mineral.

Bastar, Chanda and parts of Durg districts are covered by red sandy soils which have developed from acidic parent material like granite, quartzite, sandstone etc. These soils are rich in coarse and medium sand. Usually kaolinite or illite is present. They are usually coated with red haematite or yellow limonite. They have a moderately deep *A* horizon of pale reddish brown to red coloured coarse sand to loamy sand. The *B* horizon is dark reddish brown loam or sandy clay loam.

Red and yellow soils occur in the eastern part of Madhya Pradesh. These gently sloping and upland soils have developed from granite gneiss and quartzite schist complexes. The fairly deep, yellow, brown and reddish coloured soils are coarse to medium in texture in the A horizon and finer in texture in the *B* horizon. These well drained acidic soils contain iron concretion.

10.11. SOILS OF KARNATAKA

Medium to deep black soils occur in Gulbarga, Bijapur, Belgaum, and parts of Dharwar, Raichur and the Chickmagalur districts. These dark greyish brown silty clay to black clay soils were fromed from varieties of rock sike basaltic trap and Dharwar schist. They are plastic and sticky when wet. Cracks appear on the surface of land when they dry. They are rich in the smectite group of clay minerals. They are moderately rich in humus and nitrogen and lime and magnesia and soluble salts. Lime nodules are present in the sub soil. These soils are alkaline in reaction.

Mixed red and black soils occur at Raichur, Dharwar, belgaum, Bellary and Chitradurga districts. They were formed from Granite and Granite-Gneiss containing lime and albite. Red soils usually occur at the elevated areas and Black soils occur at the bottom land. Fine clay and plant nutrients are washed down from red soils occurring at the elevated areas to black soils occurring at the bottom land. Consequently, red soils are coarser in texture and poorer in plant nutrients than the black soils. Black soils possess a high moisture retention capacity and are therefore more suitable for growing Rabi crops.

Red sandy soils occur at Mysore, Mandya, Bangalore, Kolar, Tumkur, Hassan, Chickmagalur, Shimoga, Dharwar and Bellary districts. These red, yellow to brown stony and coarse gravelly to clay loam soils (Black coloured clay loam soils also occur in Mandya and Mysore districts are open in texture, well drained, low in humus, total nitrogen, total and available phosphorus and lime.

Moderately well drained red sandy soils occur in the Bangalore district and are loamy sand to sandy loam in texture. They have originated from granite. A thick layer of ferruginous gravel and quartz occurs in the sub oil. These soils are poor in organic matter and acidic in reaction.

Moderately deep to deep gently sloping, yellowish red, dark red and brown coloured sandy clay loam and clay loam soils occur in the Bangalore district. They contain a fair amount of humus and are slightly acidic to alkaline in reaction. Loamy red soils are predominant in the plantation districts of Shimoga, Hassan and Kadur. Some of these shallow pale brown to greyish brown sandy loam soils occur at the uplands; some occur at the lowlands which contain enough potash, phosphorus and nitrogen for growing paddy.

Laterite soils occur in the western part of Shimoga, Bangalore, Kadur, Kolar and Mysore districts. Some of them are well drained brown clay loam soil occurring on hill tops. Some of them are shallow yellowish brown loam soil. Bases have been washed down from these soils and they are still not very acidic in reaction. They are poor in phosphorus.

Laterite soils of the North Kanara district are coarse, poor in lime and phosphorus but fairly high in humus, nitrogen and potash. Laterite soils of Managlore are fairly fertile but strongly acidic in reaction. The Laterite soils of Pattukkottai are also fairly fertile, but are almost neutral in reaction.

Alluvial soils which occur along the West coast in the north and south Kanara districts are reddish or greyish sand and are well drained. But poorly drained heavy-textured alluvial soils occur along the banks of the rivers Netravati and Gurpur.

10.12. SOILS OF KERALA

Laterite soils of Malabar have developed from basalt and gneiss under hot humid conditions. In laterite, the process of laterization *i.e.* decomposition of the mineral colloid to release iron and aluminium hydrous oxide or hydroxide and silicic acid resulting in base unsaturation is complete, whereas in lateritic soil, this process is considered incomplete. Lateritic soils are generally loamy in texture, without the gravelly vesicular nodules which characterise true laterite.

Laterite soils are usually very deep and red in colour. The red colour at the surface gradually changes to reddish yellowish and finally yellow in the deeper layers. Ferruginous gravel is found both in laterite soil and lateritic soil. Hydrous oxides of iron and aluminium dominate the clay, so the cation exchange capacity of the clay is even lower than the soil, as the soil contains a considerable amount of humus. Lateritic soils contain 50 per cent or more exchangeable calcium. Magnesium is the next important exchangeable cation. The surface soil contains less than 25 per cent exchangeable calcium 25 per cent of the metallic cations 50 of the exchangeable hydrogen. Hence the soil is distinctly acidic in reaction. Laterite and lateritic soils are usually poor in nitrogen, available phosphorus and lime. They may contain adequate amounts of total phosphorus and potash. Water soluble phosphorus when added to these soils are converted to water insoluble phosphorus.

In Kerala, there are two types of alluvium; coastal alluvium and inland alluvium in depressional lands. In central Kerala, the width of coastal alluvium tracts increase, whereas in the north and south, they are comparatively narrow. Alluvial soils of Kuttanand occur at low lying areas which were once a part of the sea. These low lying areas were later filled up by silt carried by Pampa and other rivers and deposited in these low lying areas. The coastal alluvium is coarse sand to loamy sand to sandy loam of a light grey colour, low in moisture and nutrient retention capacity, and acidic in reaction. The alluvium is usually fertile on the banks of rivers.

Excessive amounts of organic matter accumulated in the low lying areas, and decomposed in the humid climate of Kerala to form Peaty soils, which are locally known as Kari soils. They have been found in taluks of Ambalapuzha, Vaikom and Shertallai rivers. They remain submerged in water during the rainy season after which they are puddled, and paddy is transplanted on these heavy black coloured highly acidic soils, containing 10-40 per cent organic matter. The Soil pH may be as low as 4.0 which is due to anaerobic decomposition of organic matter. Sometimes soil contain ferrous and aluminium sulphate. Soluble salts accumulate in kari areas. They are relatively poor in phosphorus.

10.13. SOILS OF TAMIL NADU

Major portion of Tamil Nadu is covered by red sandy soil and red loam soils. Red sandy soils have developed from acidic parent material like granite, gneiss, quartzite, sandstone etc. Sand particles are coated with red coloured haematite or yellow coloured limonite which is responsible for the various shades of red and yellow of these soils, which usually contain ferruginous gravel containing iron, aluminium and silica. These sandy, loamy sand and sandy loam soils are heavily leached and therefore poor in basic elements and plant nutrients. Their pH ranges from 6.6 to 8.0. Majority of red loam soils are loamy is texture. The red colur of soils is due to the coating of ferric oxides on soil particles. Calcium is the important exchangeable cation. They are neutral to slightly alkaline in reaction. Black soils of Tamil Nadu which are either shallow (3 to 4 ft. deep) or deeper, are of very heavy texture, have a high moisture retention capacity are rich in lime and alkaline in reaction. They contain low amounts of nitrogen but sufficient amounts of phosphoric acid and potash.

Mixed red and black soils occur in Coimbatore, Madurai, Ramanathapuram and Tirunelveli districts. Black soils are dominated by beidellite, while red soils are dominated by kaolinite. So the cation exchange capacity of the black soils is much higher than that of the red soils but the cation exchange capacity of red soils is maximum at an intermediate depth. Black soils contain almost the same amount of nitrogen.

Laterite soils occur in the Chinglepet and Thanjavur districts. These were formed from varieties of parent materials in humid climate. Paddy is grown in lower elevation and tea, cinchona, rubber and coffee are grown at the higher elevation. They are rich in humus and plant nutrients and strongly acidic in reaction. Soil acidity increases with elevation.

Delatic alluvium occurs in the Thanjavur district and a belt of coastal alluvium covers Tamil Nadu from Madras to Kanyakumari. Alluvial soils are most extensive and most fertile. They are usually up to several feet deep at higher elevations. These soils consist of alternate layers of silt, clay and sand of varying thickness. The texture of the surface soil is usually loamy. The Cauvery alluvium is poor in humus, nitrogen and phosphorus, but rich in potash and lime. These soils possess a low cation exchange capacity and are alkaline in reaction. Soils developed from Cuddalore sandstone are loamy in texture and deep to light red, yellow and light yellow and even greyish white in colour and deficient in humus, nitrogen, phosphorus and lime. These soils are of a low cation exchange capacity and neutral to moderately alkaline in reaction. The soils which have developed from archaean schist are found only in a small area to the west of Budalur. The profile show brownish grey loam or clay at the top one foot surface soil and brown sub soil. Soils are usually sandy below three feet. Surface and sub soils are usually very hard. They are poor in humus, nitrogen, phosphorus and lime but rich in potash. These slightly alkaline soils are free from soluble salts and are of low cation exchange capacity.

The profile characteristics of coastal alluvial soils formed from recent marine deposits are similar to Cauvery alluvium in their alternate layers of clay, silt and sand but exhibit influence of sea as indicated by the presence of shells and bleached sand. They are poor in nitrogen and available phosphorus but rich in potash and lime. Some of them also contain salts.

Peaty soils which occur mainly on the south east coast of Tamil Nadu are usually coloured blue due to the presence of ferrous iron. They contain varying amounts of organic matter.

10.14. SOILS OF ANDHRA PRADESH

Black soils occur in the districts of Aeilabad, Nizamabad, Medak, Karimnagar, Warangal, Khammam, Guntur, Mehboobnagar and Hyderabad. Mixed red and black soils occur mainly at Kurnool, Ananthapur, Cuddapah, and Krishna districts.

Flue cured tobacco is grown on black soil. These are deep (120 cms) highly clayey soils of high moisture retention capacity but of poor structure and permeability. They have been tenatively grouped as follows:

(*i*) Dry black soils which include most of the area under tobacco.
(*ii*) Wet land black soils where tobacco is grown in rotation with paddy.
(*iii*) Tank bed soils of low lying area which remain in inundated during the monsoon and are cultivated late in the season.
(*iv*) Saline land black soils.
(*v*) High land black soils of the Deltaic region which are comparatively lighter in texture and more fertile.

Soils of the Central Tobacco Research Institute, Rajahmundri and the Farm of Cigarette Tobacco Research Sub-Station, Guntur are uniformly clayey in texture down the profile where calcium carbonate increases with depth. These soils are low in organic carbon and medium in total nitrogen, contain soluble salts and are alkaline in reaction. Calcium and maganesium are the important exchangeable cations. These highly clayey soils have a high moisture retention capacity. The Soils of Nizamabad district are sandy clay loam to clay loam and alkaline in reaction.

Soil of Ananthapur and Kurnool are mostly black clays which enclose small areas of red and mixed black soil.

Red sandy soils occur in Chittoor, Ananthapur, Cuddapah, Nellore, Kurnool, Mehboobnagar, Guntur, Hyderabad, Warangal, Khammam, east and west

Godavari, Visakhapattanam and Krishna districts. The important clay minerals are kaolinite and illite which are coated with red haematite or yellow limonite. *A* horizon is pale reddish brown to red coloured coarse sand, fine sand of loamy sand. *B* horizon is dark redddish brown loam or sandy clay. Deltaic alluvium occurs in the east and west Godavari, Krishna and Guntur districts. It represents the heterogenous sediments brought by rivers Godavari and Krishna and deposited by them near their mouths, where they joun the Bay of Bengal. These rivers drain the black soils which have developed from basaltic traps, so the delta soils are largely composed of dark silty clay. the soils are usually well drained. Godavari and Krishna deltaic alluvium contain black fertile mud of a high moisture and nutrient retention capacity. A variety of crops like paddy, sugarcane, banana and tobacco are grown on these soils.

10.15. SOILS OF ORISSA

Major parts of Orissa are covered by red and yellow soils. They are usually well drained and deep soils of sandy loam texture at the surface and clay loam at the sub surface. Their colour is pale brown to reddish brown. These acidic soils contain a moderate amount of humus. Laterites of Orissa have been found capping hills and in Plateau. Large areas in Khurda and Balasore are occupied by laterite. They are granular clay laom soils that contain gravels.

Deltaic Alluvium occur in Cuttack, Balasore, and Mayurbhunj districts. Mahanadi delta alluvium is light brown and ligh yellow sandy loam, loam, and silty loam. They contains enough potash but a low amount of Phosphorus. Marshy soils occur in the coastal areas. These soils contain varying amounts of organic matter. They are generally blue due to the presence of ferrous iron.

10.16. SOILS OF WEST BENGAL

Alluvial soils of the Coochbehar district have developed from sediment laid down by Tista, Mahananda and Jaldhaka rivers. These new alluvial soils have no genetic profile. They are greyish brown to dark brown in colour, loam to silt loam in texture and are of a low cation exchange capacity. Illite in the important clay mineral.

The river Ganga brought the products of weathering of rocks constituting the Himalayas and deposited them in the districts of West Dinajpur, Murshidabad, Nadia, Hooghly and 24-Parganas. The Damodar and Ajoy rivers have brought the products of weathering of rocks in Chhota Nagpur and Santhal Parganas of Bihar and deposited these in the Burdwan, Bankura, Midnapore and Hooghly districts. Alluvial soils have developed from silt laid down by these rivers.

Alluvial soils of West Bengal have been grouped into two broad families called Ganga alluvium, which has originated from the material deposited by the river Ganges and Vindhyan alluvium, which has originated from the material deposited by the river Damodar.

The Ganga alluvium comprises of the following soils:

(*i*) Ganga riverine lands that have an immature profile. These soils contain plenty of greyish white fine sand and lime.

(*ii*) Ganga flat lands that have a slightly mature profile. The soil texture varies from sand to clay, sub-soil contain more clay than the Surface soil. Sub-soil contains lime at many places.

(*iii*) Ganga low lands that have flat clayey surface soil and lighter textured sub-soil which often contains lime concretions at many places.

(*iv*) Ganga Uplands occur over older alluvial fans, alluvial plains and terraces. They have better developed profiles. A moderate amount of clay and high amount of lime have accumulated.

Vindhyan Alluvium includes the following soils:

(*i*) Riverine lands that have a stratified, immature soil profile. These soils contain yellowish brown sand.

(*ii*) Flat land soils that are usually poorly drained, as indicated by mottling in the lower horizon. They contain brown coarse sand. Some clay has been illuviated in the lower horizons.

(*iii*) Upland have a mature soil profile. Some clay, sesquioxide has been illuviated in the *B* horizon.

These alluvial lands are fairly fertile and support the growth of many varieites of crops.

The Southern parts of 24-Parganas and Midnapore districts have been covered by deltaic alluvium which is usually poorly drained lighter coloured silty clay. The Lower horizons are mottled. Swampy vegetation of mangroves occur at the mouth of the river Ganges, which is responsible for a large accumulation of organic matter in the soil, which has been formed from the deposit brought by tidal currents. The original deltaic branches of the Ganges have been choked up because the head water has been cut off. Consequently these lands were converted into numerous tidal flats. These lands have been bunded to prevent the ingress of the sea water, while sluice gates have been provided in order to allow the accumulated rain water to drain off. Coastal alluvial soils which occur in canning, are saline and contain a high amount of neutral salts like sodium chloride and are thus almost neutral in reaction. They are usually medium to heavy textured and occur in depressions. In the Sundarbans and other places in West Bengal, peaty soils usually bluish in colour due to the presence of ferrous iron are found. They contain large amounts of organic matter.

Red and yellow soils of the Burdhaman district, are pale brown to dark brown loamy sand at the surface and brownish yellow and yellowish brown sandy loam to clay loam at the sub surface. These soils are slightly acidic in reaction and contain low amounts of humus and nitrogen. Red soils of the Burdhaman district have a moderately deep *A* horizon of pale reddish brown to red coloured coarse sand, fine sand or loamy and a *B* horizon of dark reddish brown loam or sandy clay loam of varying depth. These soils are acidic in reaction and low in humus.

Laterite soils also occur in the Midnapore, Bankura, Burdhaman and Birbhum districts. These soils have been heavily leached and are consequently acidic, and of a poor fertility.

The Darjeeling district is mainly covered by Brown Hill soils are usually highly acidic in reaction and of a high exchange capacity due to their high humus content. The surface soil is a mixture of well decomposed humus and inorganic material. Its colour gradually turns to the colour of the parent rock.

Tarai soils of the Jalpaiguri district have originated from the material brought from elevations of more than 3000 metres above sea level at the Himalayas by hilly rivers like Tista, Mahananda, Torsa and Jaldhaka and deposited by them at an elevation of 50 to 150 metres above sea level along the foothills. They are naturally fertile but poorly drained soils.

10.17. SOILS OF ASSAM

Alluvial soils of Assam have developed from the deposition of the products of weathering of rocks consituting the Himalayas by the river Brahmaputra. They are acidic in reaction. Generally those on the hills and old alluvium are more acidic than the new alluvial soils along the river bank. Soil acidity increases with the increase in rainfall and fineness of the soil texture. Soils on the northern bank of the Brahmaputra are less acidic than those on the southern bank. In the Surma valley, soils on Tillas and Bheel are markedly acidic in reaction. Soils of low lying inundated areas are less acidic. The new alluvial soils are often neutral in reaction. Soils of the Brahmaputra valley are of a coarser texture, while soils of the Surma valley are of finer texture. Alluvial soils of Assam contain moderate amounts of organic matter and nitrogen and high amounts of phosphorus and potash. Soils of the Surma valley contain more humus than those of Brahmaputra valley. The soils of Brahmaputra valley are richer in total magnesium. The five following local soil types have been recognised in the Assam valley:

(*i*) a layer of micaceous sand skirting the base of the Himalayas in Kamrup and Mangaldai.

(*ii*) the silt of the plateau rising suddenly out of the lower plains in Tezpur, Bishnath, Sibsagar and Golaghat, sandy soils of Nowgong and Jorhat.

(*iii*) heavy soils, chiefly of very fine silt, comprising north Lakhimpur and Dibrugarh.

(*iv*) a series of sand ridges running more or less across the lime of the valley of Jorhat, extending east and west particularly from Titabar to Jhamzie river and

(*v*) a series of low hills skirting the base of the Naga Hills and separated by rolling land.

Red soils occur in Karbianglong and north Cachar Hill districts. These soils contain high amounts of organic matter and nitrogen and are of a finer texture.

10.18. SOILS OF NORTH EASTERN HILL REGIONS OF INDIA

Farmers of the North Eastern Hill regions of India have been practising shifting cultivation. They clear jungles by burning them and grow crops there for a few years. Thereafter they abandon the land and burn forests, and grow crops in a new area. Burning causes severe loss of nutrients especially nitrogen and sulphur. They do not follow any soil conservation practices. Consequently fertile top soil is being eroded at an increasing rate. These soils have been tentatively classified by Govinda Rajan and Gopal Rao (1978) as discussed below:

Most of the land area of Nagaland is covered by red loam soil except for a very small area bordering the Sibsagar and Lakhimpur districts of Assam covered by Laterite soil. The soils of the Twensang district are low in organic carbon and therefore in available nitrogen, available phosphoric acid and available potash. Soils of Kohima, Mokokchung, Phek and Wokha districts have a high organic carbon and content therefore a high available nitrogen and low available phosphoric acid content. Soils of Mon and Zunheboto district are medium in nitrogen. Soils of all the districts contain medium amounts of organic carbon and therefore, available nitrogen soils of all the districts except those of organic cabon and therefore Soils of the Zunheboto district have a medium quality available phosphoric acid. Soils of Kohima, Phek, Wokha and Zunheboto district have medium amounts of available potash and those of the rest of the districts are low in available potash.

Almost the whole of Manipur is covered by red and yellow soils except for a small area on the Assam border that is covered by laterite soil. Manipur soils are moderately rich in organic carbon and therefore in available nitrogen, low in phosphoric acid content and medium in potash content.

Practically the whole of Mizoram is covered by red and yellow soil, Mizoram soils contain a high amount of organic carbon and are therefore high in available nitrogen, have a low to medium amount of available phosphoric acid and a medium amount of available potash. Tripura is covered by red and yellow soils. Soils of Tripura contain medium to high amounts of organic carbon and therefore the same amounts of an available nitrogen, low to medium amounts of available phosphoric acid and low amounts of available potash.

The Garo Hills are covered by laterite soils and alluvial soils, Jaintia Hills are covered by red and yellow soils. Laterite, red and yellow soils and alluvial soils cover the Khasi Hills. Almost the entire soils of Meghalaya contain low amounts of available phosphoric acid and a medium of potash. The soils of Khasi and Jaintia Hills contain high amounts of organic carbon and hence a high amount of available nitrogen.

The whole of Lohit district, two-thirds of the Siang and Tirap district and half of Kameng district is covered by red loam soil. The rest of the areas in the Siang, Subansiri and Kamang districts are covered by brown hill soils. Laterite soils have been found in the Tirap district. Soils of all the districts contain a high amount of organic arbon and hence of available nitrogen. Soils of the Kameng district contain medium amounts of available phosphoric acid, and the soils of the remaining districts contain low amounts of available phosphoric acid. Soils of all the district except those of Siang district contain low amounts of available potash. Soils of Siang district contain medium amounts of available potash.

SELECTED READINGS

1. Agarwal, R.R. and Gupra, R.N. (1968), *Saline and Alkali Soils of India*, I.C.A.R New Delhi.
2. Arakeri, H.R. and Donahuc, R.L. (1985), *Principles of Soil and Water Conservation and Water Management*. Oxford & IBH Publishing Co., 66 Janpath, New Delhi.
3. Chetri, M.S. and Tandon, U. (1980), *Fertility status of Soils of Nagaland, Agriland* VII, 7, 6-10.
4. Daji J.A. (1968), *A textbook of Soil Science*, Media Promoters & Publishing, Pvt. Ltd., Bombay.
5. Govinda Rajan, S.V. & Gopala Rao, H.G. (1978), *Studies on Soils of India*, Vikas Publishing House Pvt. Ltd., Ansari Road, New Delhi.
6. Mandal, S.C. *et al.* (1982), *Acid soil of India & Their Management*. I.C.A.R. New Delhi.
7. Prasad, R.N. *et al.* (1981), Soil Fertility Management in N.E.H. Region *I.C.A.R. Research Bulletin No. 9.*
8. Ray Choudhury, S.P. & Govinda Rajan, S.V. (1971), Soils of India, *I.C.A.R. Tech. Bul. No. 25.*
9. Tamhane, R.V. *et al.* (1964): *Soils, Their Chemistry and Fertility in Tropical Asia*, Prentice Hall of India, Pvt. Ltd., New Delhi.

11
MINERAL NUTRITION OF CROPS

The growth of crops depends upon a number of soil and climatic factors, of which the cultivator can control the supply of nutrients and water to crops. He may increase the availability of nutrients in the soil by modifying the soil conditions or may add the nutrients in the form of manures and fertilizers. Crops absorb a large number of elements from the soil. About ninety elements have been found in crops, all of which are not essential to them. Then the question arises which elements should be considered as essential to crops. The answer is that an element is considered essential for crops only if it fulfils the three following criteria of essentiality as suggested by D.I. Arnon and takes an active part in the growth and development processes.

11.1. CRITERIA OF ESSENTIALITY OF ELEMENTS

(*i*) The crop cannot grow normally and complete its life cycle in the absence of an essential element.

(*ii*) The deficiency of an essential element can be corrected by supplying only that particular essential element to the crop.

(*iii*) The concerned essential element must play a direct role in plant metabolism.

Sixteen elements fulfil the three above-mentioned criteria of essentiality of elements: carbon, oxygen, hydrogen, nitrogen, phosphorus, potassium, magnesium, sulphur, iron, manganese, zinc, copper and Molybdenum are essential both for higher plants and micro-organisms. Calcium and Boron are essential to higher plants but not for all micro-organism according to the criteria of essentiality of elements.

A few elements like rubidium, strontium, nickel, chromium and arsenic stimulate the growth of certain plants or have other beneficial effects. Silicon and aluminium are present in plants in appreciable amounts but are not essential to them because they do not fulfil the three criteria of essentiality of elements. So they are called balast elements.

11.2. FUNCTIONS OF ESSENTIAL ELEMENTS

Carbon, Oxygen and Hydrogen

Carbon, oxygen and hydrogen are the major constituents of all organic compounds, of which crops are made and they are concerned with different metabolic reactions which are vital for the growth of plants. Carbohydrates fats and proteins are oxidized to liberate energy which is essential for all living organisms.

Nitrogen

Nitrogen is a major structural constituent of the cell. It is also an essential constituent of metabolically active compounds like proteins, nucleic acids, chlorophyll, enzymes, adenosine di and triphosphates (A.D.P. and A.T.P.) hormones etc. Chlorophyll is essential for photosynthesis; nucleic acids are essential for transfer of genetic information. Enzymes are biological catalysts. A.D.P.and A.T.P. are essential for energy transfer. Hormones are regulators of plant growth.

When excessive amounts of nitrogen all supplied to crops they become succulent when they are adversely affected by unfavourable factors like deficiency of water, high temperature, insect pests and diseases, lodging etc. Whenever nitrogen is deficient in soil, flower buds turn pale and are shed prematurely, fruits are poor in quality, size, and weight.

Phosphorus

Phosphorus is a constituent of the cell nucleus, and metabolically active compounds like nucleic acids, adinosine di and triphosphate etc. Nucleic acids are essential for the transfer of genetic information and A.D.P. and A.T.P. are involved in energy transfer, A.D.P. is involved in basic reactions of photosynthesis.

Phosphorus is essential for cell division and development of the meristematic tissues at the growing points and for root growth. It offsets the harmful effects of excess nitrogen in plants.

Potassium

Potassium plays a vital role in the formation of amino acids and proteins from ammonium ions, which are absorbed by roots form the soil It is also responsible for the transfer of carbohydrates, proteins etc. from the leaves to the roots. It also plays a vital role in the uptake of other elements particularly nitrogen, phosphorus and calcium. Potassium regulates the permeability of the cellular membrane. It increases the hydration of protoplasm. It activates a number of enzymes, e.g. alcohol dehydrogenase and its deficiency decreases photosynthesis. Potassium increases the resistance of crops to hot and dry conditions and insect pests and diseases. It increases the stiffness of straw in cereals and therefore the lodging of cereals is reduced. It improves the quality of fruits and grains.

Calcium

Calcium also regulates the permeability of the protoplasm and decreases the hydration of the protoplasm. Calcium is a structural component of chromosomes where it possibly binds the DNA to the protein. chromosomes are responsible for hereditary characters of crops. Chromosomes become abnormal when calcium is deficient because mitosis is abnormal in the roots of calcium deficient plants and is not often followed by cell wall formation, resulting in two nuclei within the same cell. It also activates enzymes e.g. Lipase. When calcium is applied to acid soils, the availability of nitrogen, phosphorus and molybdenums is increased. Crops cannot take enough potassium and magnesium from calcareous soils.

Magnesium

Magnesium is a constituent part of chromosomes which are the bearers of hereditary characters. It is an essential constituent of polyribosomes involved in protein synthesis. It is a constituent of chlorophyll. It affects the translocation of phosphorus in plants and activates enzymes e.g. pyrophosphatase.

Sulphur

Some proteins contain sulphur. Sulphur activates enzymes which participate in the dark reactions of photosynthesis. It affects the development of tissue. Xylem and collenchyma fibres thicken, resulting in a hard and woody stem.

Iron

Iron is a structural part of both enzymes and their activators. It is a constituent of prophyrin compounds which include cytochromes, ferredoxin, haemoglobin etc. The cytochromes participate in respiration during which electrons flow along the cytochromes bridge between the inner layer of the protoplasm and the outer layer of protoplasm and the anions flow in the opposite direction when iron is alternately oxidized to the ferric state and reduced to the ferrous state. By being a structural constituent of cytochrome, B_6 and ferredoxin iron play direct role in photosynthesis and nitrogen fixation by Rhizobium bacteria found within the root nodules of leguminous plants, which, in turn, contain a red substance called leghaemoglobin.

Manganese

Manganese is a constituent of nitrite reductase and hydroxilamine reductase, which reduce nitrite and hydroxilamine respectively to ammonia, which is later converted to amino acids and proteins. It also activates enzymes. Manganese deficiency reduces photosynthesis even when the chlorophyll content is not reduced.

Copper

Copper is a constituent of several oxidation and reduction enzymes; photosynthesis is reduced when copper is deficient even when their chlorophyll content is not reduced. Copper deficiency adversely affects respiration.

Zinc

Zinc is a constituent of enzymes. When zinc is deficient, photosynthesis is reduced even when the chlorophyll content is not reduced. When it is deficient, the auxin, concentration is reduced and flowering and fruit formation is reduced and delayed.

Molybdenum

Molybdenum is a constituent of the enzyme nitrate reductase which reduces nitrate to nitrite. It is essential for nitrogen fixation by free living bacteria e.g. Azotobacter, algae and symbiotic bacteria.

Boron

Boron is required for proper development of tissues, particularly the vascular elements. Boron deficiency causes the necrosis of tissues and sterility and malformation of the reproductive organs. It is also involved in the translocation of sugars in plants.

Chlorine, Cobalt and Sodium

Chlorine is involved in the evolution of oxygen in the primary reactions of photosynthesis.

Cobalt is essential for micro-organisms e.g. rhizobium bacteria. It is a structural component of vitamin B_{12}. Sodium maintains a high internal osmotic concentration in plants, which enables them to withstand desiccation when the soil solution contains high amounts of salts.

11.3. SYMPTOMS OF DEFICIENCY OF NUTRIENTS

When crops cannot absorb any nutrient in sufficient quantities, their metabolism changes, which is indicated by specific hunger signs exhibited by crops. These hunger signs are called deficiency symptoms which appear first on older leaves of the whole plant when nitrogen, phosphorus, potassium, magnesium, molybdenum and zinc are deficient. These deficiency symptoms appear first on new leaves and buds when calcium, boron, manganese iron and sulphur are deficient.

Nitrogen

Nitrogen deficient plants have a stunted and spindly growth, the stem is errect but lateral buds do not develop. So tillering is reduced in cereals. Leaves make an acute angle with the main stem. First, older leaves become pale green or yellowish green. This starts from the apex of older leaves and moves towards the base. These pale green or yellow green areas often develop brilliant tints of lemon, yellow or orange or less frequently reddish purple. Later younger leaves are affected if the deficiency continues for a sufficiently long period when older leaves become completely dry. This is called firing.

Phosphorus

Growth of phosphorus deficient plants is severely restricted. They are thin, erect and spindly. Lateral buds do not develop and so tilering is reduced in cereals. The bluish-green older leaves become bronzed or develop reddish brown or purple tints. These effects appear first on the apex of older leaves and then gradually spread towards the base. The severely affected apical portions of leaves may become rolled. Leaves may be shed prematurely and flowering and fruiting may be delayed.

Potassium

Potassium deficient plants have a stunted and bushy growth. Pale green older leaves develop chlorosis between veins and light grey to bronze, reddish brown or brown colouration along the leaf apex and the apical margin. The leaf tip and apical margin of the leaf become scorched and necrotic. Thin brown roots are poorly developed; the small leaflets of the potato are crinkled and curved downwards.

Calcium

The first symptoms of calcium deficiency is chlorosis of young leaves, followed by distortion of the growing points of the stem. In crops belonging to the genus Brassica, young leaves forming the terminal bud do not expand, but become hooked. The leaves near them become cupped and the growing point dies. In wheat and barely, the upper parts of younger leaves fail to unroll and appear thread-like.

Magnesium

The normal green colour of the foliage fades, resulting in chlorosis between veins, leading to chlorotic stripping in the parallel veined leaves and chlorotic mottling in the reticulate veined leaves. These chlorotic areas later develop brilliant tints of orange, red, purple or mauve. The area between the veins may turn brown and necrotic if a severe deficiency continues for long time.

Sulphur

The most common symptom is the fading of the normal green colour of younger leaves followed by chlorosis. The young leaves of sulphur deficient plants *i.e.* cereals exhibit chlorotic striping between veins. Brassica crops are susceptible to sulphur deficiency the symptoms of which are curled leaves and a non-development of apical bud older leaves exhibit wrinkled and inwardly raised areas. They may be tinted with orange or red colour and shed prematurely. The plant becomes brittle.

Iron

Chlorosis appears first at the basal part of younger leaves and later spreads towards the apical portion. Veins remain green. A continuous deficiency may result in the total bleaching of leaves.

Manganese

Small chloritic patches appear between the veins of middle leaves. Later, these chlorotic patches unite, resulting in a chlorotic striping between the veins in parallel veined leaves, and chlorotic mottling in reticulate veined leaves. The chlorotic areas of leaves turn reddish-brown and necrotic.

Copper

A copper deficiency results in first young leave exhibiting chlorosis and drying, and a distortion of the terminal leaves. The shoot apex may die prematurely, resulting in the development of several auxiliary buds. Apices of young leaves often dry up. Young leaves of cereals may turn severely chlorotic, fail to unroll and wither. In species of Brassica e.g. cauliflower and cabbage, young leaves develop fine chlorotic mottling between the veins; mottled areas often develop white necrotic patches, particularly along the veins and leaf margins.

Zinc

The commonest symptoms of zinc deficiency are chlorosis between veins, reduction in the sizes of young leaves, which are often clustered and a bronzing, purple, violet reddish brown or brown colouration of the foliage.

Molybdenum

In most plants with reticulate venation, the first symptoms of Molybdenum deficiency is chlorotic mottling between the veins of old or middle leaves , or all over the surface when nitrogen is supplied to the crops in the form of nitrates. Crops like cauliflower, a species of Brassica, exhibit chlorotic mottling and a cupping of the middle leaves. A drying of severely affected leaves begins from the margin and covers the whole leaf, leaving only the petioles. Later, younger leaves are also affected and fail to expand fully; ultimately the growing point becomes necrotic and further growth ceases.

Boron

The commonest symptoms of boron deficiency exhibited by crops include necrosis of the softer tissue, particularly phloem and the death of the growing points. In heart rot of sugar beat, young leaves at the centre of the crown fail to expand fully and become curled. The petiole and the basal part of the mid rib turn brown or black and become brittle. The young terminal leaves of the growing points and later, the entire crown become necrotic.

11.4. NUTRIENT TOXICITY SYMPTOMS

Iron

Plant growth is reduced, Roots become thick and brown. Tiny brown spots appear on lower leaves of paddy plant. These spots begin to develop from the apex and then spread towards their base. Entire leaf become purplish if the toxicity is severe.

Aluminium

Usually roots are affected first and more severely than the shoot. Roots become thick and brown and their tips are enlarged. The paddy crop display orange yellow chlorosis of leaves which may become necrotic if aluminium toxicity is very severe.

Manganese

Young leaves become chlorotic; Minutex brown spots often appear on chlorotic leaves, leaf sheaths and lower parts of stems of cereal crops. The paddy plant is stunted. Tillering is reduced. Leaves especially the lower ones, develop brown spots on their veins and sheaths.

Copper

Young leaves develop interveinal chlorosis. Margins of older leaves are rolled. Older leaves are coloured reddish orange or pink..Ultimately they dry up. Roots become reddish brown and necrotic.

Zinc

First young leaves develop interveinal chlorosis. Later these interveinal chlorotic areas turn reddish brown and dry up. Reddish brown colouration often appear first near the base of the leaf and later spreads towards the apex. The margin of the affected leaf is rolled. Roots become brown and necrotic.

Boron

Interveinal chlorosis of young leaves begins from their base and then spreads towards their apex. Then diffuse brown or reddish brown colour appears in the leaf margins and chlorotic areas. Such colour is often very prominent in the interveinal areas and along veins which is often bluish green. The colouration intensifies and spreads to cover the entire interveinal area which become dry and papery.

Yellowish white discolouration of the tip of older leaves of paddy plant is observed six weeks after the paddy crop has been transplanted. Then their tips and margin become yellow. Then eliptical large dark brown spots appear in the discoloured area. Ultimately the entire leaf become light brown and dry up.

Molybdenum

Molybdenum toxicity symptoms are brilliant golden yellow or blue tints, Reddish or golden yellow tints appear on shoots of potato and tomato while blue tints appear on shoots of barley and maize.

Chlorine

Chlorine toxicity symptoms are reduction in the number and size of leaves, bronzing and yellowing of the folliage and browning and scorching of leaf margins.

11.5. FORMS AND BEHAVIOURS OF NUTRIENTS IN SOIL

Nitrogen

Nitrogen basically occurs in the soil in complex organic forms like proteins which are decomposed by proteinase enzymes produced by micro-organisms to amino acids, that are further decomposed to ammonium compounds. Ammonium ions may replace basic cations from the clay and humic micelle.

Ammonium compounds are oxidised to nitrites by nitrosomonas bacteria, which are further oxidised to nitrates by nitrobacter. Plants take up nitrogen from the soil in the form of ammonium and nitrate ions. Ammonia is volatilized from hot, alkaline waterlogged soil when nitrates are reduced to nitrites, which are further reduced to oxides of nitrogen and nitrogen gas. Nitrates are leached down if added to paddy fields.

Phosphorus

Phosphorus originates in the soil, both in the inorganic form *i.e.* apatite mineral, as well as in the organic form *i.e.* phytin, inositol, nucleic acids etc. Microorganisms decompose these organic phosphorus compounds to release the phosphorus in the simple inorganic form *i.e.* orthophosphate, H_2PO_4 ions. Carbon dioxide is produced from the decomposition of organic matter and respiration of roots and micro-organisms. It reacts with water to form carbonic acid, the reacts with tricalcium phosphate to form water soluble monocalcium phosphate:

$$Ca_3(PO_4)_2 + 4H_2CO_3 \longrightarrow \underset{\text{Monocalcium Phosphate}}{Ca(H_2PO_4)_2 + Ca(HCO_3)_2}$$

Water soluble monocalcium phosphate reacts with aluminium and ferric ions in strongly acidic soil to form water insoluble aluminium and ferric hydroxy phosphate.

$$2Al^{+++} + \underset{\text{Monocalcium Phosphate}}{Ca(H_2PO_4)_2} + 2H.OH \rightleftharpoons \underset{\text{Aluminium Hydroxy Phosphate}}{2Al\begin{cases}OH\\OH\\OH\end{cases}} + Ca^{++} + 2H^{+}$$

$$2Fe^{+++} + \underset{\substack{\text{Monocalcium Phosphate}\\ \text{Soluble}}}{Ca(H_2PO_4)_2} + 2H.OH \rightleftharpoons \underset{\text{Ferric Hydroxy Phosphate Insoluble}}{2Fe\begin{cases}OH\\OH\\H_2PO_4\end{cases}} + Ca^{++} + 2H^{-}$$

Water soluble monocalcium phosphate reacts with calcium ions and calcium carbonate to form insoluble tricalcium phosphate in strongly alkaline soils.

$$\underset{\text{Monocalcium Phosphate}}{Ca(H_2PO_4)_2} + 2Ca^{++} + 4\,OH^{-} \rightleftharpoons \underset{\text{Tricalcium Phosphate}}{Ca_3(PO_4)_2} + 4\,H.OH$$

$$\underset{\substack{\text{Monocalcium Phosphate}\\ \text{Soluble}}}{Ca(H_2PO_4)_2} + CaCO_3 \rightleftharpoons \underset{\text{Tricalcium Phosphate Insoluble}}{Ca_3(PO_4)_2} + 2CO_2 + 2H_2O$$

Plants assimilate phosphorus in the form of H_2PO_4 ions.

Potassium

Potassium basically occurs in primary minerals like orthoclase, mica etc. in the soil. These mineral decompose to release the potassium contained them in the water soluble form.

$$\underset{\text{Orthoclase or Microcline}}{2\,KAl\,Si_3O_8} + H_2CO_3 \longrightarrow K_2CO_3 + Al_2(OH)_2Si_4O_{10} + 2Si\,O_2$$

$$K_2CO_3 \rightleftharpoons 2K^+ + CO_3^{--}$$

When excessive amounts of potassium ions are present in the soil solution, they may be lost by leaching. Crops assimilate potassium in excessive amounts but the crop yield does not increase in proportion to the amount of potassium absorbed by the root. This is known as luxury consumption of potassium. Potassium ions are fixed by certain clays like illite.

Calcium and Magnesium

Calcium and Magnesium occur in primary minerals like calcite, dolomite, hornblende etc. Which are decomposed by organic acids to release calcium and magnesium in the form of ions. Usually calcium and magnesium ions are held by clay and humic micelle is exchangeable cations which are replaced by hydrogen ions as shown below:

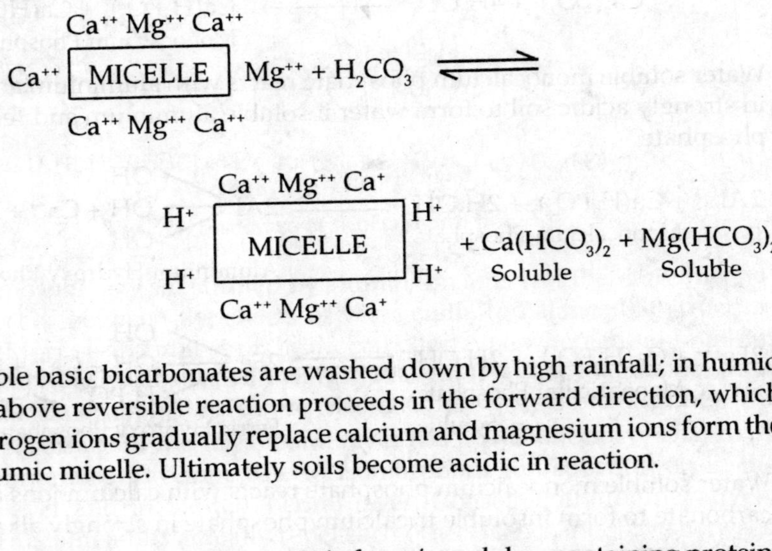

These soluble basic bicarbonates are washed down by high rainfall; in humid region the above reversible reaction proceeds in the forward direction, which means hydrogen ions gradually replace calcium and magnesium ions form the clay and humic micelle. Ultimately soils become acidic in reaction.

Sulphur

Sulphur mainly occurs in the organic form *i.e.* sulphur containing proteins which are decomposed by micro-organisms to sulphides, that are further oxidized to sulphate. The soil solution contain HSO_3^{--}, SO_3^{--}, SO_4^{--} ion.

Micronutrients

The micronutrients occur in primary minerals which decompose during the formation of soils resulting in the formation of oxide and sulphides and of iron, manganese and zinc. Boron, molybdenum and chlorine occur as borate and borosilicate, molybdate and chloride respectively. Some of the micronutrients, also occur in the organic form in the surface soil. Micronutrients usually occur in the soil solution in the ionic form.

All micronutrients except molybdenum are extremely soluble in a strongly acidic soil, and thus become toxic to crops. Boron has been leached form strongly acidic sandy soil. When the pH is increased, Iron, manganese, zinc and copper are precipitated as hydroxides. So the availability of all micronutrients except molybdenum decrease, as the pH of the soil increases.

Iron, manganese and copper exist in the lower valent form in poorly aerated waterlogged soils. They are oxidized to the higher valent form is well aerated soils. The higher valent forms of micronutrients are less soluble than the lower valent forms, at the most common pH values of soil e.g. ferric hydroxide precipitates at a pH of 3.0 whereas ferrous hydroxide precipitates at a pH at 6.0.

Manganese and zinc also occur as integral parts of some silicate clays like those belonging to the smectite group. When large amounts of water soluble phosphate fertilizers are added to the soil, the crop takes up less iron and zinc and more molybdenum. Iron, Manganese, zinc and copper combine with organic compounds to form chelates which protect them from certain harmful reactions, e.g. precipitation of water soluble phosphate by ferric ions. At the same time, they remain available for the growth of the crop.

11.6. CHELATES

The word "chelate" means claw in Greek. Chelates are organic compounds in which certain metallic cations are bound with the organic molecule to form the organic compounds of cyclic structure of the metal atom and the oxygen atom as shown below.

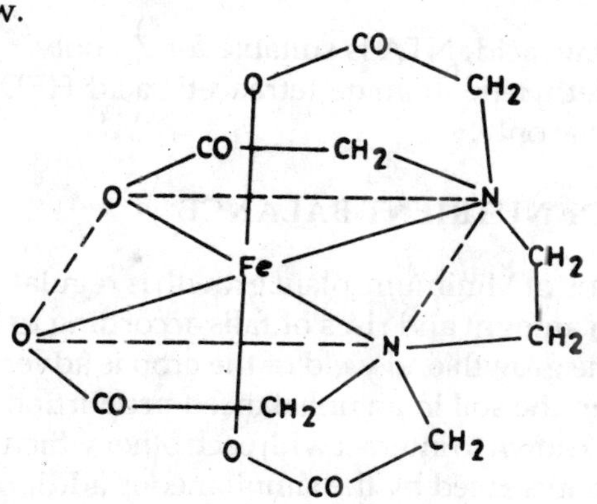

Structural formula of a common iron chelate ferric ethylenediamine tetraacetate (Fe-EDTA)

The chelated nutrients dissociate slightly. The chelated nutrient cations remain soluble at a much higher pH than in their unchelated forms. Iron, manganese, copper and zinc can be chelated. If iron salt is added to a soil of alkaline reaction the iron is precipitated as illustrated in the undermentioned equation.

$$\underset{\text{Soluble}}{Fe^{+++} + 3(OH)^-} \rightleftharpoons \underset{\text{Insoluble}}{Fe(OH)_3}$$

If iron chelate is applied to the soil, then it remains in the soluble form and therefore available to plant roots. Hence the primary function of the chelate is to keep the micronutrient cation in the soluble form at a higher pH than at would remain in its unchelated form.

Chelates vary in their stability. Except in soils of very high pH, iron chelates tends to before stable than those of copper and zinc which in turn are more stable than that of manganese. Thus iron is more strongly attracted by the chelating agents than other micronutrient cations. Consequently if a zinc chelate is applied to a soil containing significant quantity of iron in the soluble form, then the iron chelate is formed as shown below:

$$\text{Zn-chelate} + Fe^{+++} \rightleftharpoons \text{Fe-chelate} + Zn^{++}$$

As the iron-chelate is more stable than the zinc chelate, the above-mentioned reaction proceeds in the forward direction.

Chelates of other micronutrients have also been used successfully because the replacement of iron in the soil is slow enough to allow absorption of the micronutrients by the plant roots.

The undermentioned synthetic chelating agents have found to be effective for ions indicated below.

(*i*) Ethylene diaminetetra acetic acid, EDTA is suitable for iron, zinc, manganese and copper.
(*ii*) Ethylenediamine di-o-hydroxyphenyl acetic acid E DDHA is suitable for iron only.
(*iii*) Diethylenetriamine penta acetic acid, DTPA is also Suitable for iron only
(*iv*) Nitrolo triacetate acid, NTA is suitable for Zn only.
(*v*) Hydroxyethylethylene diamine tetraacetic acid HEDTA is suitable for iron and zinc only.

11.7. ESSENTIALITY OF NUTRIENT BALANCE

According to Liebig's Law of Minimum, plant growth is regulated by the factor present in the minimum amount and rises or falls according to the increase or decrease of this. Thus the growth and yield of the crop is adversely affected if any nutrient is present in the soil in an unbalanced proportion.

Several nutrients are known to interact with each other which means that the yield of the crop may be increased by the simultaneous addition of these two nutrients increase the yield more than the sum of the increase in the yield of the crop due to the separate addition of two nutrients to the soil.

The utilisation of a nutrient by the crop to increase its yield depends on the availability of some other nutrient. Fir example, liming of acid soils, increases the availability of molybdenum to crops growing in them. The utilisation of phosphate by crops enables them to take up more molydbenum from the soil. Crops absorb calcium and boron in a definite ratio. The assimilation of zinc by crops depends on their proper utilization of copper.

If excessive amounts of some nutrient(s) is/are applied to the soil or is/are present in it, the absorption of some other nutrient(s) is/are decreased. For example, excessive amount of calcium in the soil decreases the absorption of magnesium by crops. Similarly the excessive amount of magnesium in soils decreases the absorption of calcium more than that of potassium. When liberal doses of potassic fertilisers are added to the soil, the absorption of calcium by the crop, is decreased more than that of magnesium. Some micronutrients catalyze some specific biochemical reactions in crops which may be poisoned by the presence of excessive amounts of some specific micronutrients as mentioned below:

1. Crops cannot assimilate molybdenum if excessive amounts of copper and sulphur are present in the soil.
2. The presence of excessive amount of zinc, manganese or molybdenum induces iron deficiency.
3. Crops cannot assimilate iron, zinc and copper if excessive amounts of phosphate are present in the soil.
4. Addition of excessive amounts of nitrogenous fertilizers encourage deficiencies of copper and zinc.
5. Crops cannot absorb sufficient amounts of manganese if excessive amounts of potassium and sodium are present in the soil.
6. Crops cannot assimilate a sufficient amount of boron from calcareous soil.
7. Crops cannot absorb manganese in sufficient qualities, if an excessive amount of iron, copper or zinc are present in the soil.

Fodder grasses that have been raised on phosphate deficient soils may contain so little phosphorus that the health of the cattle to grazing on them is adversely affected. If forage grasses have been grown in the presence of adequate nitrate, but absence of adequate molybdenum, they will accumulate toxic amounts of nitrate.

11.8. NUTRIENT ABSORPTION BY THE ROOT

Soil nutrients occur as the constituents of easily decomposable minerals and ions in the soil water, as well as exchangeable ions absorbed on the surface of the clay and humic particles. The soil water containing nutrient ions is known as the soil solution. Whenever nutrient ions are absorbed on the surface of the root from the soil solution, these move from the surface of the clay and humic particles to the soil solution in the root zone, since they are in equilibrium. The nutrient ions also move from the soil solution away from the root zone to the soil solution in the root zone by mass flow and diffusion. Mass flow of nutrients means the movement of nutrient ions with the flow of water. Diffusion of nutrients means the movement of nutrients ions from the zone of higher nutrient ion concentration to the zone of lower nutrient ion concentration. Roots also rapidly grow towards the surrounding areas of higher nutrient ion concentration where they intercept the nutrient ions. This phenomenon is known as the root interception. So the nutrient ions reach the surface of the root. The cytoplasmic membrane adjacent to the cell wall of the root is known as the plasmalemma, and is impermeable to ions. The plasmalemma contains some amphoteric compounds which combine with the nutrient ions to form the carrier-ion complex that passes across the root membrane and reaches the vacuole where the carrier-ion complex breaks up and the ions are released. The carrier then returns to the outer surface of the plasmalemma where it combines with another ion of the same species, carried it across the plasmalemma and leaves it in the vacuole. This process is continuous infinitely. Different carriers are required to transport different ions. For example different carriers are required to transport nitrate and orthophosphate ions. Calcium, magnesium and potassium also need different carriers to transport then across the plasmalemma to the vacuole.

11.9. SOIL FERTILITY VERSUS PRODUCTIVITY

The fertility of the soil is its capacity to supply proper amounts of different nutrients in the proper proportion for the growth of crops. This means that soil fertility is concerned with the amounts of plant nutrients present in the soil.

Soil productivity is the capacity of the soil to produce specific crops under specified systems of management. A productive soil must be fertile and well-aerated, and also requires proper amounts of water either as rain or by irrigation.

11.10. MANURES AND FERTILISERS AND SOIL AMENDMENTS OR AMELIORENT AND SOIL CONDITIONERS

Manures are usually bulky in nature, and are formed from the decomposition of waste organic materials by. They contain small amounts of almost all plant nutrients. Manures are added to soils to improve their productivity. Fertilisers are usually concentrated inorganic materials usually containing high amounts of one or two nutrients in a form usable by plants.

Soil Amendments/Ameliorents bring about certain improvements in the condition of certain problem soils, just to make them more suitable for plant growth. For example, lime and gypsum are added to acid and alkaline soils respectively to control soil acidity and soil alkalinity respectively.

Materials which are applied to the soil improve or maintain its physical condition have been designated as the soil conditioners. Various synthetic organic polymers with a long flexible molecule are effective in stabilizing the soil structure by bridging across adjoining particles in a manner similar to the naturally occurring polysaccharide.

An American chemical company Montsanto produced vinyl acetate-maleic anhydride and hydrolysed polyacrylonitrile that were released in the market under the trade name of Krillium. These materials can stabilize soil structure whenever they are added @ 0.01 to 0.1 per cent by weight to the soil. Later polyvinyl alcohol (PVA) and poly acrylamide (PAM) have come into use and have shown promise as soil conditioners.

The uncharged polyvinyl alcohol can be manufactured with a range of molecular weight. Uncharged polymers can diffuse in the soil pores more easily than the charged ones because the former is absorbed by weak hydrogen bonds of van der Waal forces. Lower their molecular weight, more they diffuse into peds. Soil conditioners cannot be extensively used since they are very costly.

11.11. FARM YARD MANURE

Farm Yard Manure (FYM) is the material resulting from the decomposition of the mixture of animal dung and urine soaked litter by micro-organisms.

Loss of Plant Nutrients During the Collection of Animal Dung and Urine and Preparation of Farmyard Manure

(*i*) In India, a major portion of cattle dung is used as fuel by farmers, after being converted into dung cake. Urine is completely soaked into the earthern floor of the cattle shed, thus wasting a major portion of the cattle dung and urine.

(*ii*) The mixture of cattle dung and urine soaked litter is usually left exposed to hot sun and heavy rain when urea of urine and dung is rapidly decomposed to ammonia and is lost as shown in the equation.

$$\begin{matrix} NH_2 \\ | \\ CO \\ | \\ NH_2 \end{matrix} + 2H_2O \xrightarrow[\text{Enzyme}]{\text{Urease}} (NH_4)_2CO_3 \rightarrow 2NH_3 + CO_2\uparrow + H_2O$$

Cattle dung is rapidly decomposed by high temperatures, when the complex organic form of nutrients contained in them are converted to simple inorganic forms of nutrients which are washed down by high rainfall.

How to Reduce Loss of Nutrients From Cattle Dung and Urine

(*i*) By using a better method of collection of dung and urine. Farmers should be provided with other sources of cheap fuel. A Biogas plant should be provided to them at a subsidized rate. The floor of the cattle shed should be cemented to prevent from soaking in the urine. Sufficient quantity of waste organic material should be used for soaking the urine.

(*ii*) By the use of chemical preservatives. Urea of urine and dung is decomposed by the urease enzyme to ammonium carbonate, which is very unstable and decomposes to ammonium carbonate, carbon dioxide and water. So gypsum or single super phosphate should be spread evenly over the litter. These chemical preservatives reduce the loss of ammonia by converting ammonium carbonate to ammonium sulphate as shown in the following equations. If gypsum is used, ammonium sulphate and calcium carbonate are formed.

$$\begin{matrix} NH_2 \\ | \\ CO \\ | \\ NH_2 \end{matrix} + 2H.OH \xrightarrow[\text{Enzyme}]{\text{Urease}} (NH_4)_2CO_3$$

$$(NH_4)_2CO_3 + \underset{\text{Gypsum}}{CaSO_4} \longrightarrow (NH_4)_2SO_4 + CaCO_3$$

when manure + heap dries up, the ammonium sulphate reacts with calcium carbonates to re-form ammonium carbonates, which are then decomposed to ammonia, carbon dioxide and water. If a single super phosphate is used, then ammonium sulphate and insoluble tricalcium phosphate are formed, as shown in the following equation:

$$2CaSO_4 + Ca(H_2PO_4)_2 + 2(NH_4)_2CO_3 \longrightarrow Ca_3(PO_4)_2 + 2(NH_4)_2SO_4 + 2H_2O + 2CO_2$$

Single super phoshate Tricalcium Phosphate

Thus single super phosphate is better than gypsum because tricalcium phosphate does not react with ammonium sulphate and there is no loss of ammonia.

(*iii*) A better method of preparation of farm yard manure (FYM) : The manure pit should be about 0.9 metres deep, 1.8 to 2.4 metres wide and of any suitable length. The length of the manure pit should be about 5 meters if the farmer has 2 pairs of animals. The manure pit should be provided with a roof to protect the FYM from the hot sun and heavy rain. The mixture of cattle dung and urine soaked litter should be directly taken to the manure pit and evenly spread uniformly at its bottom. If necessary, water should be uniformly added to it. The daily collection of the mixture of cattle dung and urnie soaked litter is spread evenly over the previous layer. This is continue till the manure heap rises about 30 cms above the ground. It is then watered thoroughly and mud plastered. The farm yard manure is ready in about six months. In India, Farm Yard Manure contains (Percentages on dry matter) about 0.32 per cent nitrogen, 0.05 per cent phosphorus (P), 0.25 per cent potassium (K), 1.20 per cent calcium (Ca) and 0.33 per cent magnesium (Mg).

Use of Farm Yard Manure

Partially rotten farm yard manure should generally be applied 3 to 4 weeks before sowing the crops. This farm yard manure FYM will decompose in moist soil to improve the soil structure and release the nutrients contained in it, in the soluble form for growth of the crop. If it is applied too long before sowing the crop, the nutrients are lost by leaching by rain water. Well rotted farm yard manure should be thoroughly worked into the soil just before the crop is sown. The application of Farm Yard Manure and fertilisers to young vegetable and fruit plants have given the best results. As farm yard manure contains a relatively low amount of phosphorus, it should be used in conjunction with single super phosphate (Bone meal to acid soils) as a basal dose and nitrogenous fertilisers should be used as a top dressing.

Direct Use of Cowdung as Manure

Since fresh cowdung contains very little available plant nutrients, it should not be applied to the field just before sowing because micro-organisms assimilate all the available plant nutrients in the soil, and crops are therefore adversely affected by the lack of available nutrients. However, cowdung may be applied to the field about one month before sowing seeds or transplanting crops, to allow it to decompose. It should be thoroughly mixed up with the soil by plaughing and harrowing. It will decompose in about a month to improve the soil structure and to release the nutrients contained in it, in the soluble form for the growth of the crop.

11.12. COMPOST

Compost or artifical or synthetic farm-yard manure is formed from the decomposition of plant waste materials by micro-organisms.

Principles of Compose Manufacture

Waste plant materials contain nutrients in the form in which plant roots cannot absorb them. Waste plant materials are decomposed by micro-organisms when complex organic form of nutrients like proteins are converted to their respective simple inorganic form in which plant roots can absorb them. The plant waste materials are decomposed by microorganisms in the presence of sufficient available nitrogen, phosphorus, potassium and other essential elements which are used to synthesize cell protoplasm. If the waste plant material contains 1.2 per cent nitrogen on a dry weight basis, no extra soluble nitrogen is to be added to it. Usually, however, plant waste materials contain about 0.5 per cent nitrogen, so about 0.7 parts of soluble nitrogen should be added to every 100 parts of the plant waste material, so that it is decomposed by micro-organisms to release the nutrients crops need for them growth.

Processes of Compost Manufacture

(*i*) *Indore process:* The manure pit is provided with a roof to protect it from hot sun and heavy rain. It should be about one metre deep, two metres wide and five metres in length if the farmer possesses two pairs of animals. About 25 kg of dry waste plant material is spread under each pair of animals as bedding. The next morning, the mixture of cattle dung and urine soaked litter is taken to the manure pit and is spread uniformly at the bottom. A slurry made of 4.5 kg cattle dung + 3.5 kg of urine earth + 4.5 kg of fungus inocculum + 0.5 kg of wood ash in 4 gallons of water, is uniformly applied over it. The fungal inocculum is the material taken from an old compost heap which contains the micro-organisms. Water is uniformly sprinkled over this material. In this way, the compost pit is gradually filled up day-by-day when the layer of the mixture of cattle dung and urine soaked litter alternate with the layer or slurry and water till it rises about 30 cms above the ground. The composting material is dug out twice, 15 and 30 days after the manure pit has been filled up, water is sprinkled over it, it is thoroughly mixed and replaced in the pit. It is taken out of the pit after one month, water is sprinkled over the composting material which is thoroughly mixed, and a heap is made above the ground below a shed. It is ready in 3 to 4 months.

The main advantage is that well decomposed manure is obtained in a relatively shorter period. The disadvantage is the considerable labour cost due to turning and watering. Some organic matter and nitrogen are lost.

(*ii*) *Bangalore method*. A roof is provided over the manure pit whose dimensions are about the same as in the Indore method. About 18 kg of waste plant material is spread as bedding under each pair of animals at the cattle shed for the night. The next morning the mixture of cattle dung + urine soaked litter is taken to the manure pit, which is filled in sections of 60 to 120 cms at a time. A mobile wooden or bamboo partition is placed across the manure pit at a distance of 60 to 120 cms from one end. The mixture of cattle dung and urine soaked litter is spread uniformly in this section. Water is uniformly applied to it. Everyday a mixture of cattle dung and urine soaked litter is spread over the previous layer in the pit. When this section has been filledup, the temporary partition is removed and a similar section is made next to it. This second section is filled up in the same manner. The manure trench is filled up section wise in this manner. Then it is thoroughly watered and plastered with mud. The manure is ready in about six months.

Advantage: The cost of labour is saved because it is not stirred. The material slowly decomposes and therefore less organic matter and nitrogen is lost.

Disadvantage: It takes longer time it is not well decomposed.

(iii) Coimbatore method : A roof is provided over the manure pit. It is three feet deep and four feet wide and its length depends upon the quantity of material available. Waste plant material is spread uniformly at the bottom of the pit. It is throughly moistened with a suspension of cowdung in water (@ 10 kg cowdung in 50 litres of water) and about half to one kilogram of powdered bone meal is sprinkled over it. Then it is thoroughly watered. Again a layer of waste plant material is placed over the previous layer, it is moistened with the suspension of cowdung in water and powered bone meal is uniformly applied to it. In this way, the manure trench is filled up until the material in it rises about 60 cms about the ground level. The composting material is taken out of this manure pit after two months, water is applied to it. It is thoroughly mixed and a rectangular heap is made below a shed and left undisturbed. Occasionally, water is sprinkled over it. It is ready in about 6-7 months.

Well prepared compost usually contains 0.8 to 1.0 per cent of nitogen, has all the properites of farm yard manure and can be used the same way.

11.13. GREEN MANURE

Green manure is the fresh green plant material which is mixed up with the soil by ploughing. This material is decomposed by soil micro-organisms to release the nutrients contained in it for the growth of crops. It the crop is grown on the field itself and then ploughed the process is called green manuring. If green plant material is brought from elsewhere and ploughed, the process is called green leaf manuring.

Important Green Manuring Crops

Green manuring crops must grow rapidly on varieties of soil and must decompose quickly enough to release the nutrients they contain in the soluble form, for the growth of crops. A few most important green manure crops are as follows:

(i) Sunhemp: Crotalaria juncea is most suitable for loamy soils. It is sown in May or June when the monsoon breaks. It grows very fast and to a height of one to two metres. It can grow even on poor soils and adds about 20 to 25 tonnes of fresh green plant material per hectare to the soil. Sunhemp is the most outstanding green manure crop. It is well suited to almost all parts of India and

accommodates the growing seasons of sugarcane, potato and garden crop, irrigated wheat in north India and second season paddy crop in South India.

(*ii*) *Dhaincha:* Sesbania bispinosa, it can tolerate waterlogging and salinity and dry conditions if the germination has been good. It grows up to a height of 1.5 to 1.8 metres in water logged paddy fields within a very short period of time. It is best suited for loamy or clayey soils where it adds about 10 to 20 tonnes of green material per hectare. It is mainly grown as a green manure crop in Assam, Bihar, West Bengal, Crissa and Tamil Nadu.

(*iii*) *Sesbania speciosa:* It can also tolerate waterlogging and dry periods. It is suitable for loamy or clayey soils. It grows very slowly during the earlier stages. It is also suitable for paddy field and adds about 25 to 20 tonnes/hectare of green plant materials.

(*iv*) *Mung (Phaseolus ausreus) is a pulse crop:* The variety Mung Type 1 which is sown in the first week of July matures and yields about 3 to 4 quintals of seed per hectare by the first week of September. After picking the pods, the crop is ploughed down as green manure and wheat may be sown in the first week of November.

(*v*) *Pillipesara, (Phaeolus trilobus):* It is a minor pulse crop. It is mainly suitable for clayey soil where it grows very fast, completely covers the surgace and is quickly decomposed. About 30 kg of seed is sown in an hectare field where 10 kg of nitrogen and 30 kg of phosphorus (P_2O_5) has been applied. It adds about 8 to 10 tonnes of green material per hectare.

(*vi*) *Gaur (Cyamopsis tetragonoloba):* It is a good fooder cum green manuring crop in the dry regions of north-western India where it grows even on poor soils.

(*vii*) Other pulse crops like horse gram, cowpea etc. can also be grown as green manure crops.

Important Green Leaf Manuring Plants

Some plants are grown on field bunds and waste land. Their green foliage is added to the soil. Therefore organic matter and plant nutrients are added to the soil from outside sources. A few most common plants are:

(*i*) Gliricida maculata grows on most soils to a height of about four and half metres. One cutting should be taken in June when the monsoon breaks, and another one in December. Each plant supplied about 22 to 23 kg of green material containing about 0.48 per cent nitrogen.

(*ii*) *Pongamia glabra.* Each plant supplies about 130 kgs of fresh green material containing about 0.56 per cent nitrogen.

(*iii*) Sesbania speciosa is grown around the borders of paddy fields and irrigation channels. Sesbania speciosa seedlings planted on the field bunds four inches apart around the paddy field of hectare, produces about 3000 kgs of green leaves per hectare.

Technique of Green Manuring

Usually the field in ploughed, a little fertiliser *i.e.* 10 kgs of nitrogen and about 30 kgs of phosphorus (P_2O_5) per hectare is added to the soil. The fields is again ploughed and about 50 to 60 kgs sunhemp or Dhaincha seeds are sown in June when the monsoon breaks. These grow rapidly and are ploughed after 5 to 6 weeks, just before the flowering stage when they are the most succulent. The main crop is sown after another two months when the green plant materials have been thoroughly decomposed. Dhaincha and sunhemp add about 75 kgs of nitrogen per hectare to the soil.

Advantages and Disadvantages of Green Manuring

Advantages: (*i*) Leguminous green manure crops add nitrogen to the soil.

(*ii*) Green manure crops usually deep rooted. They take nutrients from the deeper layers and leave them on the surface in the form of decaying organic matter. In green leaf manuring, nutrients are brought from elsewhere and added to the field.

(*iii*) Green manure, while it decomposes, improves the soil structure.

Disadvantages: (*i*) It is difficult to practice green manuring under rainfed condition.

(*ii*) It is not convenient to include a green manure crop in crop rotation in areas where intensive farming is practised.

Economics of Green Manuring

Wheat, paddy and sugarcane have responsed to green manuring under irrigated conditions in India. The results of experiments conducted in different parts of India have shown that it is more remunerative to grow legumes for grain and/or fodder and to apply moderate amounts of fertilisers to the succeeding crops, then to grow them for green manuring only.

Nowadays, agricultural research has advanced enough to produce a large number of short duration high yielding cereal crops like paddy ('TN.1', 'TN.3', 'T.65') wheat ('Sonara 64', 'Lerma Roja', 'S. 227'), many hybrids of maize ('Ganga 2', 'Ganga 3') and Sorghum ('CSH. 1', CSH. 2'). Prices are also increasing steeply. So farmers are trying to grow as many crops as possible every year if irrigation water is available. Green manure crops cannot be accommodated in these multiple cropping rotation systems. The opportunity cost of green manure in wheat rotation has been assessed to be equivalent to 40 to 50 quintals per hectare of paddy, maize or pearl millet.

Green manuring may be practised under special circumstances *e.g.* to reclaim alkali soils. Actually it is a very costly source of nitrogen and therefore can not complete with relatively cheap nitrogenous fertiliser which, when applied with crop residues, will increase the yields of crops and sustain the fertility of the soil.

11.14. OIL CAKE

Oil cake is the residue left after extraction of oil from oil seeds. When oil cakes are added to the soil, they are quickly decomposed by micro-organisms.

Oil cakes are rich in nitrogen, phosphorus and potash. They also supply micronutrients to the soil. For example, mustrad cake contains about 5.2 per cent N, 1.8 per cent P_2O_5 and 1.2 per cent K_2O. Usually the nitrogen content varies from 2.5 to 8.0 per cent, phosphoric acid from 1.0 to 2.5 per cent and potash from 0.5 to 3.0 per cent. Some oil cakes like groundnut oil cake decompose quickly and make available the nutrients contained in it for plant growth. Others, for example, castor cake, decompose relatively slowly. Usually oil cakes take more time to decompose if they still contain some oil. Groundnut cake contains about 7.3 per cent N, 1.6 per cent P_2O_5, and 1.3 per cent K_2O while castor cake contains 4.4 per cent N 1.8 per cent P_2O_5 and 1.4 per cent K_2O. Sesame or till cake contains about 6.2 per cent N, 2.1 per cent P_2O_5 and 1.3 per cent K_2O. Oil cakes are also called concentrated organic manures.

11.15. NIGHT SOIL, SEWAGE AND SLUDGE

Night Soil

Night soil is human excreta, which is very rich in plant nutrients. Agricultural chemists have found that fresh human faeces contain about 1.5 per cent N, 1.1 per cent P_2O_5, 0.5 per cent K_2O and 1.0 per cent CAO. Although human urine, which contains about 0.6 per cent N, 0.1 per cent P_2O_5, 0.5 per cent K_2O and 0.3 per cent CaO, can be directly applied to the field, hygeinically fresh human faeces cannot be directly applied to the field. This material may be dried in shallow beds after it is covered with soil for about eight to ten days, when it is known as poudrette which contains about 1.1 per cent N, 0.7 per cent P_2O_5 and 0.9 per cent K_2O. Poudrette may also be prepared in trenches of about twelve feet length three feet width and one foot deep. Night soil is uniformly spread in a thin layer at the bottom of the pit and is covered with another layer of soil The trench is filled in this manner. The material dries in the trench in about eight to ten months when it contains about 1.8 per cent N, 1.9 per cent P_2O_5 and 1.1 per cent K_2O. In the Nasik system, organic refuse is added to human excreta. This mixture is then deposited in trenches.

Sewage and Sludge

Human excreta is flushed by water through underground sewage pipes. This known as the sewerage system. Human faeces and urine mixed with water is known as raw or crude sewage, which consists mainly of a colloidal solution of human excreta rich in the available forms of plant nutrients. Its application to the field therefore increases crop yields. The colloidal material in raw sewage may choke the soil pores and nake the soil sewage sick if it is continuously applied to the field. Therefore, its continuous application to the field should be avoided.

The sewage is treated in either of the following to make it more suitable as a manure.

(*i*) Septic tank method,
(*ii*) Filter bed method,
(*iii*) Activated sludge process.

All these processes work on the principle of biological oxidation of faecal material either aerobically or anaerobically; a part of it is converted to methane, hydrogen and ammonia, another part remains in solution and a third part settles down to the bottom which is known as sludge, and the liquid portions known as effluent. In the septic tanks method, the sewage is left undisturbed in the septic tank, where the faecal material is anaerobically decomposed to evolve considerable amounts of methane, hydrogen and ammonia, although considerable amounts of colloidal faecal materials remain in the effluent. In the Filter Bed Method, the sewage is allowed to percolate through filters where the faecal material aerobically decomposes. A part of it is converted to gas, another part remains in solution, and a third part is thrown down as the sludge. The effluent is almost entirely clear. In the Activated Sludge Process, air is forced through the sewage, and almost all the faecal material is oxidised, and only a little settles down to the bottom as sludge, leaving the effluent as clear as water.

The effluent obtained from the filter bed method and the activated sludge process can safely be directly applied to the field, but the septic tank effluent should be carefully used because it still contains lots of objectionable colloidal materials and pathogens. The effluent is usually rich in ammoniacal nitrogen in the septic method while the effluent obtained by the activated sludge process is usually rich in nitrates. The sludge is rich in nitrogen and phosphorus and poor in potassium.

Sullage

Sullage is the liquid which passes in the open drains, mainly containing human urine and kitchen washing. It can be directly used as manure. It does not contain large amounts of solid or colloidal material as in Sewage.

11.16. CONCEPTS OF BIOLOGICAL GAS AND USE OF SLURRY

Principles

Animal dung is anaerobically decomposed in biogas plants to liberate large volumes of combustible gases like methane and hydrogen, the mixture of which can be used as a cooking gas. The residue left after anaerobic decomposition of cattle dung is richer in nutrients than the original animal dung, because large amounts of cellulosic material in the cattle dung has already decomposed to yield the combustible gases.

Bio-gas Plant

The bio-gas plant has been diagramatically represented in the Fig. 11.1. The size varies with amount of cowdung available. If 50 kgs of dung is available from three animals, the bio-gas plant consists of a masonary fermentation tank or a digester, 3.6 metres deep and with an internal diameter of 1.5 metres in its lower two third portion. A steel drum of diameter 1.5 metres which acts as the gas holder is inverted over the fermentation tank. It is held by three iron chains that pass over three pulleys which have been fitted at the top of three iron posts (Only two have been shown in Fig. 11.1). A small masonary tank has been built by the side of the fermentation tank. A pipe joins the bottom of the small tank with the bottom of the digester or the fermentation tank. A small outlet has been provided at one side of the fermentation tank for overflow of the digested cattle dung. Similarly, an outlet is also present for the gas, provided with a stop cock or a regulator.

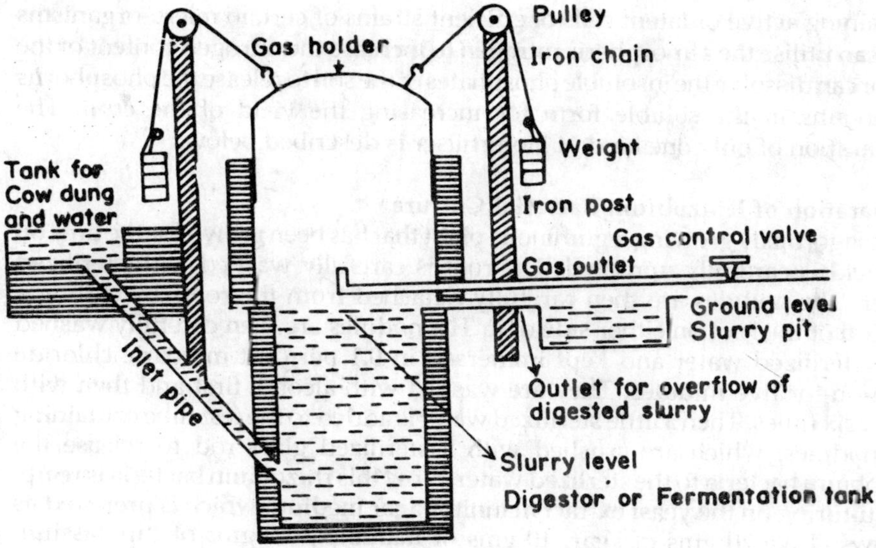

Fig. 11.1. Bio-gas Plant

To begin the operation of the bio-gas plant, 50 kgs of dung are mixed with 50 litres of water. Then this slurry and 50 kgs of slurry from another bio-gas plant is poured in the small tank. This 50 kgs slurry from another gas plant serves as the inoculum to start the fermentation process in the fermentation tank or digester. These materials go down into the fermentation tank and begin to decompose within about ten days, when another mixture of about 250 kgs of cattle dung and about 250 litres of water are added to the inlet tank. As more gas is evolved, further quantities of the mixture of cattle dung and water is added to the inlet tank every alternate day, until the fermentation tank or digester is filled up. The gas is collected in the collecting drum which slowly rises. This gas collecting drum is suitably counterposed with weights to facilitate the collection of gas. When the gas collecting drum is completely filled, it rises to its maximum. The gas may be used for cooking, by opening the gas stop cock.

Gas production is maximum in the summer. The optimum temperature for maximum gas production is 30°C to 35°C.

Use of Digested Slurry as Manure

The digested slurry that overflows through the outlet is richer in plant nutrients than the original dung. It contains about 2.0 to 2.4 per cent nitrogen on oven dry basis.

11.17. BIOFERTILISER

The process of mixing the cultures of certain micro-organisms with the seed or the soil is known as inoculation. The Microbial inoculant of seeds of certain crops or soils are known as biofertilisers, which are actualy preparations containing active or latent cells of efficient strains of certain micro-organisms that can utilise the atmospheric nitrogen to increase the nitrogen content of the soil or can dissolve the insoluble phosphates of the soil to release the phosphorus it contains in the soluble form for increasing the yield of the crop. The preparation of only one kind of biofertiliser is described below:

Preparation of Rhizobium Bacterial Culture

A vigorously growing leguminous plant that has been grown for the time on the field is carefully uprooted. The root is carefully washed with sterilized water. The nodules are then carefully detached from the roots by cutting a portion of the root on either side of it. The nodules are then carefully washed with sterilized water and kept immersed in 0.1 per cent mercuric chloride solution for five minutes. They are washed with alcohol first and then with water six times. Then a little sterilized water is added to the test tube containing the nodules, which are crushed with a sterilized glass rod to release the rhizobium bacteria to the sterlized water. Now this rhizobium bacteria is ready for culturing on the yeast extract mannitol-agar medium which is prepared as follows: Take 20 gms of agar, 10 gms of mannitol, 0.5 gms of dipotassium hydrogen phosphate, 0.2 gm of magnesium sulphate, 0.1 gm of sodium chloride, 3 gms of powered calcium carbonate, 1 gm of yeast extract and 1000 ml of distilled water in an Erlenmeyer flask, plug its mouth with cotton wool and sterilize it in the autoclave, at a pressure of 15 pounds per square inch for 15 minutes. Then pour it into a few sterilized test tubes. Then a little water containing the rhizobium becteria is added to each of these test tubes with the help of sterilized platinum loop. Thoroughly mix the bacteria with the culture medium by gently rotating the test tubes. Than pour the contents into one of these test tubes containing the bacteria in agar medium into a sterilized petridish. Pour a little of the sterilized agar medium to these empty test tubes containing drops of agar medium containing the bacteria. Shake it gently and pour its contents into another sterilized petri dish. Maintain the temperature of the agar medium at 45°C and spread the contents of the petri dish uniformly over it, by gently rotating it in a clockwise manner. When the agar medium has solidified, invert it. Put the petridish and the test tube containing the bacteria in a slanting position in the incubator at a temperature of 30°C. Usually bacterial colonies appear after four to five days. They resemble shining waterdrops.

Preparation of Rhizobium Bacterial Culture in Soil

Thoroughly mix 1000 gms of soil, 10 gms of sugarcane powder, 10 gms of legume hay powder, 0.9 gms of potassium sulphate, and 120 ml water. The soil is low in lime, also add 0.5 gms of powdered calcium carbonate to it.

Put 400 gms of the above mixture in each tin. Then sterilize tins containing the mixture for 2 hours, at 9 kgs pressure for two consecutive days and again on the fourth day.

Take 150 mls of the yeast extract-mannitol-medium in a 500 ml conical flask; plug it with cotton wool and sterilize it at 6.8 kgs of pressure for half an hour. This sterilized medium is then inoculated with the rhizobium bacterial culture from the slant of one test tube where the rhizobium bacteria have grown for 24 hours. Shake the flask in the shaker for 24 to 72 hours, depending upon the generation time of the organisms. Then add 100 ml of the bacterial culture from the flask to the soil contained in the tin. Mix it thoroughly with the soil. Incubate the soil in the tin for a week at 28°C.

Use of this Bacterial Culture in Soil

Dissolve 150 gms of gur (raw indeginous sugar) in 1000 ml of water. Boil for half an hour and the cool. Add one tinfull of the soil culture to this gur solution and mix thoroughly. This gur solution containing the soil culture is then thoroughly mixed with the seed meant for sowing one hectare of land. The seeds are then dried in the shaker for an hour and shown in this field.

It India, the nitrogen content of the soil was increased when the seeds of the leguminous crops that had been inoculated with the Rhizobium bacteria of the concerned cross inoculation group were sown. Fox example, the agricultural scientists of the Indian Agricultural Research Institute have found that the nitrogen content of the soil increased by 120 kg per hectare per year when barseem (Trifoliom alexandrinum) which had been inoculated with the Rhizobium bacteria of the concerned cross-inoculation group (Rhizobium trifolii), was grown.

11.18. NITROGENOUS FERTILISERS

Urea

Manufacture: Ammonia reacts with carbon dioxide under a pressure of about 400 atmospheres, at a temperature of about 160 to 200°C to form urea.

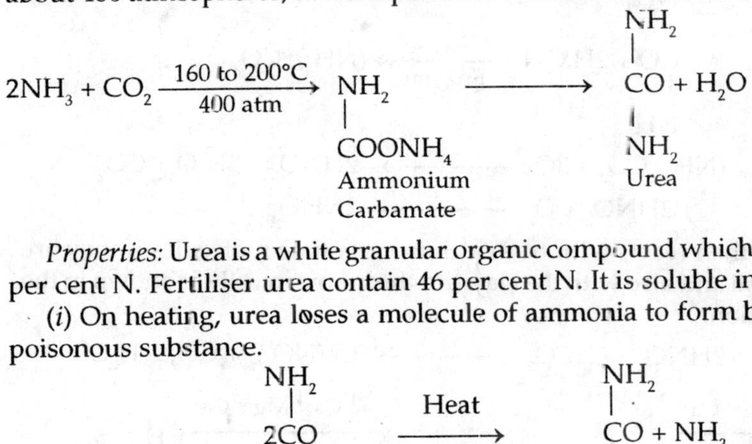

Properties: Urea is a white granular organic compound which contains 46.67 per cent N. Fertiliser urea contain 46 per cent N. It is soluble in water.

(*i*) On heating, urea loses a molecule of ammonia to form biuret which is poisonous substance.

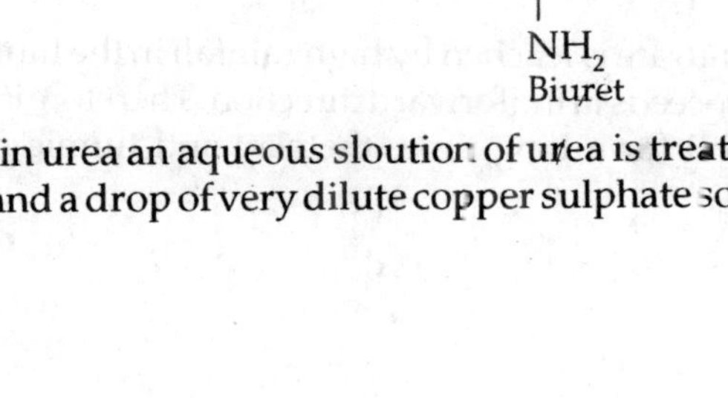

For detecting biuret in urea an aqueous sloution of urea is treated with sodium hydroxide solution and a drop of very dilute copper sulphate solution is added.

A violet colouration is produced if biuret is present in it.

(*ii*) Urea reacts with nitrous acid to produce nitrogen gas

$$\begin{array}{l} NH_2 \\ | \\ CO + 2HNO_2 \longrightarrow 2N_2 + CO_2 + 3H_2O \\ | \\ NH_2 \end{array}$$

(*iii*) When urea is boiled with dil. acid, alkali or water under pressure or when it is treated with the enzyme urease it decomposes to produce ammonia.

$$\begin{array}{l} NH_2 \\ | \\ CO + 2H.OH \xrightarrow[\text{Enzyme}]{\text{Urease}} (NH_4)_2CO_3 \longrightarrow 2NH_3 + CO_2 + H_2O \\ | \\ NH_2 \end{array}$$

Urea slowly decomposes to ammonia as shown in the above equation at a temperature of 80 to 90°F and relative humidity of 50-75 per cent. This decomposition of urea is accelerated in the presence of monocalcium phosphate. Hence urea should not be mixed with single super phosphate.

Reaction in Soil

When urea is added to the soil it is decomposed by the enzymes urease produced by micro-organisms to ammonium carbonate which is oxidized to nitric acid as shown below:

$$\begin{array}{l} NH_2 \\ | \\ CO + 2H.OH \xrightarrow[\text{Enzyme}]{\text{Urease}} (NH_4)_2\,CO_3 \\ | \\ NH_2 \end{array}$$

$$(NH_4)_2CO_3 + 3O_2 \longrightarrow 2HNO_2 + 3H_2O + CO_2$$

$$2HNO_2 + O_2 \longrightarrow 2NHO_3$$

This nitric acid then reacts with free calcium carbonate molecules if they are present in the soil and also with the clay and the humic micelle as shown in the following equation:

$$2HNO_3 + CaCO_3 \longrightarrow Ca(NO_3)_2 + CO_2 + H_2O$$

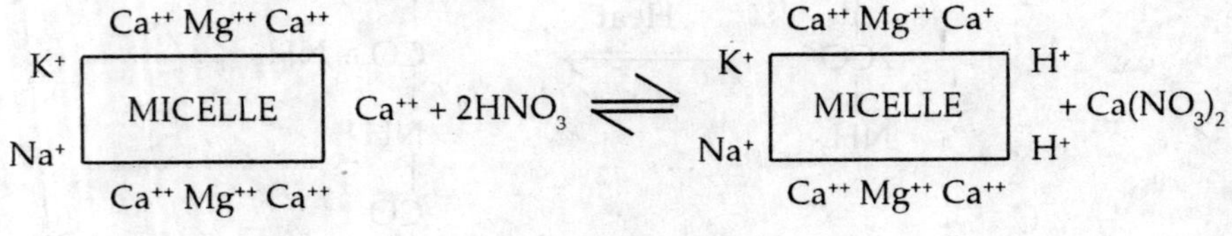

This calcium nitrate is leached by high rainfall in the humid region when the above reaction proceeds in the forward direction. Therefore if urea is continuously applied to the soil for a long time, the clay and humic micelle is gradually saturated with hydrogen ion and ultimately the soil becomes acidic in reaction if there is insufficient calcium carbonate. This soil acidity can be neutralized by adding 80 kgs of limestone for each 100 kg of urea. If rainfall is not high then calcium nitrate accumulates in the soil and is assimilated by crops. It may be applied at the sowing time or as top dressing to most soils, but should not be allowed to come in contact with the seed.

Ammonium Sulphate

Manufacture from coal: Coal contains about one per cent nitrogen. When it is heated in a closed retort in the absence of air ammonia and ammonia salts are formed. Slaked lime is then added to it. When it is reheated ammonia is evolved, which is absorbed in 60 per cent sulphuric acid, obtained by product from industries.

$$\text{Coal} \xrightarrow{\text{Heat}} NH_3$$

$$NH_3 + CO_2 + H_2O \longrightarrow (NH_4)_2CO_3$$

$$(NH_4)_2CO_3 + Ca(OH)_2 \longrightarrow 2NH_3 + CaCO_3 + 2H_2O$$

$$2NH_3 + H_2SO_4 \longrightarrow (NH_4)_2SO_4$$

(ii) Synthetic process: When Nitrogen and Hydrogen in the ratio of 1 : 3 are passed over heated (500°C) platenized asbestos or powdered iron at a pressure of 250 atmospheres, ammonia is formed which is passed through a suspension of gypsum in water through which carbon dioxide is also passes.

$$N_2 + H_2O \xrightarrow[250\ \text{atm.}]{500°C} 2NH_3$$

$$2NH_3 + CO_2 + H_2O \longrightarrow (NH_4)_2CO_3$$

$$(NH_4)_2CO_3 + CaSO_4 \longrightarrow (NH_4)SO_4 + CaCO_3$$

Calcium carbonate is removed by filtration

An aqueous solution of ammonium sulphate is evaporated when crystals of ammonium sulphate is formed which is cried and bagged.

Properties: Fertiliser ammonium sulphate is white to yellowish grey in colour. It contains about 20.5 per cent Nitrogen and 23.5 per cent Sulphur. It is neutral to litmus. It reacts with monocalcium phosphate of single super phosphate, potassium chloride (Muriate of Potash) and calcium carbonate in the presence of moisture as shown in the following equation.

$$(NH_4)_2SO_4 + Ca(H_2PO_4)_2 \longrightarrow 2(NH_4)H_2PO_4 + CaSO_4$$

$$(NH_4)_2SO_4 + 2KCl \longrightarrow 2NH_4Cl + K_2SO_4$$

$$(NH_4)_2SO_4 + CaCO_3 \longrightarrow (NH_4)_2CO_3 + CaSO_4$$

$$(NH_4)_2CO_3 \xrightarrow{\text{heat}} 2NH_3 + CO_2 + H_2O$$

Ammonium sulphate, on heating, decomposes as shown in the equation below:

$$(NH_4)_2SO_4 \xrightarrow{\text{Heat} < 300°C} NH_4.HSO_4 + NH_3$$

$$(NH_4)_2SO_4 \xrightarrow{\text{Heat} > 300°C} 4NH_3 + N_2 + 3SO_2 + 6H_2O$$

Reaction in Soils

When ammonium sulphate is applied to the soil, ammonium ions replace basic cations from the clay and humic micelle as shown in the following equation.

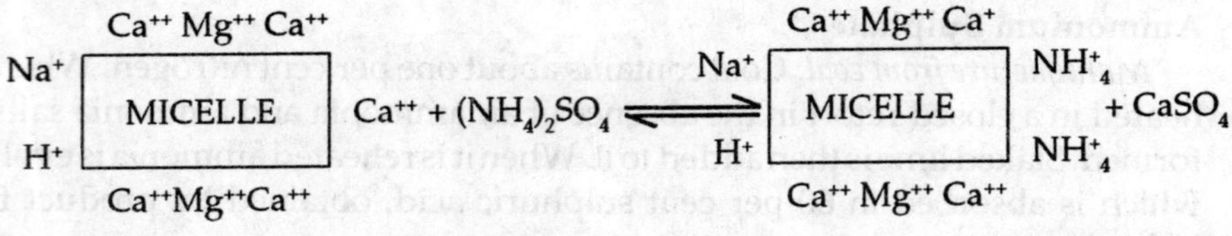

These absorbed ammonium ions are oxidized to nitrate as shown below:

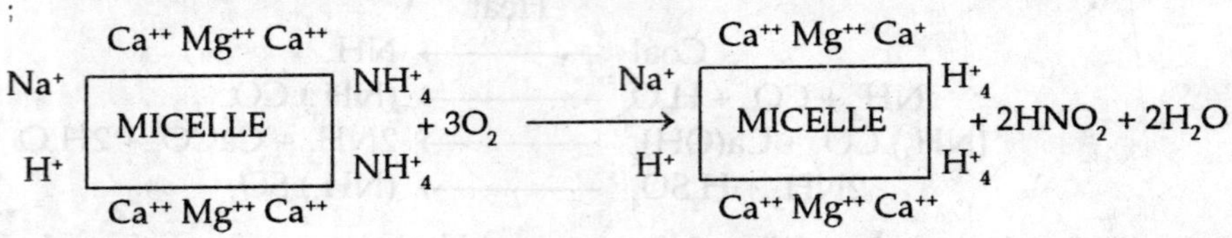

$$2HNO_2 + O_2 \longrightarrow HNO_3$$

This nitric acid then reacts with the calcium carbonate and the clay and humic micelle as shown below:

$$CaCO_3 + 2HNO_3 \longrightarrow Ca(NO_3)_2 + CO_2 + H_2O$$

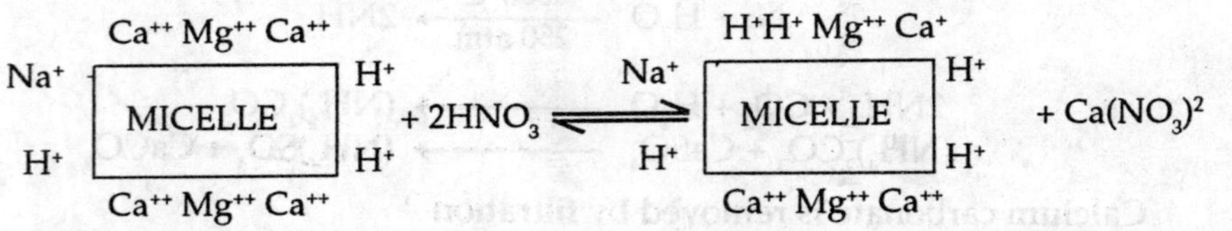

In the humid region, calcium nitrate is leached by high rainfall; the above reaction proceeds in the forward direction when the clay and humic micelle is gradually saturated with hydrogen ions; ultimately the soil becomes acidic in reaction, if it does not contain calcium carbonate. Soil acidity is caused by adding 100 kgs of ammonium sulphate can be neutralized by adding 110 kg of limestone. Some ammonium ions are fixed by certain clay minerals like illite, vermiculate etc.

Calcium Ammonium Nitrate (Nitrochalk OR A.N.L.)

Manufacture : When a mixture of Nitrogen and Hydrogen in the ratio of 1 : 3 is passed over heated (500°C) platinized asbestos at a pressure of about 250 atmospheres, ammonia is formed. $N_2 + 3H_2 \longrightarrow 2NH_3$. A part of this ammonia is mixed with air and passed over a red hot platinum gauge and nitric acid is formed.

$$4NH_3 + 5O_2 \longrightarrow 4NO + 6H_2O + 215 \text{ K.cal. of energy}$$

$$4NO + 2O_2 \longrightarrow 4NO_2$$

$$4NO_2 + O_2 + 2H_2O \longrightarrow 4HNO_3$$

Summing up

$$4\,NH_3 + 8O_2 \longrightarrow 4HNO_3 + 4H_2O$$

or

$$NH_3 + 2O_2 \longrightarrow HNO_3 + H_2O$$

This nitric acid then reacts with the remaining ammonia to form ammonium nitrate. $NH_3 + HNO_3 \longrightarrow NH_4\,NO_3$

This ammonium nitrate solution still contains some nitric acid which is neutralized by adding limestone.

$$2HNO_3 + CaCO_3 \longrightarrow Ca(NO_3)_2 + H_2O + CO_2$$

$$Ca(NO_3)_2 + NH_4NO_3 \longrightarrow Ca(NO_3)_2\,NH_4NO_3$$

Calcium Ammonium Nitrate

Properties: Calcium ammonium nitrate is a brown or light grey. It contains 20.5 per cent Nitrogen, one-fourth of which is ammoniacal, and three-fourths, nitrate. Calcium ammonium nitrate containing 25 per cent nitrogen is now being manufactured.

Reaction in soil: It reacts with soil as the ammonium sulphate does, but the soil does not become acidic in reaction even if there is no reserve calcium carbonate as shown in the following equation.

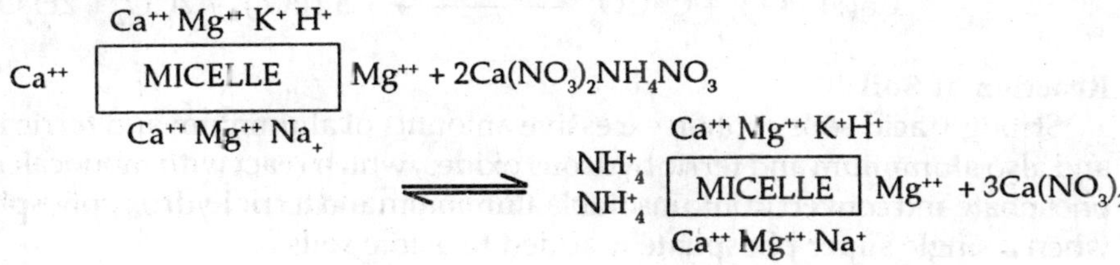

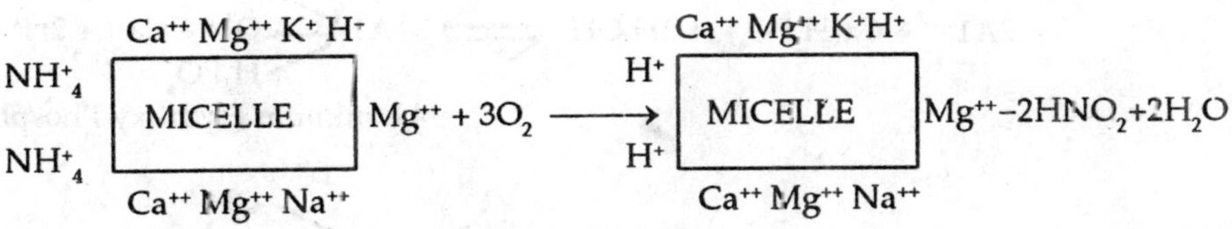

$$2NHO_2 + O_2 \longrightarrow 2\,HNO_3$$

This nitric acid reacts with calcium carbonate present in the soil:

$$2HNO_3 + CaCO_3 \longrightarrow Ca(NO_3)_2 + CO_2 + H_2O.$$

If the soil does not contain free calcium carbonate this nitric acid reacts with the clay and humic micelle.

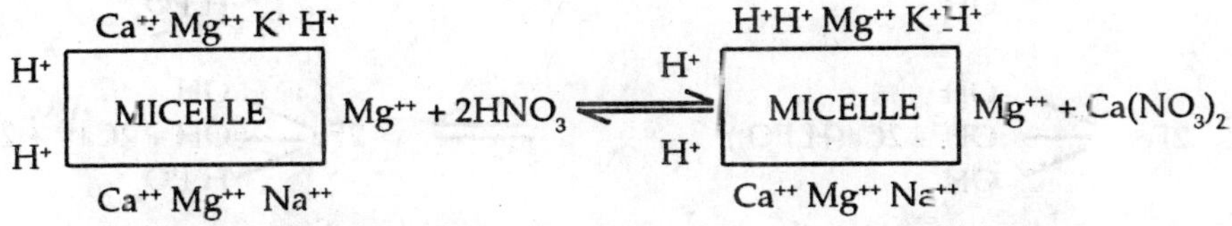

Since the soil contains enough calcium nitrate, the above reaction does not proceed in the forward direction. Hence the soil does not become acidic in reaction. Calcium nitrate is assimilated by crops.

11.19. PHOSPHATIC FERTILIISERS

Single Super Phosphate

Manufacture: Calculated amount of sulphuric acid is added to finely ground rock phosphate in a pit or den where the following reaction takes place:

$$\underset{\text{Flourapatite}}{[Ca_3(PO_4)_2]_3CaF_2} + 7H_2SO_4 \longrightarrow \underset{\substack{\text{Mono. Cal. Phos.} \\ \text{Single Super Phosphate}}}{3Ca(H_2PO_4)_2} + \underset{\text{Gypsum}}{7CaSO_4} + 2HF$$

The temperature rises to about 100°C when hydroflouric acid (HF) evaporates. The mixture of gypsum (about 50 per cent), Mono calcium phosphate (about 27 per cent) and dicalcium phosphate (about 3 per cent) is ground to powder and is bagged for marketing as Single Super Phosphate.

Properties: Single super phosphate is an ash coloured powder containing about 16 per cent water soluble P_2O_5, 20 per cent Ca and 12 per cent S. It should not be mixed with lime because water soluble monocalcium phosphate is gradually converted to water insoluble tricalcium phosphate as shown in the following equation:

$$Ca(H_2PO_4)_2 + CaCO_3 \longrightarrow Ca_3(PO_4)_2 + 2CO_2 + 2H_2O$$

Reaction of Soil

Strongly acid soils contain excessive amounts of aluminium and ferric ions and also aluminium and ferric hydrous oxides, which react with monocalcium phosphate and convert in into insoluble aluminium and ferric hydroxy phosphate when a single super phosphate is added to acidic soils

$$2A1^{+++} + Ca(H_2PO_4)_2 + 2H.OH \rightleftharpoons A1\left<\begin{array}{l} OH \\ OH + Ca^{++} + 2H^{+} \\ H_2PO_4 \end{array}\right.$$

Aluminium Hydroxy Phosphate

$$2Fe^{+++} + Ca(H_2PO_4) + 2H.OH \rightleftharpoons 2Fe\left<\begin{array}{l} OH \\ OH + Ca^{++} + 2H^{+} \\ H_2PO_4 \end{array}\right.$$

Ferric Hydroxy Phosphate

$$2A1\left<\begin{array}{l} OH \\ OH + 2Ca(H_2PO_4) \\ OH \end{array}\right. \rightleftharpoons 2A1\left<\begin{array}{l} OH \\ OH + 2Ca^{++} + 2(OH)^{-} \\ H_2PO_4 \end{array}\right.$$

$$2Fe\left<\begin{array}{l} OH \\ OH + 2Ca(H_2PO_4)^{2} \\ OH \end{array}\right. \rightleftharpoons 2Fe\left<\begin{array}{l} OH \\ OH + 2Ca^{++} + 2(OH)^{-} \\ H_2PO_4 \end{array}\right.$$

This above process of conversion of water soluble phosphate to water in soluble phosphate to known as fixation of phosphrous.

Strongly alkaline soils contain excessive amounts of calcium ions and calcium carbonate, which react with the monocalcium phosphate and convert it into water insoluble tricalcium phosphate, as shown in the following equations:

$$Ca(H_2PO_4)_2 + 2Ca^{++} + 4\,(OH)^- \rightleftharpoons Ca_3(PO_4)_2 + 4H.OH$$

$$Ca(H_2PO_4)_2 + 2CaCO_3 \longrightarrow \underset{\text{Tricalcium Phosphate}}{Ca_3(PO_4)_2} + 2CO_2 + 2H_2O$$

The most phosphate is available to crops in soils of pH 6.0 to 7.0.

Bone Meal

Bones have been used as manures for time immemorial. Bone meals are of two kinds *i.e.* (*i*) raw bone meal and (*ii*) steamed bone meal. Raw bone meal contains about 25 per cent P_2O_5 and 4 per cent nitrogen, which is in the slow acting organic form. Steamed bone meal contains about 25 to 30 per cent total phosphorus (P_2O_5) and about 1-2 per cent N. It contains about 25 per cent citrate soluble phosphorus (P_2O_5). Steamed bone meal can be readily ground to powder because it is more brittle. Bone meal is applied to the soil few days before sowing the crop. Boen meal is mainly used on acidic soils. Almost all the phosphorus is converted into a readily available form if bones are charred, although charring of bones destroys almost half the nitrogen.

Triple Superphosphates

Triple superphosphate is manufacured by treating rock phosphate with phosphoric acid:

$$[Ca_3(PO_4)_2]_3CaF_2 + 14H_3PO_4 \rightleftharpoons 10Ca(H_2PO_4)_2 + 2HF$$

Triple superphosphate (also called concentrated super phosphate) contains 43 per cent water soluble P_2O_5. It reacts with the soil the same way as single superphosphate.

Basic Slag

Basic slag is a by product of the steel industry, where the original iron ores contain appreciable amounts of phosphorus. It is a greyish black powder with a very high specific gravity. Some chemists consider it to be the silicate and phosphate of lime of the formula $(CaO)_5P_2O_5SiO_2$. It contains 8 to 12 per cent P_2O_5 Basic slag produced in Tata Iron & Steel Company contains 3 to 4 per cent of citrate soluble P_2O_5.

11.20. POTASSIC/POTASH FERTILISERS

Muriate of Potash and Sulphate of Potash

Potash fertilisers include muriate of potash and sulphate of potash, which are manufactured from the following minerals:

(*i*) Sylvinite or sylvite KCl — It contains 63.1% K_2O
(*ii*) Carnalite KCl $MgCl_26H_2O$ — It contains 17.1% K_2O
(*iii*) Kainite KCl $MgSO_4.3H_2O$ — It contains 18.9% K_2O
(*iv*) Langbeinite $K_2SO_4.MgSO_4$ — It contains 26.6% K_2O

Muriatic acid and muriates are the old names of hydrochloric acid and its salts respectively.

Manufacture of Muriate of Potash

Sylvinite or Sulvite is dissolved in hot water and sodium chloride is added to it. The temperature of the solution is gradually increased and more sylvinite is gradually added to it, until a hot saturated solution of potassium chloride is formed. This is suddenly cooled and crystals of potassium chloride separate, which are removed, dried and bagged.

Manufacture of Sulphate of Potash

Langbeinite is dissolved in hot water to which magnesium sulphate is also added. The temperature of the solution is gradually increased and more Langbeinite is dissolved until a saturated solution is formed. When it is suddenly cooled crystals of magnesium sulphate separate, because these are less soluble than potassium sulphate. These are removed by filtration. When the mother liqour is evaporated, crystals of potassium sulphate separate. These are dried and bagged. Muriate of Potash is a reddish or dirty white crystalline material which contains about 61 per cent K_2O.

Sulphate of Potash is a dirty white powder which contains about 52 per cent K_2O.

Reaction in Soil

When potassium chloride and potassium sulphate are added to the soil, they are assimilated by plants. Potassium ions replace other ions from the clay and humic micelle as shown below:

$$\mathrm{Ca^{++}}\ \left[\begin{array}{c}\mathrm{Ca^{++}\ Mg^{++}\ K^{+}\ H^{+}} \\ \text{MICELLE} \\ \mathrm{Ca^{++}\ Mg^{++}\ H_{+}}\end{array}\right]\ \mathrm{Mg^{++}} + \mathrm{K_2SO_4} \underset{K^{+}}{\overset{K^{+}}{\rightleftharpoons}} \left[\begin{array}{c}\mathrm{Ca^{++}\ Mg^{++}\ K^{+}\ H^{+}} \\ \text{MICELLE} \\ \mathrm{Ca^{++}\ Mg^{++}\ H^{+}}\end{array}\right]\ \mathrm{Mg^{++}} + \mathrm{CaSO_4}$$

$$\mathrm{Ca^{++}}\ \left[\begin{array}{c}\mathrm{Ca^{++}\ Mg^{++}\ K^{+}\ H^{+}} \\ \text{MICELLE} \\ \mathrm{Ca^{++}\ Mg^{++}\ Na^{++}}\end{array}\right]\ \mathrm{Mg^{++}} + 2\mathrm{KCl} \underset{K^{+}}{\overset{K^{+}}{\rightleftharpoons}} \left[\begin{array}{c}\mathrm{Ca^{++}\ Mg^{++}\ K^{+}\ H^{+}} \\ \text{MICELLE} \\ \mathrm{Ca^{++}\ Mg^{++}\ Na^{++}}\end{array}\right]\ \mathrm{Mg^{++}} + \mathrm{CaCl_2}$$

Some potassium ions are also fixed by clay minerals like illite and vermiculite.

Use of Muriate of Potash and Sulphate of Potash

Muriate of potash should be added to acidic and clayey soils. Sulphate of potash should be added to sandy to loamy soil and paddy fields, alkaline and calcareous soils and highly leached acidic soils which are deficient in sulphur. Sulphate of potash is more expensive than muriate of potash thus sulphate of potash should be applied to the vegetable and fruit to improve their quality.

Other Sources of Potash

Other sources of potash include wood ash, tobacco stem and water hyacinth.

Wood ash contains about 5-6 per cent potash in the form of potassium carbonate, 1-2 per cent phosphorus (P_2O_5) and 25-30 per cent lime. Wood ash therefore may be used to neutralise soil acidity. Finely ground tobacco stem contains 2-3 per cent of nitrogen, 6-10 per cent of readily available potash Dry water hyacinth contains about 1 per cent N, 4 per cent K and a small quantity of phosphorus. The ash of paddy husk, sugarcane bagasse and groundnut shell also contains a fair amount of potash and some phosphorus.

11.21. COMPLEX FERTILISERS

Straight Versus Complex Fertilisers

Straight fertilisers supply one of the fertiliser elements, either nitrogen or phosphorus of potash e.g. urea, needed for crop growth. Complex fertilisers supply more than one fertiliser element needed for crop growth. When they supply any two of the fertiliser element needed for crop grwoth, they are called incomplete complex fertilisers, *e.g.* Diammonium Phosphate. When they supply all the three fertiliser elements for crop growth, they are called complete complex fertiliser *e.g.* NPK complex fertilisers.

Mono and Diammonium Phosphate

Mono Ammonium Phosphate is manufactured by slowly neutralising orthophosphoric acid with ammonia gas as shown in the following equation:

$$NH_3 + H_3PO_4 \longrightarrow \underset{\text{Mono Ammonium Phosphate}}{NH_4H_2PO_4}$$

A considerable amount of heat is generated in this reaction. Consequently the liquid mono ammonium phosphate is thickened. When this is cooled, the mono ammonium phosphate crystallises out. The mother liquor is drained and the crystals are dried and bagged.

Diammonium phosphate is produced, when an excess of ammonium gas is passed through orthophosphoric acid in order to maintain the pH of the slurry at 5.8 to 6.0 as shown below:

$$2NH_3 + H_3PO_4 \longrightarrow \underset{\text{Diammonium Phosphate}}{(NH_4)_2HPO_4}$$

Both mono and diammonium phosphate are soluble in water.

Mono ammonium phosphate is a light grey material. Diammonium phosphate is a dark brown material. They absorb a little moisture from the atmosphere and are acidic fertilisers. The acidity caused by the application of every 100 pounds of mono ammonium phosphate can be neutralized by adding about 55 pounds of powdered limestone to the soil. The acidity caused by adding every 100 pounds of Diammonium phosphate can be neutralized by adding about 77 pounds of powdered limestone to the soil. Mono and Diammonium phosphate can be used in alkaline and calcareous soils.

Mono Ammonium Phosphate contain 11.0 per cent N and 48.0 per cent P_2O_5.

Diammonium Phosphate contains 21.0 per cent N and 54.0 per cent P_2O_5.

Nitrophosphate

This is manufactured by treating apatite with a mixture of nitric and sulphuric acids. They react together as shown below:

$$[Ca_3(PO_4)_2] CaF_2 + 12HNO_3 + 4H_2SO_4 \longrightarrow 6H_3PO_4 + 4CaSO_4 + 6Ca(NO_3)_2 + 2HF$$

Liquid ammonia is then added to the slurry and they react together as shown below:

$$6H_3PO_4 + 4.CaSO_4 + 6.Ca(NO_3)_2 + 2HF + 13NH_3 \longrightarrow \underbrace{5CaHPO_4 + NH_4H_2PO_4 + 4CaSO_4 + 12NH_4NO_3 + CaF_2}_{\text{Nitro Phosphate}}$$

Nitrophosphate is a granulated fertiliser containing a stabilizer which prevents the reversion of citrate soluble phosphate to insoluble tricalcium phosphate. It contains nitrogen and phosphorus in various proportions. If it is necessary, potassium salts are also added to it. Three types of nitro phosphates (Trade Name SUPHALA) 20:20:2, 18:18:9 and 15:15:15 are manufactured.

The First type contains 20% N, 20% P_2O_5 and 2% K_2O
The Second type contains 18% N, 18% P_2O_5 and 9% K_2O
The Third type contains 15% N, 15% P_2O_5 and 15% K_2O

Usually about one-third of the phosphate contained in these is water soluble and the rest of the two thirds is citrate soluble. The water soluble phosphate is readily available and hence useful in the early stages of crop growth, while the citrate soluble phosphate helps the crop to grow and attain maturity. Nitrophosphate contains ammoniacal and nitrate nitrogen.

11.22. MIXED FERTILISERS

A mixed fertiliser means a mixture of two or more straight fertilisers. For example, ammonium sulphate and single super phosphate may be thoroughly mixed to get a mixed fertiliser. Mixed fertilisers are marketed as wheat fertiliser mixtures, paddy fertiliser mixtures etc.

Advantages of Mixed Fertilisers

1. Two or more fertiliser elements are added together to make a mixed fertiliser to be applied to the field. Less labour is therefore required for the application of a mixed fertiliser.
2. The fertiliser elements can be more uniformly applied to the field especially when they are required in small quantities.
3. Mixed fertilisers can be easily drilled in the filed because they are in a good physical condition.
4. When a mixed fertiliser has been prepared on a scientific basis and compensates for the deficiency of nutrients in the soils of the region, the farmer is assured of the right proportion of plant nutrients in the soil.

Disadvantages of a Mixed Fertiliser

1. The use of mixed fertiliser does not permit application of individual fertiliser elements, which may be required by the crop at specific times.

2. The illiterate farmer cannot effectively control the quantity of plant food present in the mixture. He has to rely entirely on the grade advertised by the firms.

3. The firms charge for mixing the fertilisers.

Types to Mixed Fertilisers

Mixed fertilisers are of two kinds:

1. *Open formula mixed fertiliser:* In this case, the manufacturing firms disclose the names and quantities of the straight fertilisers that are constituents of the mixed fertilzer.

2. *Close formula mixture:* In this case, the manufacturing firms do not disclose the names and quantities of the straight fertilisers that are constituents of the mixed fertiliser.

Materials Used to Make Mixed Fertilisers

The following materials are required to make the mixed fertiliser:

(*i*) Straight fertilisers are the primary materials used for making the mixed fertiliser. They supply the fertiliser element required in the proper amount to the soil.

(*ii*) Some low grade organic materials are added @ 100 pounds per ton to the mixed fertiliser in order to prevent caking, and to maintain them in a good physical condition. These organic materials (peat, paddy husk etc.) are called conditioners.

(*iii*) Some basic material like Dolomite is added to neutralize acidity if the fertiliser mixture produce acids.

(*iv*) Some materials like finely ground ash, soil etc. are added to make up the difference between the weight of the fertiliser necessary to supply the plant nutrients and the desired quantity of fertiliser mixture.

Incompatibilities in Fertiliser Mixture

A detailed knowledge of the properties of fertilisers is essential for preparing the fertiliser mixture. The following kind of fertilisers should not be mixed at all or only mixed just before use:

(*i*) Ammonium sulphate, ammonium nitrate etc. should not mixed with basically reactive fertilisers like basic slag, rock phosphate etc., because they may decompose to liberate ammonia gas.

$$(NH_4)_2SO_4 + CaCO_3 \longrightarrow (NH_4)_2CO_3 + CaSO_4$$
$$(NH_4)_2CO_3 \longrightarrow 2NH_3 + CO_2 + H.OH$$

(*ii*) Fertilisers like single super phosphate, ammonium phosphate etc. which contain water soluble phosphorus should not be mixed with fertilisers then contain free lime because water soluble phosphorus would be converted to water insoluble phosphorus as shown below:

$$Ca(H_2PO_4)_2 + 2CaCO_3 \longrightarrow Ca_3(PO_4)_2 + 2CO_2 + 2H_2O$$

(*iii*) Some hygroscopic fertilisers like urea and sulphate of potash tend to form lumps after mixing. Therefore, they should be mixed only just before application.

(*iv*) If urea is to be mixed with a phosphatic fertiliser, it should be mixed with mono or diammonium phosphate, or ammoniated single super phosphate or ammoniated triple super phosphate.

(*v*) Urea should not be mixed with ammonium nitrate because the mixture will readily liquify.

Granulation of Mixed Fertiliser

Recently the granulated mixed fertilisers are being manufactured to decrease the separation of the constituent straight fertilisers and to prevent them from caking. There are more conveniently applied in the field and are difficult to adulterate. Most urea is either prilled or pelleted in sizes too large to form stable dry formulations with most other fertilisers. At present, a granular form of urea is also available.

11.23. SOIL TEST AND FERTILISER RECOMMENDATION

Soil nutrients are gradually withdrawn from the soil by the crops growing on it. Therefore the soil nutrient reserve would be at least partially withdrawn by the time the crop is harvested. The farmer must know the accurate level of depletion of the soil nutrients, before the next crop is sown, so that he can compensate for their deficiency after then removal by the previous crop and leaching, as well as be able to supply the amount of nutrient needed by the next crop in the form of fertiliser and manures.

What is a Soil Test

The available nutrient contents of soils and a few other soil conditions are mainly determined in the soil testing laboratories by specific analytical methods called soil tests.

What is Available Nutrient

Soils contain plant nutrients in different forms, which are not equally responsible for increasing the crop yield. That form of each soil nutrient which is most responsible for increasing the yields of crops is known as the available form of the nutrient.

To find the available form of each of the plant nutrients, soil scientists have grown the crop on a large number of soils and analyzed them for all the forms of the plant nutrients, the amounts each different form, and the yield of the crop. These have been tabulated. A thorough study of this table has revealed that the yield of the crop steadily increases if the amount of a particular form of each plant nutrient in the soil. This particular form of each plant nutrient is therefore called the available form of that nutrient.

Levels of Available Nutrients in the Soil

If high yields of crops from a particular field are to be obtained, the soil must contain the available form of all these chemical elements in quantities sufficient for a good yield of crops. Some soils may not contain sufficient amounts of available nutrients. Soil scientists have chemically analyzed a large number of soils and determined the quantities of available nutrients that should be present in the soil for getting a high yield. They have also produced crops on those soils and worked out a relationship between the amount of available nutrients present in the soil and the yields of crops that have been obtained from them. Thereafter they have concluded that the yield of crops will be low if soils contain a certain amount of the available nutrients, which is termed a 'low amount'. They have further concluded that a medium yield of crops is obtained if the soils contain a particular amount of nutrients, called 'a medium amount'. Also, high yields of crops are obtained if soil contain available nutrients above a certain quantity called 'high amount'. These low, medium and high amounts of available fertiliser nutrients are represented in Table 11.1.

TABLE 11.1. Rating chart used for fertiliser recommendation

Nutrient	*Anal. Method*	*Low*	*Medium*	*High*
Available N kg/ha	Alkaline Permanganate	<280	280 to 560	>560
Available P_2O_5 kg/ha	Bray (soil pH <6) Olsen (soil pH>6)	<22.4	22.4 to 56	>56
Available K_2O kg./ha	N. Am. Ac (pH 7)	<112	112 to 280	>280
Org. Carbon %	Walkely & Black	<0.5	0.5 to 0.75	>0.75
Total Nitrogen %	Kjeldahl	<0.03	0.03 to 0.06	>0.06
	Acidic	*Normal*	*Tending to be alkaline*	*Alkaline*
pH (1 : 2 Soil : Water)	<6	6 to 8	8 to 8.5	>8.5
Soluble Salts 1 : 2 Soil : Water Elect. Conductivity m. mhos/cm at 25°C	Normal <1	Critical for Germination 1 to 2	Critical for Growth of Salt sensitive crops 2 to 3	Serve injury to most crops >3

Collection of Soils Samples

Roots absorb nutrients from a 0 to 20 cm depth of soil under average are conditions. Since the properties of soils vary from place to place, soil samples collected from 25 spots which are uniformly scattered over a one acre field. At each spot a "V" shaped hole of 20 cms is made. One uniform slice of soil is cut from this vertical surface of the hole with a sharp implement and put in a clean bucket. In a similar way, soil samples are collected from each of the 25 spots and put in the same bucket. The entire quantity of soils is spread on a cloth or paper and thoroughly mixed. A round heap is made of the soils. The heap is divided into four equal parts and any two diagonally opposite heaps are chosen, while the other two are discarded. The soils of the two chosen soil heaps are thoroughly mixed and another heap is made with them. This second heap is divided into four equal parts and any two diagonally opposite heaps are discarded. The soils of the remaining two heaps are thoroughly mixed and another third heap of soil is made. This process is continued till about a kilogram of soil is left. This is put in a clean cloth bag and sent to the soil testing laboratory along with the following information.

1. Name of the farmer, village and district.
2. Plot number of his field.
3. Whether the land is level or sloping.
4. Whether the land is well drained or poorly drained.
5. Yield of the previous crop and manures and fertilisers applied to it.
6. The next crop the farmer wishes to grow.

Analysis of the Soil Sample and Fertiliser Recommendation

The available nitrogen is estimated in the soil by the alkaline permanganate method. The available phosphoric acid is estimated by Bray's method in acidic soils and by Olsens method in alkaline soils. The organic carbon and total nitrogen are estimated by Walkey-Black's and Kjeldahle's methods respectively. Morgan's reagent or normal neutral ammonium accetate is used for extracting potassium from the soil. The soil sample is analyzed in the soil testing laboratory to determine if it contains a low, medium or high amount of fertiliser elements. Its organic carbon and soluble salt contents and pH are also determined. If the soil contains a low amount of any one, two, or all three of the available nitrogen, phosphorus acid and potash, the soil chemist in charge of the Soil Testing Laboratory will ask the farmer to apply the full dose of nitrogen and/or phosphoric acid and/or potash in the form of fertiliser necessary for growing the next crop. If the soil contains a medium amount of one or all the fertiliser elements, the application of a half dose of the concerned nutrient in the form of fertiliser will do. The soil chemist in charge may ask the farmer to apply a part of nitrogen and potash (if they are in low amount) in the form of Farm Yard Manure or composit, if the soil contains a low (less than 0.50 per cent) or medium (0.50-0.75 per cent) amount of organic carbon. The soil pH should be adjusted to suit the optimum pH requirement of the next crop to be grown on the filed by adding lime if the soil is too acidic. Gypsum may be applied to the soil if the soil pH is more than 8.5.

11.24. PRINCIPLE OF FERTILISER APPLICATION

Frequency of Fertiliser Application

Usually larger quantities of fertilisers are added to clayey soils at longer intervals, than to the sandy soils because clayey soils are richer in humus than sandy soils and both clay and humus have a high capacity to retain nutrient ions, by a phenomenon called base exchange. These absorbed nutrient ions are not lost by leaching and can be gradually taken up by the roots. If a heavy dose of water soluble fertiliser is applied to a sandy soil, most of it will be leached down by high rainfall in the humid region.

Quantity of Fertiliser to be Applied

Cultivated soils where grain crops are raised have been placed in Low, or Medium or High categories, according to their content of available Nitrogen and/or Phosphoric acid and/or Potash. If the soils are low in one or more of the concerned nutrients, application of that nutrient in a full dose to the soil, will increase the crop yield. If the nutrients are present in a medium amount, only half the dose of that nutrient is applied to the soil. Different crops require different amounts of nutrients. Vegetables, and sugarcane require very high amounts. A requisite amount of nutrients must be given to the crop if the soil contains low quantities of that nutrient. The nutrient requirement of dwarf varieties of wheat and paddy are more than those of the tall varieties of wheat and paddy less nitrogen is given to the tall varieties, because they will lodge if they as given the dose of nitrogen meant for dwarf varieties.

Time of Application of Fertiliser

The rate of assimilation of nitrogen by crops is equal to their rate of growth. Crops require less nitrogen immediately after germination because they grow less at that time. The crops demand for nitrogen increases from the early growth stage to the flowering stage, when their growth rate increases. Therefore nitrogenous fertilisers should be applied in split doses at least twice; once just before sowing or planting the crop and then about one and half to two months after sowing or planting. The crops utilize about two-thirds of their phosphatic and potash requirements during their early growth period. So then entire quantity of phophate and potash may be applied as a basal dose just before sowing the crop.

Kind or Fertilisers to be Applied

Nitrate should not be applied to sandy soil, especially under conditions of high rainfall, because they are readily leached. So nitrate fertilisers are usually not applied to paddy fields where ammoniacal and amide fertiliser (Urea) should be applied. Ammoniacal and amide fertilisers can be continuously applied to neutral and alkaline soils, but not to acidic soils because they increase soil acidity. Water soluble single super phosphate is best suited for application to neutral soils. If it is added to strongly acidic or alkaline soil, the water soluble monocalcium phosphate contained in it is precipitated as water insoluble aluminium and ferric hydroxy phosphate in strongly acidic soil. and as water insoluble tricalcium phosphate in strongly alkaline soil. Hence lime and gypsum should be added to strongly acidic soil and strongly alkaline soil respectively, in order to adjust the soil pH to approximately 6.8, at which most of the nutrients are made available to crops. Citrate soluble bone meal or basic slag should be applied to acidic soils. They will continue to release phosphorus for the growth of the crop during the growing season. Sodium meta and pyro phosphates should be applied to calcareous soil because they do not react with calcium carbonate.

Placement of Fertilisers

Plant nutrients should be placed in the root zone, so that roots can assimilate them quickly. The fertilisers should come in minimum contact with the soil in order to reduce the fixation of nutrient ions like potassium, ammonium and phosphate. Acidic soils contain aluminium and ferric ions, which precipitate water soluble phosphate as water insoluble phosphate. Phosphate is also fixed by aluminium, iron oxide and kaolinite. Phosphate ions also move very slowly from their place of application. Hence single super phosphate or diammonium phosphate should be placed in a band at a distance of about two inches from the rows of seeds. In order to prevent weeds from using the fertilisers, nitrogenous fertilisers should also be placed in a band near rows of widely spaced crops. The up take of phosphate from the band by the crop increases if the fertiliser band also contain nitrogenous fertiliser. Roots of closely spaced crops like wheat evenly permeate the soil, so nitrogenous fertilisers may also be uniformly scattered over the surface.

Ammoniacal fertilisers like ammonium sulphate diammonium phosphate etc. should be placed in the reducing zone (about nine inches deep) of paddy soil where they remain as ammonium ions which are assimilated by the crop.

If they are placed in the surface oxidizing zone, they are oxidized to nitrate which more down to the reducing zone where they are reduced to oxides of nitrogen and nitrogen gas, which is lost to the atmosphere.

11.25. METHODS OF APPLICATION OF FERTILISERS

Broadcasting localized placement and spraying of fertiliser solution on crops are the three main methods of fertiliser application.

Broadcasting

Broadcasting of fertilisers means the uniform spreading of fertiliser over the entire field. It proves effective when the roots uniformly permeate the soil, when heavy doses of nitrogenous and potassic fertiliser are to be applied, and when citrate soluble phosphatic fertilisers like bone meal and basic slag are to be applied to the acidic the soil. In India, cultivators broadcast the fertilisers on the field and incorporate them in the soil by ploughing. Thereafter the field is thoroughly prepared and seeds of the crop are sown. Broadcasting of fertilisers (usually nitrogenous) over the standing crop is known as top dressing.

The only advantage of broadcasting is that fertilisers may be more conveniently applied in the hilly regions by this method. Fertilisers are also broadcast on grasslands. Broadcasting of fertilisers should be discouraged due to the following disadvantages.

(*i*) Most of the plant nutrients are assimilated by weeds.

(*ii*) Since fertilisers come in contact with a large volume of soil, most of the nutrients, especially phosphates, are fixed up.

(*iii*) Nutrients cannot be fully assimilated by the roots of widely spaced crops.

Localized Placement

The localized placement of fertilisers means their application very near the seed or the plant.

Drill placement of fertilisers means the placement of fertilisers in the soil with the help of drills, which usually have separate seed box and fertiliser hoppers with drills attached, so that the seed and fertilisers are put in same furrow. The fertilisers, if placed a little below the seed, will not harm the seed and the tender roots of the young seeding. Roots will also develop deeper. Drill application saves the cost of labour.

Plough sole placement of fertilisers means the application of fertilisers in a continuous band at the bottom of the furrow when the field is being ploughed, with a furrow turning plough. This method encourage the development of a deeper root system, because fertilisers have been placed in the moist sub soil where roots develop.

Band Placement

The fertiliser is placed by the side of the crop, either along the row of crops or in the hill near each plant. This is called Band Placement, where the fertilisers are placed to one or both the side of the row of the crop plant at a distance of 2.5 cm to 7.5 cm from it.

Spot Placement

This means the placement of fertilisers at the base of each pair of crop plants.

Advantages of Localized Placement

(*i*) The fixation of nutrients, especially phosphate, is considerably reduced because the fertiliser comes in minimum contact with the soil.

(*ii*) The nutrients are near the root hairs which can easily assimilated them, yet are not injured by the high concentration of fertilisers.

(*iii*) Plough sole placement of fertilisers encourages a deeper root system because the fertiliser is placed in the moist deeper layers where roots readily extend to assimilate the nutrients.

(*iv*) In the fertiliser band, the high ion concentration facilitates their movement in the roots.

(*v*) The residual effect of bands of fertiliser is more than that of the broadcasted fertilisers.

Spraying of Fertiliser Solutions

A dilute solution of nitrogenous and micro nutrient fertilisers are sprayed on the crops and leaves readily absorb the nutrients.

11.26. EFFECT OF FERTILISERS ON SOILS AND CROPS

When water soluble fertilisers are added to the soil, the soluble salt content of the soil is increased. The salt index of a fertiliser is a measure of the tendency of the fertiliser to increase the soluble salt content of the soil. The salt index per unit of plant nutrient of a few common fertilisers are given below:

Fertiliser	*Salt index per unit of plant nutrient*
Sodium Nitrate	6.06—Salt Index per unit of N.
Potassium Nitrate	5.34—Salt Index per unit of N.
Ammonium Sulphate	3.25—Salt Index per unit of N.
Ammonium Nitrate	2.99—Salt Index per unit of N.
Mono Ammonium Phosphate	2.45—Salt Index per unit of N.

Urea	1.62—Salt Index per unit of N.
Diammonium Phosphate	1.61—Salt Index per unit of N.
Diammonium Phosphate	0.64—Salt Index per unit of N.
Mono Ammonium Phosphate	0.48—Salt Index per unit of N.
Super Phosphate	0.39—Salt Index per unit of CaO
Triple super phosphate	0.21—Salt Index per unit of CaO
Potassium Magnesium Sulphate	1.97—Salt Index per unit of K_2O
Potassium Chloride (Muriate of Potash)	1.94—Salt Index per unit of K_2O
Potassium Nitrate	1.58—Salt Index per unit of K_2O
Potassium Sulphate	0.85—Salt Index per unit of K_2O

Thus we find that the continuous application of a heavy dose of nitrogenous fertiliser, especially sodium nitrate, potassium nitrate, ammonium sulphate and ammonium nitrate highly increases the soluble salt content of the soil. The continuous application of potassic fertilisers also increases the soluble salt content of the soil, but this increase is much less than that caused by the application of nitrogenous fertilisers. The continuous application of phosphatic fertilisers increases the soluble salt content of the soil by a negligible amount. The salt index of nitrogenous fertiliser per unit of nitrogen decreases when their nitrogen content increases.

Whenever ammoniacal fertilisers are applied to the soil, ammonium ions replace the basic cations, especially calcium from the clay and humic micelle as shown below supposing the clay and Humic micelle have absorbed calcium . magnesium, potassium, sodium and hydrogen ions in the ratio of 50 : 20 : 5 : 5 : 20.

$$K_5^+ \overset{Ca_{50}^{++}\,Mg_{20}^{++}}{\underset{H^{20+}Na_5^+}{\boxed{\text{MICELLE}}}} + (NH_4)_2SO_4 \rightleftharpoons K^+ \overset{Ca_{49}^{++}\,Mg_{20}^{++}}{\underset{H_{20}^+(NH_4)_2^+}{\boxed{\text{MICELLE}}}} Na_5^+ + CaSO_4$$

Calcium sulphate is gradually dissolved in rain water in the humid region. The above reversible reaction proceeds in the forward direction which means that basic cation are gradually replaced from the micelle by ammonium ions which are then oxidized to nitrous, which is further oxidized to nitrate as shown below:

$$K_5^+ \overset{Ca_{49}^{++}\,Mg_{20}^{++}}{\underset{H_{20}^+(NH_4^+)_2}{\boxed{\text{MICELLE}}}} Na_5^+ + 3O_2 \rightleftharpoons K_2 \overset{Ca_{49}^{++}\,Mg_{20}^{++}}{\underset{H_{22}^+\,Na^+}{\boxed{\text{MICELLE}}}} + 2HNO_2 + 2H_2O$$

$$2HNO_2 + O_2 \longrightarrow 2\,HNO_3$$

When urea is added to the soil, it is converted to ammonium carbonate, which oxidized to nitrous acid that, in turn, is oxidized to nitric acid.

$$\begin{matrix} NH_2 \\ | \\ CO \\ | \\ NH_2 \end{matrix} + 2H.OH \xrightarrow[\text{Enzyme}]{\text{Urease}} \underset{\text{Ammonium Carbonate}}{(NH_4)_2CO_3}$$

This ammonium carbonate then may react with the clay and humic micelle and ammonium ions gradually replace the basic cations from them, as shown below:

$$\underset{H_{20}^{+}}{\overset{Ca_{50}^{++}\,Mg_{20}^{++}}{K_5^{+}\,\boxed{\text{MICELLE}}}} + Na_5^{+} + (NH_4^{+})_2CO_3 \rightleftharpoons \underset{H_{20}^{+}\,(NH_4^{+})_2}{\overset{Ca_{49}^{++}\,Mg_{20}^{++}}{K_5\,\boxed{\text{MICELLE}}}}\;Na^{++} + CaCO_3$$

The ammonium ions of ammonium carbonate and the clay and humic micelle are then oxidized to nitrous acid, that is further oxidized to nitric acid as shown below:

$$\underset{H_{20}^{+}\,(NH_4^{+})_2}{\overset{Ca_{49}^{++}\,Mg_{20}^{++}}{K_5^{+}\,\boxed{\text{MICELLE}}}}\;Na_5^{+} + 3O_2 \rightleftharpoons \underset{H_{22}^{+}Na_5^{+}}{\overset{Ca_{49}^{++}\,Mg^{20++}}{K_5^{+}\,\boxed{\text{MICELLE}}}} + 2HNO_2 + 2H_2O$$

$$(NH_4)CO_3 + 3O_2 \longrightarrow 2HNO_2 + 3H_2O + CO_2$$

$$2HNO_2 + O_2 \longrightarrow 2HNO_3$$

Ultimately the clay and humic micelle is saturated with hydrogen io[ns an]d nitric acid is also formed. So if a heavy dose of ammoniacal and amide fert[ilise]rs (urea) are continuously applied to the soil, they become acidic in reaction.

In a similar way, if sodium nitrate is continuously applied to soil, sodium ions gradually replace other cations as shown in the following equation, and the soils become alkaline in reaction.

$$\underset{K_5^{+}\,H_{20}^{+}\,Na_5^{+}}{\overset{Ca_{50}^{++}\,Mg_{20}^{++}}{\boxed{\text{MICELLE}}}} + 3NaNO_3 \rightleftharpoons \underset{K_4^{+}\,H_{20}Na_8^{+}}{\overset{Ca_{49}^{++}\,Mg_{20}^{++}}{\boxed{\text{MICELLE}}}} + Ca(NO_3)_2 + KNO_3$$

Tall varieties of wheat and paddy lodge *i.e.*, fall down, if a heavy dressing of nitrogenous fertilisers is applied to them. The dwarf varieties of wheat have responded more to the increasing doses of nitrogenous and phosphatic fertilisers than the tall varieties. For example, the yields of tall varieties of wheat are increased when the dose of nitrogenous fertilisers are increased to up to 80 kg nitrogen per hectare, whereas the yields of dwarf varieties of wheat continued to increase when the dose of nitrogenous fertilisers are increased up to 120 kg nitrogen per hectare. The application of nitrogenous, phosphatic fertilisers have increased the yields of the high yielding varieties of wheat.

Tall varieties of paddy have responded to up to 60 kg of nitrogen per hectare whereas the drarf varieties have responded to up to 120 kg of nitrogen per hectare. Paddy usually responds to the application of phosphatic fertilisers much less than to nitrogenous fertilisers. But dwarf varieties of paddy may respond to phosphatic fertilisers. High yielding paddy varieties yield more if nitrogenous, phosphatic and potassic fertilisers are applied to them.

Yields of hybrid maize have increased much more as a result of application of nitrogenous and phosphatic fertilisers than the yields of local maize. Hybrid maize has responded to the application of potassic fertilisers to potassium deficient soils.

Pulses like pigeon pea (Cajanus cajan) and green gram (phseolus aureus) have responded only to the application of phosphatic fertilisers. Lentil (Lens esculenta) and pea (Pisum sativam) have responded to the application of nitrogenous and phosphatic fertilisers whereas Gram (Cicer arietinum) has responded to the application of phosphatic fertilisers. The yields of oilseeds (Brassica sp) have increased as a consequence of the application of nitrogenous fertilisers only. The application of phosphatic fertilisers to oil seeds increases their oil content. A balanced application of Nitrogenous, phosphatic and potassic fertilisers increase the yields of potato and other vegetables. Phosphorus and potassium are essential for improving their quality.

It has been reported by Chowdhury and Bains (1967) from the Indian Agricultural Research Institute, Delhi, that the protein content of wheat grain and straw increased when increasing doses of nitrogenous fertilisers are applied to them. Similar results have been obtained for other crops. The phosphorus, thiamine and riboflavin content of wheat have also increased as a consequence of application of nitrogenous, phosphatic and potassic fertilisers, at the same place. If excessive amounts of nitrate fertilisers are applied to fodder crops, they may contain toxic amounts of nitrates.

SELECTING READINGS

1. Annonymous (1971), *Hand Book of Manures and Fertilisers*. I.C.A.R. New Delhi.
2. Asana, R.D. and Sarin M.N. (1968), *Crop Physiology in India*. I.C.A.R. Technical Bulletin No. 16.
3. Donahue, R.L. (1977), *Soils. An Introduction to Soils & Plant Growth*. Prentice Hall, Englewood cliffs, New Jersey-07652.
4. Foth, H.D. (1984), *Fundamentals of Soil Science*, 7th Ed. John Wiley & Sons, New York.
5. Jones, U.S. (1979), *Fertilisers and Soil Fertility*. Reston Publishing Co., Reston, Virginia, Prentice Hall Co, Reston, Virginia, U.S.
6. Kanwar, J.S. (1976), *Soil Fertility, Theory and Practice*. I.C.A.R. New Delhi.
7. Mariakulandai, A. and Manickam, T.S. (19), Chemistry of Manures and Fertilisers. Asia Publishing House, Bombay 38.
8. Nelson, W.L. & Tisdale, S.L. (1975), *Soil Fertility and Fertilisers*. Macmillan Publishing Co., Inc., New York.
9. Yawalkar, K.S. (1967), *Manures and Fertilisers*. Agri. Horticultural Publishing House, Nagpur-440010.

INDEX